Acoustics and Psychoacoustics

David M. Howard and Jamie A. S. Angus

ELSEVIER

AMSTERDAM · BOSTON · HEIDELBERG · LONDON · NEW YORK · OXFORD
PARIS · SAN DIEGO · SAN FRANCISCO · SINGAPORE · SYDNEY · TOKYO

Focal Press

Focal Press is an imprint of Elsevier

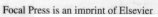

Focal Press is an imprint of Elsevier
Linacre House, Jordan Hill, Oxford OX2 8DP, UK
30 Corporate Drive, Suite 400, Burlington, MA 01803, USA

First edition 1996
Reprinted 1998, 1999, 2000
Second edition 2001
Third edition 2006
Fourth edition 2009

British Library Cataloguing in Publication Data
Howard, David M. (David Martin), 1956-
  Acoustics and psychoacoustics. – 4th ed.
  1. Psychoacoustics. 2. Music–Acoustics and physics.
  3. Sound–Recording and reproducing–Digital techniques.
  I. Title II. Angus, J. A. S. (James A. S.)
  152.1'5-dc22

Library of Congress Control Number: 2009930892

ISBN: 978-0-240-52175-6

For information on all Focal Press publications
visit our website at focalpress.com

Printed and bound in Great Britain

09 10 11 12 11   10 9 8 7 6 5 4 3 2 1

Contents

Companion website passcode: ACSTCS978235

Preface

Acoustics and Psychoacoustics continues to be adopted as a core text in courses all around the world; it has come on a long way since it was first published in 1996. We receive a number of emails from readers asking questions relating to material presented and we always try to provide appropriate answers. Such queries are always welcomed; indeed, much of the substance of the revisions made in the 2nd edition was based on this important feedback for which we are very grateful. The main change in the 3rd edition was the inclusion of the audio compact disc (CD) to provide recorded material for use in learning, and we understand from feedback received that this was a very welcome addition to the book.

In addition to minor modifications that have been made in this 4th edition, there is a major change to the final chapter which is now, we believe, better focused as the book's concluding material. It takes *Applications of acoustics and psychoacoustics* as its theme and it explores the underlying principles of devices, procedures and systems that underpin practical work in the area of acoustics and psychoacoustics. The original material is essentially retained but reorganized to enable the inclusion of new material including: new room acoustic design examples, hearing testing in practice, the principles of psychoacoustic testing, noise reducing headphones, "mosquito" units and "teen buzz" ring tones (demonstrated on a new audio on the CD—track 79). We hope not only that these practical examples will provide useful insights into existing practical applications of the subject, but also that they will trigger creative thinking in tomorrow's readers who might be responsible for the invention of new devices, procedures and systems for the future.

The musical and studio side of the field has not been neglected and additional material has been added on the following: pipe organs—to take into account the full range of acoustic harmonic synthesis that they achieve around the globe through the inclusion of what we believe to be all stop footages which are found on today's instruments (enhanced Table 5.1); unaccompanied (*a capella*) singing performance and how overall tuning can drift due to the tuning system that *a capella* singers adopt (new Section 4.5.3); and timbral descriptions that are often used in the studio and how they relate to frequency (at the end of Section 5.3.2). A new version of track 7 on the audio

CD relating to tuning systems is included which we believe provides a clearer demonstration of the differences between *just* and *equal* temperaments.

Other generally more minor changes have been made throughout the book mainly in response to reader and reviewer feedback, and we would like to thank everyone who has provided such helpful and supportive comments and suggestions. A book like this builds up a community of readers and we are humbled but happy to receive remarks that lead to improved clarity in the material offered.

Creativity is at the heart of successful work in fields such as acoustics, psychoacoustics, studio engineering, audio engineering, music technology, music composition and music performance. What we hear with our ears is an essential aspect of human communication whether by speech, music or other sounds. A basic understanding of acoustics and psychoacoustics is therefore essential if progress is to be made by tomorrow's generation, and this is the spirit in which we offer this 4th edition of *Acoustics and Psychoacoustics*.

David M. Howard (York, U.K.) and Jamie A.S. Angus (Salford, U.K.)
June 2009

Introduction to Sound

CHAPTER CONTENTS

Sound is something most people take for granted. Our environment is full of noises, which we have been exposed to from before birth. What is sound, how does it propagate, and how can it be quantified? The purpose of this chapter is to introduce the reader to the basic elements of sound, the way it propagates, and related topics. This will help us to understand both the nature of sound and its behavior in a variety of acoustic contexts, and allow us to understand both the operation of musical instruments and the interaction of sound with our hearing.

1.1 PRESSURE WAVES AND SOUND TRANSMISSION

At a physical level sound is simply a mechanical disturbance of the medium, which may be air, or a solid, liquid or other gas. However, such a simplistic description is not very useful as it provides no information about the way this disturbance travels, or any of its characteristics other than the requirement for a medium in order for it to propagate. What is required is a more accurate description which can be used to make predictions of the behavior of sound in a variety of contexts.

FIGURE 1.1 *Golf ball and spring model of a sound propagating material.*

1.1.1 The nature of sound waves

Consider the simple mechanical model of the propagation of sound through some physical medium, shown in Figure 1.1. This shows a simple one-dimensional model of a physical medium, such as air, which we call the golf ball and spring model because it consists of a series of masses, e.g., golf balls, connected together by springs. The golf balls represent the point masses of the molecules in a real material, and the springs represent the intermolecular forces between them. If the golf ball at the end is pushed toward the others then the spring linking it to the next golf ball will be compressed and will push at the next golf ball in the line, which will compress the next spring, and so on.

Because of the mass of the golf balls there will be a time lag before they start moving from the action of the connecting springs. This means that the disturbance caused by moving the first golf ball will take some time to travel down to the other end. If the golf ball at the beginning is returned to its original position the whole process just described will happen again, except that the golf balls will be pulled rather than pushed and the connecting springs will have to expand rather than compress. At the end of all this the system will end up with the golf balls having the same average spacing that they had before they were pushed and pulled.

The region where the golf balls are pushed together is known as a "compression" whereas the region where they are pulled apart is known as a "rarefaction," and the golf balls themselves are the propagating medium. In a real propagating medium, such as air, a disturbance would naturally consist of either a compression followed by a rarefaction or a rarefaction followed by a compression in order to allow the medium to return to its normal state. A picture of what happens is shown in Figure 1.2. Because of the way the disturbance moves—the golf balls are pushed and pulled in the direction of the disturbance's travel—this type of propagation is known as a "longitudinal wave." Sound waves are therefore longitudinal waves which propagate via a series of compressions and rarefactions in a medium, usually air.

1.1.2 The velocity of sound waves

The speed at which a disturbance, of either kind, moves down the "string" of connected golf balls will depend on two things:

- *The mass of the golf balls*: the mass affects the speed of disturbance propagation because a golf ball with more mass will take longer to

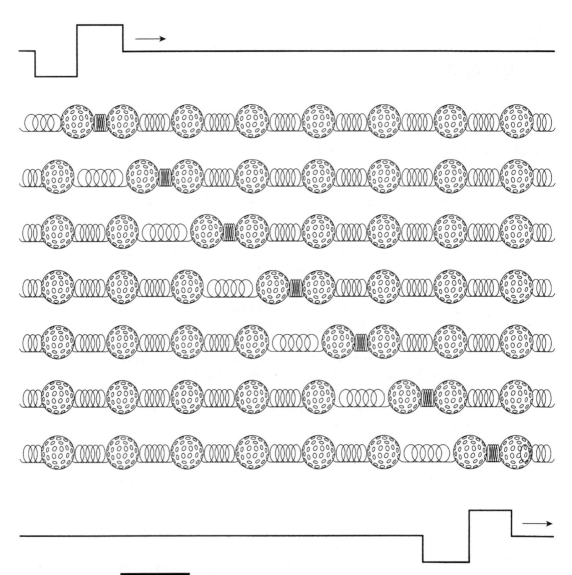

FIGURE 1.2 *Golf ball and spring model of a sound pulse propagating in a material.*

start and stop moving. In real materials the density of the material determines the effective mass of the golf balls. A higher density gives a higher effective mass and so the propagation will travel more slowly.

- *The strength of the springs*: the strength of the springs connecting the golf balls together will also affect the speed of disturbance propagation because a stronger spring will be able to push harder on the next golf ball and so accelerate it faster. In real materials the strength of the springs is equivalent to the elastic modulus of the material, which is also known as the "Young's modulus" of the material. A higher elastic modulus in the material implies a stiffer spring and therefore a faster speed of disturbance propagation.

For longitudinal waves in solids, the speed of propagation is only affected by the density and Young's modulus of the material, and this can be simply calculated from the following equation:

$$c = \sqrt{\frac{E}{\rho}} \qquad (1.1)$$

where c = the speed in meters per second (ms^{-1})
ρ = the density of the material (in kg m^{-3})
and E = the Young's modulus of the material (in N m^{-2})

However, although the density of a solid is independent of the direction of propagation in a solid, the Young's modulus may not be. For example, brass will have a Young's modulus which is independent of direction because it is homogeneous, whereas wood will have a different Young's modulus depending on whether it is measured across the grain or with the grain. Thus brass will propagate a disturbance with a velocity which is independent of direction, but in wood the velocity will depend on whether the disturbance is traveling with the grain or across it. To make this clearer let us consider an example.

This variation of the speed of sound in materials such as wood can affect the acoustics of musical instruments made of wood and has particular implications for the design of loudspeaker cabinets, which are often made of wood. In general, loudspeaker manufacturers choose processed woods, such as plywood or MDF (medium density fiberboard), which have a Young's modulus that is independent of direction.

1.1.3 The velocity of sound in air

So far the speed of sound in solids has been considered. However, sound is more usually considered as something that propagates through air, and for music this is the normal medium for sound propagation. Unfortunately air does not have a Young's modulus so Equation 1.1 cannot be applied directly, even though the same mechanisms for sound propagation are involved. Air is springy, as anyone who has held their finger over a bicycle pump and pushed the plunger will tell you, so a means of obtaining something equivalent to

---

Young's modulus is a measure of the "springiness" of a material. A high Young's modulus means the material needs more force to compress it. It is measured in newtons per square meter (N m^{-2}).

A newton (N) is a measure of force.

c is the accepted symbol for velocity, yes it's strange, but v is used for other things by physicists. Density is the mass per unit volume. It is measured in kilograms per cubic meter (kg m^{-3}).

$\sqrt{}$ is the square root symbol. It means "take the square root of whatever is inside it."

s^{-1} means per second.

EXAMPLE 1.1

Calculate the speed of sound in steel and in beech wood.

The density of steel is 7800 kg m^{-3}, and its Young's modulus is 2.1 × 10^{11} N m^{-2}, so the speed of sound in steel is given by:

$$c_{steel} = \sqrt{\frac{2.1 \times 10^{11}}{7800}} = 5189 \text{ ms}^{-1}$$

The density of beech wood is 680 kg m^{-3}, and its Young's modulus is 14 × 10^9 N m^{-2} along the grain and 0.88 × 10^9 N m^{-2} across the grain. This means that the speed of sound is different in the two directions and they are given by:

$$c_{beech \text{ along the grain}} = \sqrt{\frac{14 \times 10^9}{680}} = 4537 \text{ ms}^{-1}$$

and

$$c_{beech \text{ across the grain}} = \sqrt{\frac{0.88 \times 10^9}{680}} = 1138 \text{ ms}^{-1}$$

Thus the speed of sound in beech is four times faster along the grain than across the grain.

Pressure is the force, in newtons, exerted by a gas on a surface. This arises because the gas molecules "bounce" off the surface. It is measured in newtons per square meter (N m^{-2}).   The molecular mass of a gas is approximately equal to the total number of protons and neutrons in the molecule expressed in grams (g). Molecular mass expressed in this way always contains the same number of molecules (6.022 × 10^{23}). This number of molecules is known as a "mole" (mol).

Young's modulus for air is required. This can be done by considering the adiabatic (meaning no heat transfer) gas law given by:

$$PV^{\gamma} = \text{constant} \tag{1.2}$$

where P = the pressure of the gas (in N m^{-2})

V = the volume of the gas (in m^3)

and γ = is a constant which depends on the gas (1.4 for air)

The adiabatic gas law equation is used because the disturbance moves so quickly that there is no time for heat to transfer from the compressions or rarefactions. Equation 1.2 gives a relationship between the pressure and volume of a gas which can be used to determine the strength of the air spring, or the equivalent to Young's modulus for air, which is given by:

$$E_{gas} = \gamma P \tag{1.3}$$

The density of a gas is given by:

$$\rho_{gas} = \frac{m}{V} = \frac{PM}{RT} \tag{1.4}$$

where m = the mass of the gas (in kg)
$\quad M$ = the molecular mass of the gas (in kg mole^{-1})
$\quad R$ = the gas constant (8.31 J K^{-1} mole^{-1})
and T = the absolute temperature (in K)

Equations 1.3 and 1.4 can be used to give the equation for the speed of sound in air, which is:

$$c_{gas} = \sqrt{\frac{E_{gas}}{\rho_{gas}}} = \sqrt{\frac{\gamma P}{\left(\frac{PM}{RT}\right)}} = \sqrt{\frac{\gamma RT}{M}} \qquad (1.5)$$

Equation 1.5 is important because it shows that the speed of sound in a gas is not affected by pressure. Instead, the speed of sound is strongly affected by the absolute temperature and the molecular weight of the gas. Thus we would expect the speed of sound in a light gas, such as helium, to be faster than that in a heavy gas, such as carbon dioxide, and, in air, to be somewhere in between. For air we can calculate the speed of sound as follows.

EXAMPLE 1.2

Calculate the speed of sound in air at 0°C and 20°C.
The composition of air is 21% oxygen (O_2), 78% nitrogen (N_2), 1% argon (Ar), and minute traces of other gases. This gives the molecular weight of air as:

$\quad M$ = 21% x 16 $\times$ 2 + 78% x 14 $\times$ 2 + 1% x 18 M = 2.87 $\times$ 10^{-2} kg mole^{-1}

and

$\quad \gamma = 1.4$

$\quad R = 8.31$ J K^{-1} mole^{-1}

which gives the speed of sound as:

$$c = \sqrt{\frac{1.4 \times 8.31}{2.87 \times 10^{-2} \, T}}$$

$$c = 20.1\sqrt{T}$$

Thus the speed of sound in air is dependent only on the square root of the absolute temperature, which can be obtained by adding 273 to the Celsius temperature; thus the speed of sound in air at 0°C and 20°C is:

$$c_{0°C} = 20.1 \sqrt{(273 + 0)} = 332 \text{ ms}^{-1}$$
$$c_{20°C} = 20.1 \sqrt{(273 + 20)} = 344 \text{ ms}^{-1}$$

The reason for the increase in the speed of sound as a function of temperature is twofold. Firstly, as shown by Equation 1.4 which describes the density of an ideal gas, as the temperature rises the volume increases and, provided the pressure remains constant, the density decreases. Secondly, if the pressure does alter, its effect on the density is compensated for by an increase in the effective Young's modulus for air, as given by Equation 1.3. In fact the dominant factor other than temperature affecting the speed of sound in a gas is the molecular weight of the gas. This is clearly different if the gas is different from air, for example helium. But the effective molecular weight can also be altered by the presence of water vapor, because the water molecules displace some of the air and, because they have a lower weight, this slightly increases the speed of sound compared with dry air.

Although the speed of sound in air is proportional to the square root of absolute temperature we can approximate this change over our normal temperature range by the linear equation:

$$c < 331.3 + 0.6t \text{ ms}^{-1} \tag{1.6}$$

where t = the temperature of the air in °C

Therefore we can see that sound increases by about 0.6 ms^{-1} for each °C rise in ambient temperature and this can have important consequences for the way in which sound propagates.

Table 1.1 gives the density, Young's modulus and corresponding velocity of longitudinal waves for a variety of materials.

Table 1.1	Young's modulus, densities and speeds of sound for some common materials		
Material	**Young's modulus (N m^{-2})**	**Density (kg m^{-3})**	**Speed of sound (ms^{-1})**
Steel	2.10×10^{11}	7800	5189
Aluminum	6.90×10^{10}	2720	5037
Lead	1.70×10^{10}	11400	1221
Glass	6.00×10^{10}	2400	5000
Concrete	3.00×10^{10}	2400	3536
Water	2.30×10^{9}	1000	1517
Air (at 20°C)	1.43×10^{5}	1.21	344
Beech wood (along the grain)	1.40×10^{10}	680	4537
Beech wood (across the grain)	8.80×10^{8}	680	1138

1.1.4 Transverse and other types of wave

Once one has a material with boundaries that are able to move, for example a guitar string, a bar, or the surface of the sea, then types of wave other than longitudinal waves occur.

The simplest alternative type of wave is the transverse wave, which occurs on a vibrating guitar string. In a transverse wave, instead of being pushed and pulled toward each other, the golf ball (referred to earlier) is moved from side to side—this causes a lateral disturbance to be propagated, due to the forces exerted by the springs on the golf balls, as described earlier. This type of wave is known as a "transverse wave" and is often found in the vibrations of parts of musical instruments, such as strings, or thin membranes.

1.1.5 The velocity of transverse waves

The velocity of transverse vibrations is affected by factors other than just the material properties. For example, the static spring tension will have a significant effect on the acceleration of the golf balls in the golf ball and spring model. If the tension is low then the force which restores the golf balls back to their original position will be lower and so the wave will propagate more slowly than when the tension is higher. This allows us to adjust the velocity of transverse waves, which is very useful for tuning musical instruments.

However, the transverse vibration of strings is quite important for a number of musical instruments; the velocity of a transverse wave in a piece of string can be calculated by the following equation:

$$c_{\text{transverse}} = \sqrt{\frac{T}{\mu}}$$

where μ = the mass per unit length (in kgm^{-1})   (1.7)
and T = the tension of the string (in N)

This equation, although it is derived assuming an infinitely thin string, is applicable to most strings that one is likely to meet in practice. But it is applicable to only pure transverse vibration; it does not apply to other modes of vibration. However, transverse waves are the dominant form of vibration for thin strings. The main error in Equation 1.7 is due to the inherent stiffness in real materials, which results in a slight increase in velocity with frequency. This effect does alter the timbre of percussive stringed instruments, like the piano or guitar, and gets stronger for thicker pieces of wire. So Equation 1.7 can be used for most practical purposes. Let us calculate the speed of a transverse vibration on a steel string.

EXAMPLE 1.3

Calculate the speed of a transverse vibration on a steel wire which is 0.8 mm in diameter (this could be a steel guitar string), and under 627 N of tension.

The mass per unit length is given by:

$$\mu_{steel} = \rho_{steel}(\pi r^2) = 7800 \times 3.14 \times \left(\frac{0.8 \times 10^{-3}}{2}\right)^2 = 3.92 \times 10^{-3} kg\ m^{-1}$$

The speed of the transverse wave is thus:

$$c_{steel\ transverse} = \sqrt{\frac{627}{3.92 \times 10^{-3}}} = 400 ms^{-1}$$

This is considerably slower than a longitudinal wave in the same material; generally transverse waves propagate more slowly than longitudinal ones in a given material.

1.1.6 Waves in bars and panels

There are several different possible waves in three-dimensional objects. For example, there are different directions of vibration and in addition there are different forms, depending on whether opposing surfaces are vibrating in similar or contrary motion, such as transverse, longitudinal torsional, and others. As all of these different ways of moving will have different spring constants, they will be affected differently by external factors such as shape. This means that for any shape more complicated than a thin string, the velocity of propagation of transverse modes of vibration becomes extremely complicated. This becomes important when one considers the operation of percussion instruments.

There are three main types of wave in these structures: quasi-longitudinal, transverse shear, and bending (flexural)—the latter two are shown in Figures 1.3 and 1.4. There are others, for example surface acoustic waves, like waves at sea and waves in earthquakes, that are combinations of longitudinal and transverse waves.

Quasi-longitudinal waves

The quasi-longitudinal waves are so called because they do result in some transverse motion, due to the finite thickness of the propagating medium. However, this effect is small and, for quasi-longitudinal waves in bars and

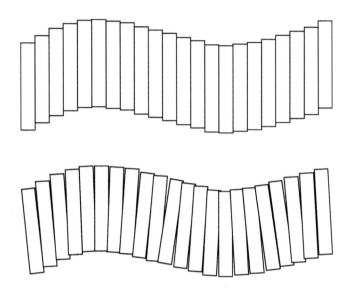

FIGURE 1.3

A transverse shear wave.

FIGURE 1.4

A bending (flexural) wave.

plates, the density and Young's modulus of the material affect the speed of propagation in the same way as pure longitudinal waves. The propagation velocity can therefore be simply calculated from Equation 1.1.

Transverse shear waves

Transverse shear waves are waves that have a purely transverse (shear) displacement within a solid. It would be helpful to have a definition and/or description of "shear." Unlike simple transverse waves on a thin string they do not rely on a restoring force due to tension, but on the shear force of the solid. Solids can resist static shear deformation and this is encapsulated in the shear modulus, which is defined as the ratio of shear stress to shear strain. The shear modulus (G) of a material is related to its Young's modulus via its Poisson ratio (v). This ratio is defined as the ratio of the magnitudes of the lateral strain to the longitudinal strain and is typically between 0.25 for something like glass to 0.5 for something like hard rubber. It arises because there is a change in lateral dimension, and hence lateral strain, when stress is applied in the longitudinal direction (Poisson contraction). The equation linking shear modulus to Young's modulus is:

$$G = \frac{E}{2(v + 1)}$$

where G = the shear modulus of the material (in Nm^{-2})   (1.8)

E = the Young's modulus of the material (in Nm^{-2})

and v = the Poisson ratio of the material

A *shear* displacement is what happens if you cut paper with scissors, or tear it. You are applying a force at right angles and making one part of the material slip, or *shear*, with respect to the other.

Thus for transverse shear waves the velocity of propagation is given by:

$$c_{\text{transverse shear}} = \sqrt{\frac{G}{\rho}}$$

(1.9)

where G = the shear modulus of the material (in Nm^{-2})

and ρ = the density of the material (in kgm^{-3})

This equation, like the earlier one for quasi-longitudinal waves, gives a phase velocity that is independent of frequency. Comparing Equations 1.1 and 1.9 also shows that the ratio of the two wave velocities is given by:

$$\frac{c_{\text{transverse shear}}}{c_{\text{quasi-longitudinal}}} = \sqrt{\frac{1}{2(c+1)}}$$

(1.10)

Thus the transverse shear wave speed is smaller than that of quasi-longitudinal waves. Typically the ratio of speeds is about 0.6 for homogeneous materials.

It is difficult to generate pure shear waves in a plate via the use of applied forces because in practice an applied force results in both lateral and longitudinal displacements. This has the effect of launching bending (flexural) waves into the plate.

Bending (flexural) waves

Bending (flexural) waves are neither pure longitudinal nor pure transverse waves. They are instead a combination of the two. Examination of Figure 1.4 shows that in addition to the transverse motion there is also longitudinal motion that increases to a maximum at the two surfaces. Also, on either side of the center line of the bar the longitudinal motions are in antiphase. The net result is a rotation about the midpoint, the neutral plane, in addition to the transverse component. The formal analysis of this system is complex, as in principle both bending and shear forces are involved. However, provided the shear forces' contribution to transverse displacement is small compared to that of the bending forces (a common occurrence), the velocity of a bending wave in a thin plate or bar is given by:

$$V_{\text{bending}} = \sqrt{\omega}\left(\frac{D}{m}\right)^{\frac{1}{4}}$$

(1.11)

where D = the bending stiffness of the plate (in Nm)

and m = the mass per unit area (in kgm^{-2})

Equation 1.11 is significantly different from that for quasi-longitudinal and transverse shear waves. In particular the velocity is now frequency dependent, and increases with frequency. This results in dispersive propagation of waves with different frequencies traveling at different velocities. Therefore waveshape is not preserved in bending wave propagation. One can hear the effect of this if one listens to the "chirp" sound emitted by ice covering a pond when hit by a thrown rock. The dispersion and the fourth root arise because, unlike quasi-longitudinal and transverse shear waves, the spatial derivatives in the wave equation are fourth order instead of second order because the bending wave is an amalgam of longitudinal and transverse waves.

A major assumption behind Equation 1.11 is that the shear contribution to the lateral displacement is small. This is likely to be true if the radius of the bend is large with respect to the thickness of the plate, that is, at long wavelengths. However, when the radius of the bend is of a similar size to the thickness, this condition is no longer satisfied and the wave propagated asymptotically approaches that of a transverse shear wave. This gives an upper limit on the phase velocity of a bending wave, which is equal to that of the transverse shear wave in the material. The ratio between the shear and bending contributions to transverse displacement is approximately:

$$\frac{contribution_{shear}}{contribution_{bending}} \approx \left(\frac{h}{\lambda_{bending}}\right)^2 \qquad (1.12)$$

where h = the thickness of the plate
and $\lambda_{bending}$ = the wavelength of the bending wave

From Equation 1.12 the contribution of the shear contribution is less than 3% when $\lambda_{bending} > 6h$. So there is an upper frequency limit. Figure 1.5 shows a comparison of the velocity of the different kinds of waves as a function of frequency for an aluminum plate that is 6 cm thick.

From Figure 1.5 we can see that both the quasi-longitudinal and transverse shear waves in bars and plates are *non-dispersive*. That is, their phase velocity is independent of frequency. This means that they also have a frequency independent group velocity and therefore preserve the waveshape of a sound wave containing many frequency components.

However, bending waves are *dispersive*. That is, their phase velocity is dependent on frequency. This means that they also have a frequency dependent group velocity and therefore *do not* preserve the waveshape of a sound wave containing many frequency components.

FIGURE 1.5

Phase velocity versus frequency for different types of wave propagation.

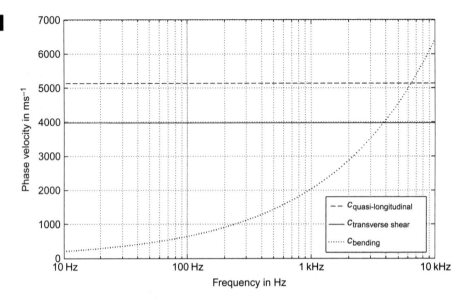

1.1.7 The wavelength and frequency of sound waves

So far we have only considered the propagation of a single disturbance through the golf ball and spring model and we have seen that, in this case, the disturbance travels at a constant velocity that is dependent only on the characteristics of the medium. Thus any other type of disturbance, such as a periodic one, would also travel at a constant speed. Figure 1.6 shows the golf ball and spring model being excited by a pin attached to a wheel rotating at a constant rate. This will produce a pressure variation as a function of time that is proportional to the sine of the angle of rotation. This is known as a "sinusoidal excitation" and produces a sine wave. It is important because it represents the simplest form of periodic excitation. As we shall see later in the chapter, more complicated waveforms can always be described in terms of these simpler sine waves.

Sine waves have three parameters: their amplitude, rate of rotation or frequency, and their starting position or phase. The frequency used to be expressed in units of cycles per second, reflecting the origin of the waveform, but it is now measured in the equivalent units of hertz (Hz). This type of excitation generates a traveling sine wave disturbance down the model, where the compressions and rarefactions are periodic. Because the sine wave propagates at a given velocity, a length can be assigned to the

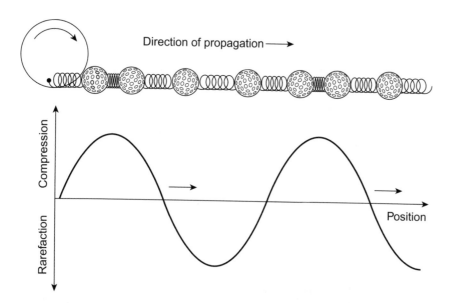

FIGURE 1.6

Golf ball and spring model of a sine wave propagating in a material.

distance between the repeats of the compressions or rarefactions, as shown in Figure 1.7. Furthermore, because the velocity is constant, the distance between these repeats will be inversely proportional to the rate of variation of the sine wave, known as its "frequency." The distance between the repeats is an important acoustical quantity and is called the wavelength (λ). Because the wavelength and frequency are linked together by the velocity, it is possible to calculate one of the quantities given the knowledge of two others using the following equation:

$$c = f\lambda \tag{1.13}$$

where c = the velocity of sound in the medium (in ms^{-1})

f = the frequency of the sound (in Hz, 1 Hz = 1 cycle per second)

and λ = the wavelength of the sound in the medium (in m)

This equation can be used to calculate the frequency given the wavelength, wavelength given the frequency, and even the speed of sound in the medium, given the frequency and wavelength. It is applicable to both longitudinal and transverse waves.

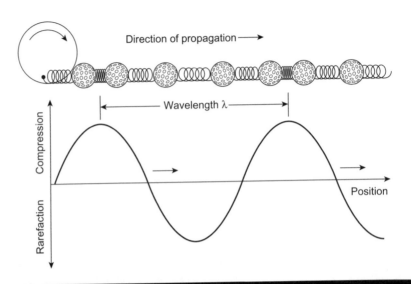

The wavelength of propagating sine wave.

EXAMPLE 1.4

Calculate the wavelength of sound being propagated in air at 20°C, at 20 Hz and 20 kHz.

For air the speed of sound at 20°C is 344 ms^{-1} (see Example 1.2); thus the wavelengths at the two frequencies are given by:

$$\lambda = \frac{c}{f}$$

which gives:

$$\lambda = \frac{344}{20} = 17.2 \, \text{m for 20 Hz}$$

and

$$\lambda = \frac{344}{20 \times 10^3} = 1.72 \, \text{cm for 20 kHz}$$

These two frequencies correspond to the extremes of the audio frequency range so one can see that the range of wavelength sizes involved is very large!

In acoustics the wavelength is often used as the "ruler" for measuring length, rather than meters, feet or furlongs, because many of the effects of real objects, such as rooms or obstacles, on sound waves are dependent on the wavelength.

A *radian* is a measure of angle, equal to 180/π degrees, which is about 57.3°.

EXAMPLE 1.5

Calculate the frequency of sound with a wavelength of 34 cm in air at 20°C.
The frequency is given by:

$$f = \frac{c}{\lambda} = \frac{344}{0.34} = 1012\,\text{Hz}$$

1.1.8 The wavenumber of sound waves

Sometimes it is also useful to use a quantity called the wavenumber that describes how much the phase of the wave changes in a given distance. Again it is a form of "ruler" that is in units of radians per meter (rad m^{-1}); most physical objects need to have a phase shift of at least a radian across their physical size before they will really interact with a sound wave.

The wavenumber of a sound wave is given by:

$$k = \frac{\omega}{c} \tag{1.14}$$

where k = the wavenumber of the wave
ω = the angular frequency of the wave
and c = the phase velocity of the wave

This is especially useful as it encapsulates any dispersive effects, and changes in wave velocity with frequency, and can be used directly to calculate various aspects of wave propagation in, and acoustic radiation from, for example, plates.

As an example, the equations for wavenumber for transverse shear and bending waves in a plate are:

$$k_{\text{transverse shear}} = \omega \left(\frac{\rho}{G}\right)^{\frac{1}{2}} \tag{1.15}$$

$$k_{\text{bending}} = \sqrt{\omega} \left(\frac{m}{D}\right)^{\frac{1}{4}} \tag{1.16}$$

Two points are of note from Equations 1.15 and 1.16. The first is that the wavenumber of lateral shear waves is proportional to frequency, just as one would expect from a non-dispersive wave. However, for a bending wave the wavenumber rises only as the square root of frequency. In both cases the coefficient is inversely proportional to the phase velocity so a low slope implies a high phase velocity. It is often helpful to plot wavenumber versus angular frequency in a dispersion diagram. Figure 1.8 shows the dispersion curves for different wave types in an aluminum plate along with the dispersion curve for sound in air.

FIGURE 1.8

Wavenumber versus frequency for different types of wave propagation.

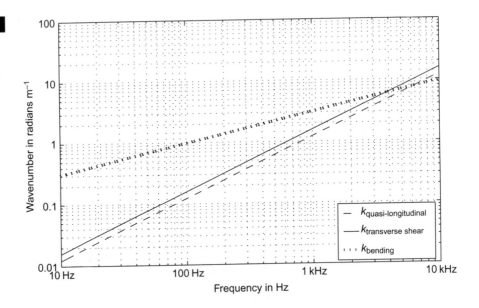

1.1.9 The relationship between pressure, velocity and impedance in sound waves

Another aspect of a propagating wave to consider is the movement of the molecules in the medium which is carrying it. The wave can be seen as a series of compressions and rarefactions which are traveling through the medium. The force required to effect the displacement—a combination of both compression and acceleration—forms the pressure component of the wave.

In order for the compressions and rarefactions to occur, the molecules must move closer together or further apart. Movement implies velocity, so there must be a velocity component which is associated with the displacement component of the sound wave. This behavior can be observed in the golf ball model for sound propagation described earlier. In order for the golf balls to get closer for compression they have some velocity to move toward each other. This velocity will become zero when the compression has reached its peak, because at this point the molecules will be stationary.

Then the golf balls will start moving with a velocity away from each other in order to get to the rarefacted state. Again the velocity toward the golf balls will become zero at the trough of the rarefaction. The velocity does not switch instantly from one direction to another, due to the inertia of the molecules involved; instead it accelerates smoothly from a stationary to a moving state and back again. The velocity component reaches its peak between the compressions and rarefactions, and for a sine wave displacement component the associated velocity component is a cosine.

Figure 1.9 shows a sine wave propagating in the golf ball model with plots of the associated components. The force required to accelerate the molecules forms the pressure component of the wave. This is associated with the velocity component of the propagating wave and therefore is in phase with it. That is, if the velocity component is a cosine then the pressure component will also be a cosine. Thus, a sound wave has both pressure and velocity components that travel through the medium at the same speed.

Air pressure acts in all directions at the same time and therefore for sound it can be considered to be a scalar quantity without direction; we can therefore talk about pressure at a point and not as a force acting in a particular direction. Velocity on the other hand must have direction; things move from one position to another. It is the velocity component which gives a sound wave its direction.

The velocity and pressure components of a sound wave are also related to each other in terms of the density and springiness of the propagating medium. A propagating medium which has a low density and weak springs

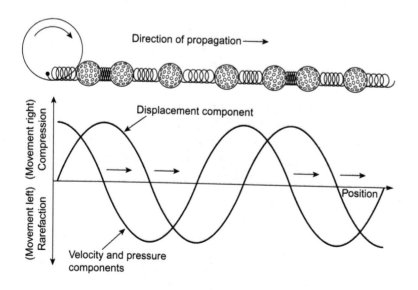

FIGURE 1.9

Pressure, velocity and displacement components of a sine wave propagating in a material.

would have a higher amplitude in its velocity component for a given pressure amplitude compared with a medium which is denser and has stronger springs. For a wave some distance away from the source and any boundaries, this relationship can be expressed using the following equation:

$$\frac{\text{Pressure component amplitude}}{\text{Velocity component amplitude}} = \text{Constant} = Z_{\text{acoustic}} = \frac{p}{U} \quad (1.17)$$

where p = the pressure component amplitude
U = the volume velocity component amplitude
and Z_{acoustic} = the acoustic impedance

This constant is known as the "acoustic impedance" and is analogous to the resistance (or impedance) of an electrical circuit.

The amplitude of the pressure component is a function of the springiness (Young's modulus) of the material and the volume velocity component is a function of the density. This allows us to calculate the acoustic impedance using the Young's modulus and density with the following equation:

$$Z_{\text{acoustic}} = \sqrt{\rho E}$$

However, the velocity of sound in the medium, usually referred to as c, is also dependent on the Young's modulus and density so the above equation is often expressed as:

> m^{-2} means per square meter.

$$Z_{\text{acoustic}} = \sqrt{\rho E} = \sqrt{\rho^2 \left(\frac{E}{\rho}\right)} = \rho c = 1.21 \times 344 \quad (1.17a)$$
$$= 416 \, \text{kg m}^{-2} \, \text{s}^{-1} \text{ in air at } 20°C$$

Note that the acoustic impedance for a wave in free space is also dependent only on the characteristics of the propagating medium.

However, if the wave is traveling down a tube whose dimensions are smaller than a wavelength, then the impedance predicted by Equation 1.17 is modified by the tube's area to give:

$$Z_{\text{acoustic tube}} = \frac{\rho c}{S_{\text{tube}}} \quad (1.17b)$$

where S_{tube} = the tube area

This means that for bounded waves the impedance depends on the surface area within the bounding structure and so will change as the area changes. As we shall see later, changes in impedance can cause reflections. This effect is important in the design and function of many musical instruments as discussed in Chapter 4.

1.2 SOUND INTENSITY, POWER AND PRESSURE LEVEL

The energy of a sound wave is a measure of the amount of sound present. However, in general we are more interested in the rate of energy transfer rather than the total energy transferred. Therefore we are interested in the amount of energy transferred per unit of time, that is, the number of joules per second (watts) that propagate. Sound is also a three-dimensional quantity and so a sound wave will occupy space. Because of this it is helpful to characterize the rate of energy transfer with respect to area, that is, in terms of watts per unit area. This gives a quantity known as the "sound intensity," which is a measure of the power density of a sound wave propagating in a particular direction, as shown in Figure 1.10.

1.2.1 Sound intensity level

The sound intensity represents the flow of energy through a unit area. In other words it represents the watts per unit area from a sound source and this means that it can be related to the sound power level by dividing it by the radiating area of the sound source. As discussed earlier, sound intensity has a direction which is perpendicular to the area that the energy is flowing through; see Figure 1.10. The sound intensity of real sound sources

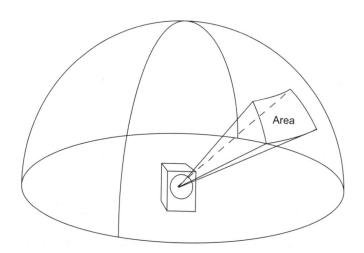

FIGURE 1.10

Sound intensity.

can vary over a range which is greater than one million-million (10^{12}) to one. Because of this, and because of the way we perceive the loudness of a sound, the sound intensity level is usually expressed on a logarithmic scale. This scale is based on the ratio of the actual power density to a reference intensity of 1 picowatt per square meter ($10^{-12}\,\text{W m}^{-2}$). Thus the sound intensity level (SIL) is defined as:

> The symbol for power in watts is W.

$$SIL = 10\log_{10}\left(\frac{I_{actual}}{I_{ref}}\right)$$

(1.18)

where I_{actual} = the actual sound power density level (in W m^{-2})

and I_{ref} = the reference sound power density level (10^{-12} W m^{-2})

The factor of 10 arises because this makes the result a number in which an integer change is approximately equal to the smallest change that can be perceived by the human ear. A factor of 10 change in the power density ratio is called the bel; in Equation 1.18 this would result in a change of 10 in the outcome. The integer unit that results from Equation 1.18 is therefore called the decibel (dB). It represents a $\sqrt[10]{10}$ change in the power density ratio, that is, a ratio of about 1.26.

EXAMPLE 1.6

A loudspeaker with an effective diameter of 25 cm radiates 20 mW. What is the sound intensity level at the loudspeaker?

Sound intensity is the power per unit area. Firstly, we must work out the radiating area of the loudspeaker which is:

$$A_{speaker} = \pi r^2 = \pi\left(\frac{0.25\ \text{m}}{2}\right)^2 = 0.049\ \text{m}^2$$

Then we can work out the sound intensity as:

$$I = \left(\frac{W}{A_{speaker}}\right) = \left(\frac{20 \times 10^{-3}\ \text{W}}{0.049\ \text{m}^2}\right) = 0.41\ \text{W m}^{-2}$$

This result can be substituted into Equation 1.19 to give the sound intensity level, which is:

$$SIL = 10\log_{10}\left(\frac{I_{actual}}{I_{ref}}\right) = 10\log_{10}\left(\frac{0.41\ \text{W m}^{-2}}{10^{-12}\ \text{W m}^{-2}}\right) = 116\text{dB}$$

1.2.2 Sound power level

The sound power level is a measure of the total power radiated in all directions by a source of sound and it is often given the abbreviation SWL, or sometimes PWL. The sound power level is also expressed as the logarithm of a ratio in decibels and can be calculated from the ratio of the actual power level to a reference level of 1 picowatt (10^{-12} W) as follows:

$$SWL = 10\log_{10}\left(\frac{W_{actual}}{W_{ref}}\right) \qquad (1.19)$$

where W_{actual} = the actual sound power level (in watts)
and W_{ref} = the reference sound power level (10^{-12} W)

The sound power level is useful for comparing the total acoustic power radiated by objects, for example ones which generate unwanted noises. It has the advantage of not depending on the acoustic context, as we shall see in Chapter 6. Note that, unlike the sound intensity, the sound power has no particular direction.

1.2.3 Sound pressure level

The sound intensity is one way of measuring and describing the amplitude of a sound wave at a particular point. However, although it is useful theoretically, and can be measured, it is not the usual quantity used when describing the amplitude of a sound. Other measures could be either the amplitude of the pressure, or the associated velocity component of the

EXAMPLE 1.7

Calculate the SWL for a source which radiates a total of 1 watt.
Substituting into Equation 1.19 gives:

$$SWL = 10\log_{10}\left(\frac{W_{actual}}{W_{ref}}\right) = 10\log_{10}\left(\frac{1 \text{ watt}}{1 \times 10^{-12} \text{ watts}}\right)$$
$$= 10\log_{10}\left(1 \times 10^{12}\right) = 120\,dB$$

A sound pressure level of one watt would be a very loud sound, if you were to receive all the power. However, in most situations the listener would only be subjected to a small proportion of this power.

sound wave. Because human ears are sensitive to pressure, which will be described in Chapter 2, and, because it is easier to measure, pressure is used as a measure of the amplitude of the sound wave. This gives a quantity which is known as the "sound pressure," which is the root mean square (rms) pressure of a sound wave at a particular point. The sound pressure for real sound sources can vary from less than 20 micropascals (20 μPa or 20×10^{-6} Pa) to greater than 20 pascals (20 Pa). Note that 1 Pa equals a pressure of 1 newton per square meter ($1\,\mathrm{N\,m^{-2}}$).

These two pressures broadly correspond to the threshold of hearing (20 μPa) and the threshold of pain (20 Pa) for a human being, at a frequency of 1 kHz, respectively. Thus real sounds can vary over a range of pressure amplitudes which is greater than a million to one. Because of this, and because of the way we perceive sound, the sound pressure level is also usually expressed on a logarithmic scale. This scale is based on the ratio of the actual sound pressure to the notional threshold of hearing at 1 kHz of 20 μPa. Thus the sound pressure level (SPL) is defined as:

> The pascal (Pa) is a measure of pressure; 1 pascal (1 Pa) is equal to 1 newton per square meter ($1\,\mathrm{Nm^{-2}}$).

$$SPL = 20\log_{10}\left(\frac{p_{\text{actual}}}{p_{\text{ref}}}\right) \tag{1.20}$$

where p_{actual} = the actual pressure level (in Pa)
and p_{ref} = the reference pressure level (20 μPa)

The multiplier of 20 has a twofold purpose. The first is to make the result a number in which an integer change is approximately equal to the smallest change that can be perceived by the human ear. The second is to provide some equivalence to intensity measures of sound level as follows.

The intensity of an acoustic wave is given by the product of the volume velocity and pressure amplitude:

> Volume velocity is a measure of the velocity component of the wave. It is measured in units of liters per second ($\mathrm{ls^{-1}}$).

$$I_{\text{acoustic}} = Up \tag{1.21}$$

where p = the pressure component amplitude
and U = the volume velocity component amplitude

However, the pressure and velocity component amplitudes are linked via the acoustic impedance (Equation 1.17) so the intensity can be calculated in terms of just the sound pressure and acoustic impedance by:

$$I_{\text{acoustic}} = Up = \left(\frac{p}{Z_{\text{acoustic}}}\right)p = \frac{p^2}{Z_{\text{acoustic}}} \tag{1.22}$$

Therefore the sound intensity level could be calculated using the pressure component amplitude and the acoustic impedance using:

$$SIL = 10\log_{10}\left(\frac{I_{\text{acoustic}}}{I_{\text{ref}}}\right) = 10\log_{10}\left(\frac{p^2}{\dfrac{Z_{\text{acoustic}}}{I_{\text{ref}}}}\right) = 10\log_{10}\left(\frac{p^2}{Z_{\text{acoustic}}\,I_{\text{ref}}}\right) \quad (1.23)$$

This shows that the sound intensity is proportional to the square of the pressure, in the same way that electrical power is proportional to the square of the voltage. The operation of squaring the pressure can be converted into multiplication of the logarithm by a factor of two, which gives:

$$SIL = 20\log_{10}\left(\frac{p}{\sqrt{Z_{\text{acoustic}}\,I_{\text{ref}}}}\right) \quad (1.24)$$

This equation is similar to Equation 1.20 except that the reference level is expressed differently. In fact, this equation shows that if the pressure reference level was calculated as

$$p_{\text{ref}} = \sqrt{Z_{\text{acoustic}}\,I_{\text{ref}}} = \sqrt{416 \times 10^{-12}} = 20.4 \times 10^{-6} \ (\text{Pa}) \quad (1.25)$$

then the two ratios would be equivalent. The actual pressure reference level of 20 μPa is close enough to say that the two measures of sound level are broadly equivalent: SIL ≈ SPL for a single sound wave a reasonable distance from the source and any boundaries. They can be equivalent because the sound pressure level is calculated at a single point and sound intensity is the power density from a sound source at the measurement point.

However, whereas the sound intensity level is the power density from a sound source at the measurement point, the sound pressure level is the sum of the sound pressure waves at the measurement point. If there is only a single pressure wave from the sound source at the measurement point, that is, there are no extra pressure waves due to reflections, the sound pressure level and the sound intensity level are approximately equivalent: SIL ≈ SPL. This will be the case for sound waves in the atmosphere well away from any reflecting surfaces. It will not be true when there are additional pressure waves due to reflections, as might arise in any room or if the acoustic impedance changes. However, changes in level for both SIL and SPL will be the equivalent because if the sound intensity increases then the sound pressure at a point will also increase by the same proportion.

This will be true so long as nothing alters the number and proportions of the sound pressure waves arriving at the point at which the sound pressure is measured. Thus, a 10 dB change in SIL will result in a 10 dB change in SPL.

These different means of describing and measuring sound amplitudes can be confusing and one must be careful to ascertain which one is being used in a given context. In general, a reference to sound level implies that the SPL is being used because the pressure component can be measured easily and corresponds most closely to what we hear.

Let us calculate the SPLs for a variety of pressure levels.

EXAMPLE 1.8

Calculate the SPL for sound waves with rms pressure amplitudes of 1 Pa, 2 Pa and 2 μPa.

Substituting the above values of pressure into Equation 1.20 gives:

$$SPL_{1Pa} = 20\log_{10}\left(\frac{p_{actual}}{p_{ref}}\right) = 20\log_{10}\left(\frac{1\,Pa}{20\,\mu Pa}\right)$$
$$= 20 \times \log_{10}(5 \times 10^4) = 94\,dB$$

1 Pa is often used as a standard level for specifying microphone sensitivity and, as the above calculation shows, represents a loud sound.

$$SPL_{2Pa} = 20\log_{10}\left(\frac{p_{actual}}{p_{ref}}\right) = 20\log_{10}\left(\frac{2\,Pa}{20\,\mu Pa}\right)$$
$$= 20 \times \log_{10}(1 \times 10^5) = 100\,dB$$

Doubling the pressure level results in a 6 dB increase in sound pressure level, and a tenfold increase in pressure level results in a 20 dB increase in SPL.

$$SPL_{2Pa} = 20\log_{10}\left(\frac{p_{actual}}{p_{ref}}\right) = 20\log_{10}\left(\frac{2\,Pa}{20\,\mu Pa}\right)$$
$$= 20 \times \log_{10}(1 \times 10^{-1}) = -20\,dB$$

If the actual level is less than the reference level then the result is a negative SPL. The decibel concept can also be applied to both sound intensity and the sound power of a source.

1.3 ADDING SOUNDS TOGETHER

So far we have only considered the amplitude of single sources of sound. However, in most practical situations more than one source of sound is present; these may result from other musical instruments or reflections from surfaces in a room. There are two different situations which must be considered when adding sound levels together.

- *Correlated sound sources*: In this situation the sound comes from several sources which are related. In order for this to happen the extra sources must be derived from a single source. This can happen in two ways. Firstly, the different sources may be related by a simple reflection, such as might arise from a nearby surface. If the delay is short then the delayed sound will be similar to the original and so it will be correlated with the primary sound source. Secondly, the sound may be derived from a common electrical source, such as a recording or a microphone, and then may be reproduced using several loudspeakers. Because the speakers are being fed the same signal, but are spatially disparate, they act like several related sources and so are correlated. Figure 1.11 shows two different situations.

- *Uncorrelated sound sources*: In this situation the sound comes from several sources which are unrelated. For example, it may come from two different instruments, or from the same source but with a considerable delay due to reflections. In the first case the different instruments will be generating different waveforms and at different frequencies. Even when the same instruments play in unison, these differences will occur. In the second case, although the additional sound source comes from the primary one and so could be expected to be related to it, the delay will mean that the waveform from the additional source will no longer be the same. This is because in the intervening time, due to the delay, the primary source of the sound will

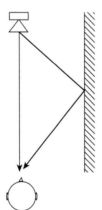
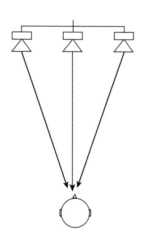

Correlation due to reflection   Correlation due to multiple sources

FIGURE 1.11

Addition of correlated sources.

have changed in pitch, amplitude and waveshape. Because the delayed wave is different it appears to be unrelated to the original source and so is uncorrelated with it. Figure 1.12 shows two possibilities.

1.3.1 The level when correlated sounds add

Sound levels add together differently depending on whether they are correlated or uncorrelated. When the sources are correlated the pressure waves from the correlated sources simply add, as shown in Equation 1.26.

$$P_{\text{total correlated}}\ (t) = P_1(t) + P_2(t) + \cdots + P_N(t) \tag{1.26}$$

Note that the correlated waves are all at the same frequencies, and so always stay in the same time relationship to each other, which results in a composite pressure variation at the combination point, which is also a function of time.

Because a sound wave has periodicity, the pressure from the different sources may have a different sign and amplitude depending on their relative phase. For example, if two equal amplitude sounds arrive in phase then their pressures add and the result is a pressure amplitude at that point of twice the single source. However, if they are out of phase the result will be a pressure amplitude at that point of zero as the pressures of the two waves cancel. Figure 1.13 shows these two conditions.

Therefore, there will be the following consequences:

- If the correlation is due to multiple sources then the composite pressure will depend on the relative phases of the intersecting waves.

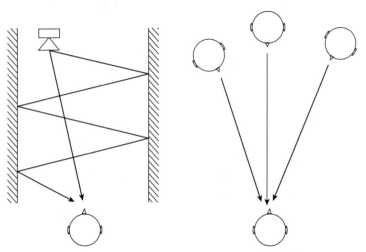

FIGURE 1.12

Addition of uncorrelated sources.

Uncorrelated reflection due to long delay   Uncorrelated multiple sources

This will depend on both the relative path lengths between the sources and their relative phases.

■ If the relative phases between the sources can be changed electronically, their effect may be different to a phase shift caused by a propagation delay. For example, a common situation is when one source's signal is inverted with respect to the other, such as might happen if one stereo speaker is wired the wrong way round compared to the other. In this case, if the combination point was equidistant from both sources, and the sources were of the same level, the two sources would cancel each other out and give a pressure amplitude at that point of zero. This cancellation would occur at all frequencies because the effect phase shift due to inverting the signal is frequency independent.

■ However, if the phase shift was due to a delay, which could be caused by different path lengths or achieved by electronic means, then, as the frequency increases, the phase shift would increase in proportion to the delay, as shown in Equation 1.27.

$$Phase\ Shift_{degrees} = 360°(Delay_{seconds} \times Frequency_{Hz}) \qquad (1.27)$$

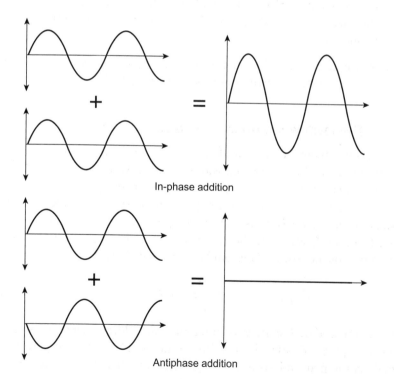

FIGURE 1.13

Addition of sine waves of different phases.

In-phase addition

Antiphase addition

As a consequence, whether the sources add in phase, antiphase, or some-where in between, they will be frequency dependent: that is, at the frequen-cies where the sources are in-phase they will add constructively whereas at the frequencies where they are in antiphase they will add destructively. At very low frequencies, unless the delay is huge, they will tend to add con-structively because the phase shift will be very small.

- If the effective phase shifts are due to a combination of delay and inversion then the result will also be due to a combination of the two effects. For example, for the case of two sources, if there is both delay and inversion, then, at very low frequencies, the sources will tend to cancel each other out, because the effect of the delay will be small. However, as we go up in frequency, the sources will add constructively when the delay causes the sources to be in-phase; this will be at a lower frequency than would be the case if the sources were not inverted with respect to each other.

- Changing the position at which the pressure variation is observed will change the time relationships between the waves being combined. Therefore, the composite result from correlated sources is dependent on position. It will also depend on frequency because the effective phase shift caused by a time delay is proportional to frequency.

As an example let us look at the effect of a single delayed reflection on the pressure amplitude at a given point (see Example 1.9).

1.3.2 The level when uncorrelated sounds add

On the other hand, if the sound waves are uncorrelated then they do not add algebraically, like correlated waves; instead we must add the powers of the individual waves together. As stated earlier, the *power* in a waveform is proportional to the square of the pressure levels, so in order to sum the powers of the waves we must square the pressure amplitudes before adding them together. If we want the result as a pressure then we must take the square root of the result. This can be expressed in the following equation:

$$P_{\text{total uncorrelated}} = \sqrt{\left(P_1^2 + P_2^2 + \cdots + P_N^2\right)} \tag{1.28}$$

Adding uncorrelated sources is different from adding correlated sources in several respects. Firstly, the resulting total is related to the power of the signals combined and so is not dependent on their relative phases. This

For coherent addition of sources, having to do everything using sines and cosines is very awkward and inconvenient. A better way is to represent the acoustic signals as complex numbers. Complex numbers are pairs of numbers based on the following form: $a + jb$ where j, or i represent the $\sqrt{-1}$. This is called an imaginary number. Consequently a is called the *real* part of the pair and b is called the *imaginary* part. To see how this works look at the figure below:

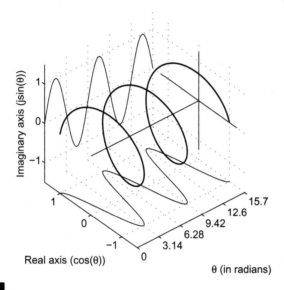

FIGURE 1.14

Here one can see that the combination of the sine and cosine in a two-dimensional complex space forms a spiral which is called a complex exponential $|a + jb|e^{arg(a+jb)} = re^{\theta} = r\cos(\theta) + jr\sin(\theta)$ where r is the radius, or modulus of the spiral and θ is the phase or rotation of the spiral. Thus $a = r\cos(\theta)$ and $b = r\sin(\theta)$. The modulus of $a + jb = |a + jb| = r = \sqrt{a^2 + b^2}$ and the argument of $a + jb = arg(a + jb) = \theta = \tan^{-1}\left(\dfrac{b}{a}\right)$

Using these simple relationships it is possible to have the following arithmetic rules:

Addition/subtraction $(a + jb) \pm (c + jd) = (a \pm c) + j(b \pm d)$, with the $\pm$ applied respectively.

Multiplication $(a + jb)(c + jd) = (ac - bd) + j(ad + bc)$. Note: $j \times j = -1$

Division $\dfrac{(a + jb)}{(c + jd)} = \dfrac{(a + jb)(c - jd)}{(c + jd)(c - jd)} = \dfrac{(ac + bd) + j(bc - ad)}{(c^2 + d^2)}$

Note: inverting the sign of the imaginary part makes the *complex conjugate* of the complex number.

Adding and subtracting the complex representation of acoustic sources, of the same frequency, naturally handles their phase differences. Multiplying the complex representation of an acoustic source by a complex number can determine the effect of a filter or propagation delay. For more details see most engineering mathematics textbooks.

EXAMPLE 1.9

The sound at a particular point consists of a main loudspeaker signal and a reflection at the same amplitude that has been delayed by 1 millisecond. What is the pressure amplitude at this point at 250 Hz, 500 Hz and 1 kHz?

The equation for pressure at a point due to a single frequency is given by the equation:

$$P_{\text{at a point}} = P_{\text{sound amplitude}} \; \sin(2\pi ft) \text{ or } P_{\text{sound amplitude}} \; \sin(360°ft)$$
$$\text{where } f = \text{ the frequency (in Hz)}$$
$$\text{and } t = \text{ the time (in s)}$$

Note the multiplier of 2π, or 360°, within the sine function is required to express accurately the position of the wave within the cycle. Because a complete rotation occurs every cycle, one cycle corresponds to a rotation of 360 degrees, or, more usually, 2π radians. This representation of frequency is called angular frequency (1 Hz [cycle per second] = 2π radians per second).

The effect of the delay on the difference in path lengths alters the time of arrival of one of the waves, and so the pressure at a point due to a single frequency delayed by some time, τ, is given by the equation:

$$P_{\text{at a point}} = P_{\text{sound amplitude}} \; \sin(2\pi f(t + \tau))$$
$$\text{or } P_{\text{sound amplitude}} \; \sin(360°f(t + \tau))$$
$$\text{where } \tau = \text{ the delay (in s)}$$

Add the delayed and undelayed sine waves together to give:

$$P_{\text{total}} = P_{\text{delayed}} \; \sin(360°f(t + \tau)) + P_{\text{undelayed}} \; \sin(360°ft)$$

Assuming that the delayed and undelayed signals are of the same amplitude this can be reexpressed as:

$$P_{\text{total}} = 2P \; \cos\left(360°f\left(\frac{\tau}{2}\right)\right)\sin\left(360°f\left(t + \frac{\tau}{2}\right)\right)$$

The cosine term in this equation is determined by the delay and frequency, and the sine term represents the original wave slightly delayed. Thus we can express the combined pressure amplitude of the two waves as:

$$P_{\text{total}} = 2P \; \cos\left(360°f\left(\frac{\tau}{2}\right)\right)$$

Using the above equation we can calculate the effect of the delay on the pressure amplitude at the three different frequencies as:

$$P_{\text{total 250Hz}} = 2P \, \cos\left(360°f\left(\frac{\tau}{2}\right)\right)$$

$$= 2P \, \cos\left(360° \times 250\text{Hz} \times \left(\frac{1 \times 10^{-3}\,\text{s}}{2}\right)\right) = 1.41P$$

$$P_{\text{total 500Hz}} = 2P \, \cos\left(360°f\left(\frac{\tau}{2}\right)\right)$$

$$= 2P \, \cos\left(360° \times 500\text{Hz} \times \left(\frac{1 \times 10^{-3}\,\text{s}}{2}\right)\right) = 0$$

$$P_{\text{total 1kHz}} = 2P \, \cos\left(360°f\left(\frac{\tau}{2}\right)\right)$$

$$= 2P \, \cos\left(360° \times 1\text{kHz} \times \left(\frac{1 \times 10^{-3}\,\text{s}}{2}\right)\right) = 2P$$

These calculations show that the summation of correlated sources can be strongly frequency dependent and can vary between zero and twice the wave pressure amplitude.

means that the result of combining uncorrelated sources is always an increase in level. The second difference is that the level increase is lower because powers rather than pressures are being added. Recall that the maximum increase for two equal correlated sources was a factor of two increases in pressure amplitude. However, for uncorrelated sources the powers of the sources are added and, as the power is proportional to the square of the pressure, this means that the maximum amplitude increase for two uncorrelated sources is only $\sqrt{2}$.

However, the addition of uncorrelated components always results in an increase in level without any of the cancellation effects that correlated sources suffer. Because of the lack of cancellation effects, the spatial variation in the sum of uncorrelated sources is usually much less than that of correlated ones, as the result only depends on the amplitude of the sources. As an example let us consider the effect of adding together several uncorrelated sources of the same amplitude.

How does the addition of sources affect the sound pressure level (SPL), the sound power level (SWL), and the sound intensity level (SIL)? For the SWL and SIL, because we are adding powers, the results will be the same whether the sources are correlated or not. However, for SPL, there will be a difference between the correlated and uncorrelated results. The main difficulty that arises when these measures are used to calculate the effect of

EXAMPLE 1.10

Calculate the increase in signal level when two vocalists sing together at the same level and when a choir of *N* vocalists sing together, also at the same level.

The total level from combining several uncorrelated sources together is given by Equation 1.28 as:

$$P_{\text{total uncorrelated}} = \sqrt{(P_1^2 + P_2^2 + \cdots + P_N^2)}$$

For *N* sources of the same amplitude this can be simplified to:

$$P_{N \text{ uncorrelated}} = \sqrt{(P^2 + P^2 + \cdots + P^2)} = \sqrt{NP^2} = P\sqrt{N}$$

Thus the increase in level, for uncorrelated sources of equal amplitude, is proportional to the square root of the number of sources. In the case of just two sources this gives:

$$P_{\text{two uncorrelated}} = P\sqrt{N} = P\sqrt{2} = 1.41P$$

combining sound sources is confusion over the correct use of decibels during the calculation.

1.3.3 Adding decibels together

Decibels are power ratios expressed on a logarithmic scale and this means that *adding decibels together is not the same as adding the sources' amplitudes together*. This is because adding logarithms together is equivalent to the logarithm of the product of the quantities. Clearly this is not the same as a simple summation!

When decibel values are to be added together, it is important to convert them back to their original ratios before carrying out the addition. If a decibel result of the summation is required, then the sum must be converted back to decibels after the summation has taken place. To make this clearer let us look at Example 1.11.

There are some areas of sound level calculation where the fact that the addition of decibels represents multiplication is an advantage. In these situations the result can be expressed as a multiplication, and so can be expressed as a summation of decibel values. In other words, decibels can be added when the underlying sound level calculation is a multiplication. In this context the decibel representation of sound level is very useful, as

EXAMPLE 1.11

Calculate the increase in signal level when two vocalists sing together, one at 69 dB and the other at 71 dB SPL.

From Equation 1.20 the SPL of a single source is:

$$SPL = 20\log_{10}\left(\frac{p_{actual}}{p_{ref}}\right)$$

For multiple, uncorrelated, sources this will become:

$$SPL = 20\log_{10}\left(\frac{\sqrt{\left(P_1^2 + P_2^2 + \cdots + P_N^2\right)}}{p_{ref}}\right)$$

$$= 10\log_{10}\left(\frac{P_1^2 + P_2^2 + \cdots + P_N^2}{p_{ref}^2}\right)$$

(1.29)

We must substitute the pressure squared values that the singers' SPLs represent. These can be obtained with the following equation:

$$P^2 = 10^{\left(\frac{SPL}{10}\right)}p_{ref}^2$$

where $p_{ref}^2 = 4 \times 10^{-10}\, N^2\, m^{-4}$

Substituting in our two SPL values gives:

$$P^2_{69dB} = 10^{\left(\frac{69}{10}\right)} \times 4 \times 10^{-10}\, N^2 m^{-4} = 3.18 \times 10^{-3} N^2 m^{-4}$$

and

$$P^2_{71dB} = 10^{\left(\frac{71}{10}\right)} \times 4 \times 10^{-10}\, N^2 m^{-4} = 5.04 \times 10^{-3} N^2 m^{-4}$$

Substituting these two values into Equation 1.29 gives the result as:

$$SPL = 10\log_{10}\left(\frac{P^2_{69dB} + P^2_{71dB}}{p_{ref^2}}\right)$$

$$= 10\log_{10}\left(\frac{3.18 \times 10^{-3} + 5.04 \times 10^{-3}}{4 \times 10^{-10}}\right) = 73.1 dB$$

Note that the combined sound level is only about 2 dB more than the louder of the two sounds and *not* 69 dB greater, which is the result that would be obtained if the SPLs were added directly in decibels.

there are many acoustic situations in which the effect on the sound wave is multiplicative, for example the attenuation of sound through walls or their absorption by a surface. To make use of decibels in this context let us consider Example 1.12.

EXAMPLE 1.12

Calculate the increase in the sound pressure level (SPL) when two vocalists sing together at the same level and when a choir of N vocalists sing together, also at the same level.

The total level from combining several uncorrelated single sources is given by:

$$P_{N\ uncorrelated} = P\sqrt{N}$$

This can be expressed in terms of the SPL as:

$$SPL_{N\ uncorrelated} = 20\log_{10}\left(\frac{p\sqrt{N}}{p_{ref}}\right) = 20\log_{10}\left(\frac{P}{p_{ref}}\right) + 20\log_{10}(\sqrt{N})$$

In this equation the first term simply represents the SPL of a single source, and the addition of the decibel equivalent of the square root of the number of sources represents the increase in level due to the multiple sources. So this equation can be also expressed as:

$$SPL_{N\ uncorrelated} = SPL_{single\ source} + 10\log_{10}(N)$$

This equation will give the total SPL for N uncorrelated sources of equal level. For example, 10 sources will raise the SPL by 10 dB, since $10\log(10) = 10$.

In the case of two singers the above equation becomes:

$$SPL_{N\ uncorrelated} = SPL_{single\ source} + 10\log_{10}(2) = SPL_{single\ source} + 3\ dB$$

So the summation of two uncorrelated sources increases the sound pressure level by 3 dB.

1.4 THE INVERSE SQUARE LAW

So far we have only considered sound as a disturbance that propagates in one direction. However, in reality sound propagates in three dimensions. This means that the sound from a source does not travel on a constant beam; instead it spreads out as it travels away from the radiating source, as shown in Figure 1.10.

As the sound spreads out from a source it gets weaker. This is not due to it being absorbed but due to its energy being spread more thinly. Figure 1.15 gives a picture of what happens. Consider a half blown-up spherical balloon which is coated with honey to a certain thickness. If the balloon is blown up to double its radius, the surface area of the balloon would have increased fourfold.

As the amount of honey has not changed it must therefore have a quarter of the thickness that it had before. The sound intensity from a source behaves in an analogous fashion in that every time the distance from a sound source is doubled the intensity reduces by a factor of four, that is, there is an inverse square relationship between sound intensity and the distance from the sound source. The area of a sphere is given by the equation

$$A_{sphere} = 4\pi r^2$$

The sound intensity is defined as the power per unit area. Therefore the sound intensity as a function of distance from a sound source is given by:

$$I = \frac{W_{source}}{A_{sphere}} = \frac{W_{source}}{4\pi r^2} \qquad (1.30)$$

where I = the sound intensity (in W m^{-2})
W_{source} = the power of the source (in W)
and r = the distance from the source (in m)

Equation 1.30 shows that the sound intensity for a sound wave that spreads out in all directions from a source reduces as the square of the distance. Furthermore this reduction in intensity is purely a function of geometry and is not due to any physical absorption process. In practice, there are sources of absorption in air, for example impurities and water molecules, or smog and humidity. These sources of absorption have greater effect at high frequencies and, as a result, sound not only gets quieter but also gets duller as one moves away from a source. The amount of excess attenuation is dependent on the level of impurities and humidity, and is therefore variable.

From these results we can see that the sound at 1 m from a source is 11 dB less than the sound power level at the source. Note that the sound intensity

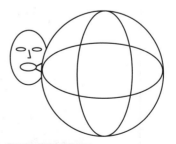

FIGURE 1.15 *The honey and balloon model of the inverse square law for sound.*

EXAMPLE 1.13

A loudspeaker radiates one hundred milliwatts (100 mW). What is the sound intensity level (SIL) at a distance of 1 m, 2 m and 4 m from the loudspeaker? How does this compare with the sound power level (SWL) at the loudspeaker?

The sound power level can be calculated from Equation 1.19 and is given by:

A milliwatt is one thousandth of a watt (10^{-3} watts).

$$SWL = 10\log_{10}\left(\frac{W_{actual}}{W_{ref}}\right) = 10\log_{10}\left(\frac{100\,\text{mW}}{1\times10^{-12}\,\text{W}}\right)$$

$$= 10\,\log_{10}\,(1\times10^{11}) = 110\,\text{dB}$$

The sound intensity at a given distance can be calculated using Equations 1.18 and 1.30 as:

$$SIL = 10\log_{10}\left(\frac{I_{actual}}{I_{ref}}\right) = 10\log_{10}\left(\frac{\dfrac{W_{source}}{4\pi r^2}}{I_{ref}}\right)$$

This can be simplified to give:

$$SIL = 10\log_{10}\left(\frac{W_{source}}{W_{ref}}\right) - 10\log_{10}(4\pi) - 10\log_{10}(r^2)$$

which can be simplified further to:

$$SIL = 10\log_{10}\left(\frac{W_{source}}{W_{ref}}\right) - 20\log_{10}(r) - 11\,\text{dB} \qquad (1.31)$$

This equation can then be used to calculate the intensity level at the three distances as:

$$SIL_{1m} = 10\log_{10}\left(\frac{100\,\text{mW}}{10^{-12}\,\text{W}}\right) - 20\log_{10}(1) - 11\,\text{dB}$$

$$= 110\,\text{dB} - 0\,\text{dB} - 11\,\text{dB} = 99\,\text{dB}$$

$$SIL_{2m} = 10\log_{10}\left(\frac{100\,\text{mW}}{10^{-12}\,\text{W}}\right) - 20\log_{10}(2) - 11\,\text{dB}$$

$$= 110\,\text{dB} - 6\,\text{dB} - 11\,\text{dB} = 93\,\text{dB}$$

$$SIL_{4m} = 10\log_{10}\left(\frac{100\,\text{mW}}{10^{-12}\,\text{W}}\right) - 20\log_{10}(4) - 11\,\text{dB}$$

$$= 110\,\text{dB} - 12\,\text{dB} - 11\,\text{dB} = 87\,\text{dB}$$

level at the source is, in theory, infinite because the area for a point source is zero. In practice, all real sources have a finite area so the intensity at the source is always finite. We can also see that the sound intensity level reduces by 6 dB every time we double the distance; this is a direct consequence of the inverse square law and is a convenient rule of thumb. The reduction in intensity of a source with respect to the logarithm of distance is plotted in Figure 1.16 and shows the 6 dB per doubling of distance relationship as a straight line except when one is very close to the source. In this situation the fact that the source is finite in extent renders Equation 1.30 invalid. As an approximate rule the nearfield region occurs within the radius described by the physical size of the source. In this region the sound field can vary wildly depending on the local variation of the vibration amplitudes of the source.

Equation 1.30 describes the reduction in sound intensity for a source which radiates in all directions. However, this is only possible when the sound source is well away from any surfaces that might reflect the propagating wave. Sound radiation in this type of propagating environment is often called the free field radiation, because there are no boundaries to restrict wave propagation.

1.4.1 The effect of boundaries

But how does a boundary affect Equation 1.30? Clearly many acoustic contexts involve the presence of boundaries near acoustic sources, or even all the way round them in the case of rooms, and some of these effects will be considered in Chapter 6. However, in many cases a sound source is placed on a boundary, such as a floor. In these situations the sound is radiating into a restricted space, as shown in Figure 1.17. However, despite the restriction

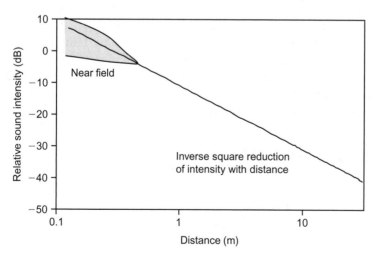

FIGURE 1.16

Sound intensity as a function of distance from the source.

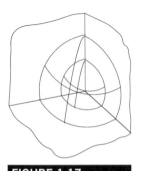

FIGURE 1.17

The inverse square law for sound at boundaries.

of the radiating space, the surface area of the sound wave still increases in proportion to the square of the distance, as shown in Figure 1.17. The effect of the boundaries is merely to concentrate the sound power of the source into a smaller range of angles. This concentration can be expressed as an extra multiplication factor in Equation 1.30. Therefore the equation can be rewritten as:

$$I_{\text{directive source}} = \frac{QW_{\text{source}}}{4\pi r^2} \qquad (1.32)$$

where $I_{\text{directive source}}$ = the sound intensity (in W m^{-2})

Q = the directivity of the source (compared to a sphere)

W_{source} = the power of the source (in W)

and r = the distance from the source (in m)

Equation 1.32 can be applied to any source of sound which directs its sound energy into a restricted solid angle which is less than a sphere. Obviously the presence of boundaries is one means of restriction, but other techniques can also achieve the same effect. For example, the horn structure of brass instruments results in the same effect. However, it is important to remember that the sound intensity of a source reduces in proportion to the square of the distance, irrespective of the directivity.

The effect of having the source on a boundary can also be calculated; as an example let us examine the effect of boundaries on the sound intensity from a loudspeaker.

From these calculations we can see that each boundary increases the sound intensity at a point by 3 dB, due to the increased directivity. Note that one cannot use the above equations on more than three boundaries because then the sound can no longer expand without bumping into something. We shall examine this subject in more detail in Chapter 6. However, it is possible to have directivities of greater than 8 using other techniques. For example, horn loudspeakers with a directivity of 50 are readily available as a standard product from public address loudspeaker manufacturers.

1.5 SOUND INTERACTIONS

So far we have only considered sound in isolation and we have seen that sound has velocity, frequency, and wavelength, and reduces in intensity in proportion to the square of the distance from the source. However, sound

---

EXAMPLE 1.14

A loudspeaker radiates 100 mW. Calculate the sound intensity level (SIL) at a distance of 2 m from the loudspeaker when it is mounted on 1, 2 and 3 mutually orthogonal boundaries.

The sound intensity at a given distance can be calculated using Equations 1.18 and 1.32 as:

$$SIL = 10\log_{10}\left(\frac{I_{actual}}{I_{ref}}\right) = 10\log_{10}\left(\frac{\dfrac{QW_{source}}{4\pi r^2}}{W_{ref}}\right)$$

which can be simplified to give:

$$SIL = 10\log_{10}\left(\frac{W_{source}}{W_{ref}}\right) + 10\log_{10}(Q) - 10\log_{10}(4\pi) - 20\log_{10}(r)$$

This is similar to Equation 1.32 except for the addition of the term for the directivity, Q. The presence of 1, 2 and 3 mutually orthogonal boundaries converts the sphere to a hemisphere, half hemisphere and quarter hemisphere, which corresponds to a Q of 2, 4 and 8, respectively. As the only difference between the results with the boundaries is the term in Q, the sound intensity level at 2 m can be calculated as:

$$
\begin{aligned}
SIL_{1\ boundary} &= SIL_{2m} + 10\log_{10}(Q) = 93\,dB + 10\log_{10}(2) \\
&= 93\,dB + 3\,dB = 96\,dB \\
SIL_{2boundaries} &= SIL_{2m} + 10\log_{10}(Q) = 93\,dB + 10\log_{10}(4) \\
&= 93\,dB + 6\,dB = 99\,dB \\
SIL_{3boundaries} &= SIL_{2m} + 10\log_{10}(Q) = 93\,dB + 10\log_{10}(8) \\
&= 93\,dB + 9\,dB = 102\,dB
\end{aligned}
$$

---

also interacts with physical objects and other sound waves, and is affected by changes in the propagating medium. The purpose of this section is to examine some of these interactions as an understanding of them is necessary in order to understand both how musical instruments work and how sound propagates in buildings.

1.5.1 Superposition

When sounds destructively interfere with each other they do not disappear. Instead they travel through each other. Similarly, when they constructively interfere they do not grow but simply pass through each other. This is because although the total pressure, or velocity component, may lie anywhere between zero and the sum of the individual pressures or velocities, the energy flow of the sound waves is still preserved and so the waves continue to propagate. Thus the pressure or velocity at a given point in space is simply the sum, or superposition, of the individual waves that are propagating through that point, as shown in Figure 1.18. This characteristic of sound waves is called linear superposition and is very useful as it allows us to describe, and therefore analyze, the sound wave at a given point in space as the linear sum of individual components.

1.5.2 Sound refraction

This is analogous to the refraction of light at the boundary of different materials. In the optical case refraction arises because the speed of light is different in different materials; for example it is slower in water than it is in air. In the acoustic case refraction arises for the same reasons, because the velocity of sound in air is dependent on the temperature, as shown in Equation 1.5.

Consider the situation shown in Figure 1.19 where there is a boundary between air at two different temperatures. When a sound wave approaches

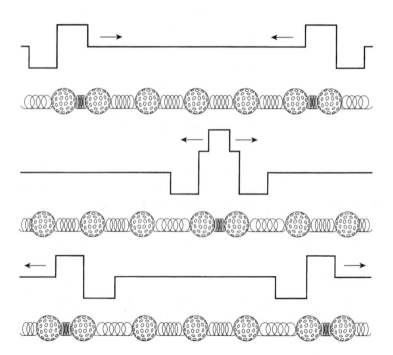

FIGURE 1.18

Superposition of a sound wave in the golf ball and spring model.

this boundary at an angle, then the direction of propagation will alter according to Snell's law, that is, using Equation 1.5:

$$\frac{\sin\theta_1}{\sin\theta_2} = \frac{c_{T1}}{c_{T2}} = \frac{20.1\sqrt{T_1}}{20.1\sqrt{T_2}} = \sqrt{\frac{T_1}{T_2}} \qquad (1.33)$$

where θ_1, θ_2 = the propagation angles in the two media

$\quad c_{T1}$, c_{T2} = the velocities of the sound waves in the two media

and T_1, T_2 = the absolute temperatures of the two media

Thus the change in direction is a function of the square root of the ratio of the absolute temperatures of the air on either side of the boundary. As the speed of sound increases with temperature one would expect to observe that when sound moves from colder to hotter air it would be refracted away from the normal direction, and that it would refract toward the normal when moving from hotter to colder air. This effect has some interesting consequences for outdoor sound propagation.

Normally the temperature of air reduces as a function of height and this results in the sound wave being bent upward as it moves away from a sound source, as shown in Figure 1.20. This means that listeners on the ground will experience a reduction in sound level as they move away from the sound, which reduces more quickly than the inverse square law would predict. This is a helpful effect for reducing the effect of environmental noise nuisance. However, if the temperature gradient increases with height then instead of being bent up the sound waves are bent down, as shown in Figure 1.21. This effect can often happen on summer evenings and results in a greater sound level at a given distance than predicted by the inverse square law. This behavior is often responsible for the pop concert effect where people living some distance away from the concert experience noise disturbance whereas people living nearer the concert do not experience the same level of noise.

Refraction can also occur at the boundaries between liquids at different temperatures, such as water, and in some cases the level of refraction can result in total internal reflection. This effect is sometimes used by submarines to hide from the sonar of other ships; it can also cause the sound to be ducted between two boundaries and in these cases sound can cover large distances. It is thought that

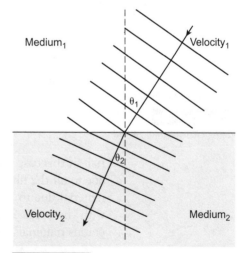

FIGURE 1.19 *Refraction of a sound wave (absolute temperature in medium₁ is T_1 and in medium₂ is T_2; velocity in medium₁ is y_{T1} and in medium₂ is y_{T2}).*

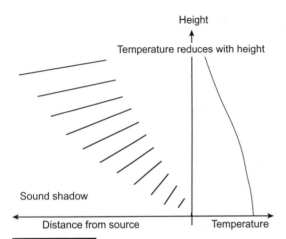

FIGURE 1.20 *Refraction of a sound wave due to a normal temperature gradient.*

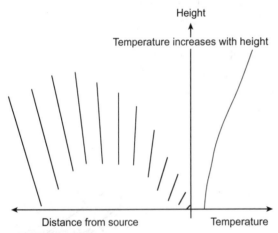

FIGURE 1.21 *Refraction of a sound wave due to an inverted temperature gradient.*

these mechanisms allow whales and dolphins to communicate over long distances in the ocean.

Wind can also cause refraction effects because the velocity of sound within the medium is unaffected by the velocity of the medium. The velocity of a sound wave in a moving medium, when viewed from a fixed point, is the sum of the two velocities, so that it is increased when the sound is moving with the wind and is reduced when it is moving against the wind. As the velocity of air is generally less at ground level compared with the velocity higher up (due to the effect of the friction of the ground), sound waves are bent upward or downward depending on their direction relative to the wind. The degree of direction change depends on the rate of change in wind velocity as a function of height; a faster rate of change results in a greater direction change. Figure 1.22 shows the effect of wind on sound propagation.

1.5.3 Sound absorption

Sound is absorbed when it interacts with any physical object. One reason is the fact that when a sound wave hits an object then that object will vibrate, unless it is infinitely rigid. This means that vibrational energy is transferred from the sound wave to the object that has been hit. Some of this energy will be absorbed because of the internal frictional losses in the material that the object is made of. Another form of energy loss occurs when the sound wave hits, or travels through, some porous material. In this case there is a very large surface area of interaction in the material, due to all the fibers and holes. There are frictional losses at the surface of any material due to the interaction of the velocity component of the sound wave with the surface. A larger surface area will have a higher loss, which is why porous materials such as cloth or rock-wool absorb sound waves strongly.

1.5.4 Sound reflection from hard boundaries

Sound is also reflected when it strikes objects and we have all experienced the effect as an echo when we are near a large hard object such as a cliff or large building. There are two main situations in which reflection can occur.

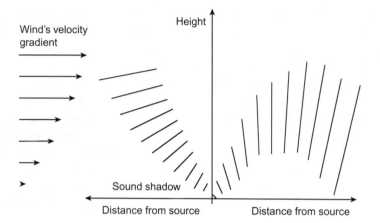

FIGURE 1.22
Refraction of a sound wave due to a wind velocity gradient.

Wind's velocity gradient

Height

Sound shadow

Distance from source

Distance from source

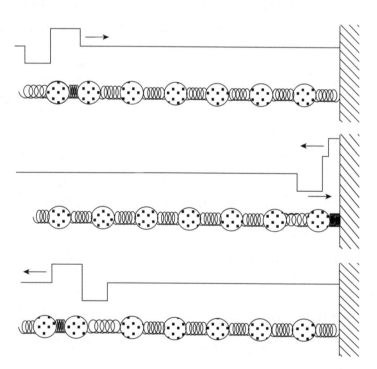

FIGURE 1.23
Reflection of a sound wave due to a rigid barrier.

In the first case the sound wave strikes an immovable object, or hard boundary, as shown in Figure 1.23. At the boundary between the object and the air the sound wave must have zero velocity, because it can't move the wall. This means that at that point all the energy in the sound is in the compression of the air, or pressure. As the energy stored in the pressure cannot transfer in the direction of the propagating wave, it bounces back in the reverse direction, which results in a change of phase in the velocity component of the wave.

Figure 1.23 shows this effect using our golf ball and spring model. One interesting effect occurs due to the fact that the wave has to change direction and to the fact that the spring connected to the immovable boundary is compressed twice as much compared with a spring well away from the boundary. This occurs because the velocity components associated with the reflected (bounced back) wave are moving in contrary motion to the velocity components of the incoming wave, due to the change of phase in the reflected velocity components. In acoustic terms this means that while the velocity component at the reflecting boundary is zero, the pressure component is twice as large.

1.5.5 Sound reflection from bounded to unbounded boundaries

In the second case the wave moves from a bounded region, for example a tube, into an unbounded region, for example free space, as shown in Figure 1.24. At the boundary between the bounded and unbounded regions the molecules in the unbounded region find it a lot easier to move than in the bounded region. The result is that, at the boundary, the sound wave has a pressure component

FIGURE 1.24

Reflection of a sound wave due to bounded–unbounded transition.

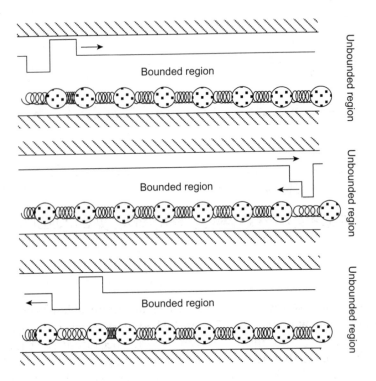

which is close to zero and a large velocity component. Therefore at this point all the energy in the sound is in the kinetic energy of the moving air molecules—in other words, the velocity component. Because there is less resistance to movement in the unbounded region, the energy stored in the velocity component cannot transfer in the direction of the propagating wave due to there being less "springiness" to act on. Therefore the momentum of the molecules is transferred back to the "springs" in the bounded region which pushed them in the first place, by stretching them still further.

This is equivalent to the reflection of a wave in the reverse direction in which the phase of the wave is reversed, because it has started as a stretching of the "springs," or rarefaction, as opposed to a compression. Figure 1.24 shows this effect using the golf ball and spring model in which the unbounded region is modeled as having no springs at all. An interesting effect also occurs in this case, due to the fact that the wave has to change direction. That is, the mass that is connected to the unbounded region is moving with twice the velocity compared to masses well away from the boundary. This occurs because the pressure components associated with the reflected (bounced back) wave are moving in contrary motion to the pressure components of the incoming wave, due to the change of phase in the reflected pressure components. In acoustic terms this means that while the pressure component at the reflecting boundary is zero, the velocity component is twice as large.

To summarize, reflection from a solid boundary results in a reflected pressure component that is in phase with the incoming wave, whereas reflection from a bounded to unbounded region results in a reflected pressure component which is in antiphase with the incoming wave. This arises due to the difference in acoustic impedance at the boundary. In the first case the impedance of the boundary is greater than the propagating medium and in the second case it is smaller. For angles of incidence on the boundary, away from the normal, the usual laws of reflection apply.

1.5.6 Sound interference

We saw earlier that when sound waves come from correlated sources then their pressure and associated velocity components simply add. This means that the pressure amplitude could vary between zero and the sum of the pressure amplitudes of the waves being added together, as shown in Example 1.9. Whether the waves add together constructively or destructively depends on their relative phases and this will depend on the distance each one has had to travel. Because waves vary in space over their wavelength then the phase will also spatially vary. This means that the constructive or destructive addition will also vary in space.

Free space is a region in which the sound wave is free to travel in any direction. That is, there are no obstructions or changes in the propagation medium to affect its travel. Therefore, free space is a form of unbounded region. However, not all bounded regions are free space. For example, the wave may be coming out of a tube in a very large wall. In this case there is a transition between a bounded and an unbounded region but it is not free space, because the wave cannot propagate in all directions.

Consider the situation shown in Figure 1.25, which shows two correlated sources feeding sound into a room. When the listening point is equidistant from the two sources (P1), the two sources add constructively because they are in phase. If one moves to another point (P2), which is not equidistant, the waves no longer necessarily add constructively. In fact, if the path difference is equal to half a wavelength then the two waves will add destructively and there will be no net pressure amplitude at that point. This effect is called interference, because correlated waves interfere with each other; note that this effect does not occur for uncorrelated sources.

The relative phases of the waves depend on their path difference or relative delays. Because of this the pattern of constructive and destructive interferences depends strongly on position, as shown in Figure 1.26. Less obviously the interference is also strongly dependent on frequency. This is because the factor that determines whether or not the waves add constructively or destructively is the relative distance from the listening point to the sources measured in wavelengths (λ). Because the shape of the amplitude response looks a bit like the teeth of a comb, the frequency domain effect of interference is often referred to as "comb filtering." As the wavelength is inversely proportional to frequency one would expect to see the pattern of interference vary directly with frequency, and this is indeed the case.

Figure 1.27 shows the amplitude that results when two sources of equal amplitude but different relative distances are combined. The amplitude is plotted as a function of the relative distance measured in wavelengths (λ). Figure 1.27 shows that the waves constructively interfere when the relative delay is equal to a multiple of a wavelength, and that they interfere destructively at multiples of an odd number of half wavelengths. As the number of

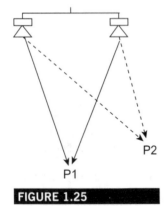

FIGURE 1.25

Interference from correlated sources.

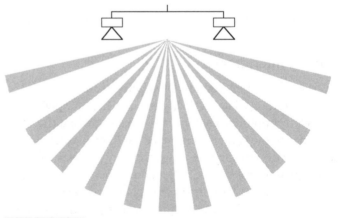

FIGURE 1.26 *Effect of position on interference at a given frequency.*

wavelengths for a fixed distance increases with frequency, this figure shows that the interference at a particular point varies with frequency. If the two waves are not of equal amplitude then the interference effect is reduced, also as shown in Figure 1.27. In fact once the interfering wave is less than one eighth of the other wave then the peak variation in sound pressure level is less than 1 dB.

There are several acoustical situations which can cause interference effects. The obvious ones are when two loudspeakers radiate the same sound into a room, or when the same sound is coupled into a room via two openings which are separated. A less obvious situation is when there is a single sound source spaced away from a reflecting boundary, either bounded or unbounded. In this situation an image source is formed by the reflection and thus there are effectively two sources available to cause interference, as shown in Figure 1.28. This latter situation can often cause problems for recording or sound reinforcement due to a microphone picking up a direct and reflected sound component and so suffering interference.

1.5.7 Standing waves at hard boundaries (modes)

The linear superposition of sound can also be used to explain a wave phenomenon known as "standing waves," which is applicable to any form of sound wave. Standing waves occur when sound waves bounce between reflecting surfaces. The simplest system in which this can occur consists of

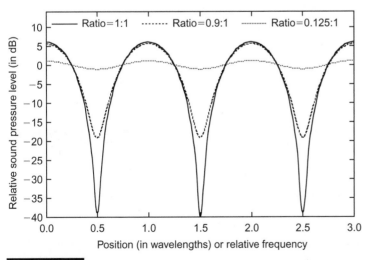

FIGURE 1.27 *Effect of frequency, or wavelength, on interference at a given position. The ratios refer to the relative amplitudes of the two waves.*

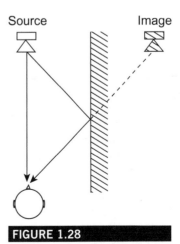

FIGURE 1.28

Interference arising from reflections from a boundary.

If we used complex numbers to represent the speaker outputs, say Ls_1 and Ls_2, then we could account for the phase shift due to the propagation delay by multiplying the speaker outputs by:

$$\theta_{delay}(k) = e^{-jkdistance}$$

multiplying *distance* by *k*, the *wavenumber*, defines the number of cycles the wave is delayed by. So if the two distances were d_1 and d_2 respectively, the result at the listener position would be:

$$S_{result}(k) = Ls_1e^{-jkd_1} + Ls_2e^{-jkd_2}$$

and one could find the magnitude and phase of the composite result by taking the argument and magnitude of the result. Note that because the phase depends on *k*, the result is frequency dependent.

EXAMPLE 1.15

Two loudspeakers are one meter apart and radiate the same sound pressure level. A listener is two meters directly in front of one speaker on a line which is perpendicular to the line joining the two loudspeakers; see Figure 1.29. What are the first two frequencies at which destructive interference occurs? When does the listener first experience constructive interference, other than at very low frequencies?

First work out the path length difference using Pythagoras' theorem:

$$\Delta_{path\ length} = \sqrt{(1m^2 + 2m^2)} - 2m = 0.24m$$

The frequencies at which destructive interference will occur will be $\lambda/2$ and $3\lambda/2$. The frequencies at which this will happen will be when these wavelengths equal the path length difference. Thus the first frequency can be calculated using:

$$\frac{\lambda}{2} = \Delta_{path\ length} \qquad \text{i.e. } \lambda = 2\Delta_{path\ length}$$

$$f_{\lambda/2} = \frac{v}{\lambda} = \frac{v}{2\Delta_{path\ length}} = \frac{344ms^{-1}}{2 \times 0.24m} = 717Hz$$

The second frequency will occur at 3 times the first and so can be given by:

$$f_{3\lambda/2} = 3 \times f_{\lambda/2} = 3 \times 717Hz = 2150Hz$$

The frequency at which the first constructive interference happens will occur at twice the frequency of the first destructive interference which will be:

$$f_\lambda = 2 \times f_{\lambda/2} = 2 \times 717Hz = 1434Hz$$

If the listener were to move closer to the center line of the speakers then the relative delays would reduce and the frequencies at which destructive interference occurs would get higher. In the limit when the listener was equidistant the interference frequencies would be infinite, that is, there would be no destructive interference.

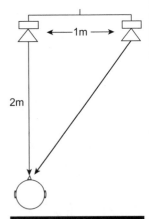

FIGURE 1.29
Interference at a point due to two loudspeakers.

two reflecting boundaries as shown in Figure 1.30. In this system the sound wave shuttles backward and forward between the two reflecting surfaces.

At most frequencies the distance between the two boundaries will not be related to the wavelength and so the compression and rarefaction peaks and troughs will occupy all positions between the two boundaries, with equal probability, as shown in Figure 1.31. However, when the wavelength is related to the distance between the two boundaries the wave keeps tracing the same path as it travels between the two boundaries. This means that the compressions and rarefactions always end up in the same position between the boundaries.

Thus the sound wave will appear to be stationary between the reflecting boundaries, and so is called a standing wave, or, more precisely, a resonant mode. It is important to realize that the wave is still moving at its normal speed—it is merely that, like a toy train, the wave endlessly retraces the same positions between the boundaries with respect to the wavelength, as shown in Figure 1.32.

Figures 1.32 and 1.33 show the pressure and velocity components respectively of a standing wave between two hard reflecting boundaries. In

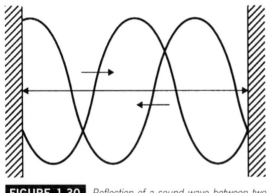

FIGURE 1.30 *Reflection of a sound wave between two parallel surfaces.*

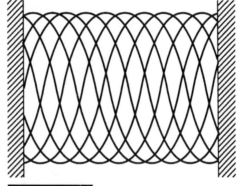

FIGURE 1.31 *A non-stationary sound wave between two parallel surfaces.*

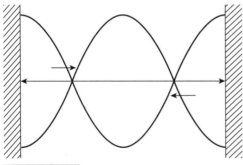

FIGURE 1.32 *The pressure components of a standing wave between two hard boundaries; this is known as a "resonant mode."*

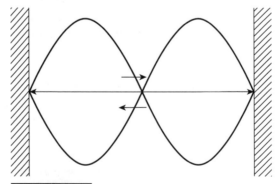

FIGURE 1.33 *The velocity components of a standing wave between two hard boundaries.*

Strictly speaking the standing waves we are discussing here should be called resonant modes. This is because it is possible to set up a standing wave without the use of two boundaries. For example, one could use two loudspeakers spaced apart and no boundaries to get a standing wave at discrete frequencies. One can do the same with one loudspeaker, or sound source, and one boundary.

this situation the pressure component is a maximum and the velocity component is a minimum at the two boundaries. The largest wave that can fit these constraints is a half wavelength and this sets the lowest frequency at which a standing wave can exist for a given distance between reflectors, which can be calculated using the following equation:

$$f_{lowest} \Rightarrow L = \frac{\lambda}{2} \Rightarrow \lambda \ 2L \Rightarrow f_{lowest} = \frac{c}{2L} \qquad (1.34)$$

where f_{lowest} = the standing wave frequency (in Hz)

L = the distance between the boundaries (in m)

λ = the wavelength (in m)

and c = the velocity of sound (in ms^{-1})

Any multiple of half wavelengths will also fit between the two reflectors, and so there is in theory an infinite number of frequencies at which standing waves occur which are all multiples of f_{lowest}. These can be calculated directly using:

$$f_n = \frac{nv}{2L} \qquad (1.35)$$

where f_n = the nth standing wave frequency (in Hz)

and n = 1, 2, ..., ∞

An examination of Figures 1.32 and 1.33 shows that there are points of maximum and minimum amplitude of the pressure and velocity components. For example, in Figure 1.31 the pressure component's amplitude is a maximum at the two boundaries and at the midpoint, while in Figure 1.33 the velocity component is zero at the two boundaries and the midpoint. The point at which the pressure amplitude is zero is called a pressure node and the maximum points are called pressure antinodes. Note that as the number of half wavelengths in the standing waves increases then the number of nodes and antinodes increases, and for hard reflecting boundaries the number of pressure nodes is equal to, and the number of pressure antinodes is one more than, the number of half wavelengths. Velocity nodes and antinodes also exist, and they are always complementary to the pressure nodes, that is, a velocity antinode occurs at a pressure node and vice versa, as shown in Figure 1.34. This happens because the energy in the traveling wave must always exist at a pressure node carried in the velocity component, and at a velocity node the energy is carried in the pressure component.

1.5.8 Standing waves at other boundaries

There are two other pairs of boundary arrangements which can support standing waves. The first, shown in Figure 1.35 and 1.36, is the case of a bounded-to-unbounded propagation boundary at both ends. An example would be a tube or pipe which is open at both ends. In this situation the pressure component is zero at the boundaries whereas the velocity component is at a maximum, as shown in Figures 1.35 and 1.36. Like the hard reflecting boundaries the minimum frequency for a standing wave occurs when there is precisely half a wavelength between the two boundaries, and at all subsequent multiples of this frequency. This means that Equation 1.35 can also be used to calculate the standing wave frequencies for this boundary arrangement.

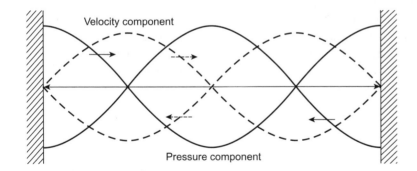

FIGURE 1.34

The pressure and velocity components of a standing wave between two hard boundaries.

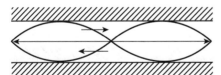

FIGURE 1.35 *The pressure components of a standing wave between two bound–unbound boundaries.*

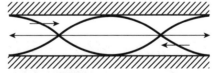

FIGURE 1.36 *The velocity components of a standing wave between two bound–unbound boundaries.*

EXAMPLE 1.16

Calculate the first two modal (standing wave) frequencies for a pipe 98.5 cm long, and open at both ends.

As this is a system with identical boundaries we can use Equation 1.35 to calculate the two frequencies as:

$$f_{1 \text{ open tube}} = \frac{nv}{2L} = \frac{1 \times 344 \text{ms}^{-1}}{2 \times 0.985 \text{m}} = 174.6 \text{Hz}$$

$$f_{2 \text{ open tube}} = \frac{nv}{2L} = \frac{2 \times 344 \text{ms}^{-1}}{2 \times 0.985 \text{m}} = 349.2 \text{Hz}$$

These two frequencies correspond to the notes F3 and F4, which differ by an octave.

The second is more interesting and consists of one hard boundary and one bound–unbound boundary, and is shown in Figures 1.37 and 1.38. In this situation there is a pressure node at the bound–unbound boundary and a pressure antinode at the hard boundary. The effect of this is to allow a standing wave to exist when there is only an odd number of quarter wavelengths between the two boundaries.

A standing wave cannot exist with even numbers of quarter wavelengths as this would require a pressure node or antinode at both ends as shown in Figures 1.32 and 1.35. The frequencies which can support standing waves exist at odd multiples of the lowest standing wave frequency. This can be expressed algebraically as:

$$f_n = \frac{(2n + 1)v}{4L} \tag{1.36}$$

where f_n = the nth standing wave frequency (in Hz)
and $n = 0, 1, 2, ..., \infty$

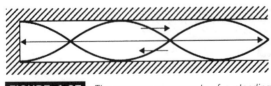

FIGURE 1.37 *The pressure components of a standing wave between mixed boundaries.*

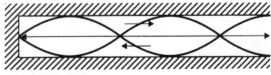

FIGURE 1.38 *The velocity components of a standing wave between two mixed boundaries.*

Figures 1.37 and 1.38 demonstrate the standing wave for $n = 2$ which is five times the lowest supported standing wave frequency. Standing waves can also occur for any type of wave propagation. A transverse wave on a string which is clamped at the ends has standing waves which can be predicted using Equation 1.36, provided one uses the propagation velocity of the transverse wave for v.

Standing waves in an acoustic context are often called the modes of a given system; the lowest frequency standing wave is known as the "first order mode," and the multiples of this are higher order modes. So the third order mode of a system is the third lowest frequency standing wave pattern which can occur in it. Standing waves are also not just restricted to situations with two parallel reflecting boundaries. In fact any sequence of reflections or refractions which returns the wave back to the beginning of its phase will support a standing wave or mode. This can happen in one, two and three dimensions and with any form of wave propagation. The essential requirement is a cyclic path in which the time of propagation results in the wave traveling around this path in phase with the previous time round. Figure 1.39 shows an example of a two-dimensional standing wave.

EXAMPLE 1.17

Calculate the first two modal (standing wave) frequencies for the same pipe as in Example 1.16 with one end closed.

As this is a system with non-identical boundaries we must use Equation 1.36 to calculate the two frequencies as:

$$f_{1 \text{ stopped tube}} = \frac{(2n + 1)v}{4L} = \frac{(2 \times 0 + 1) \times 344\,\text{ms}^{-1}}{4 \times 0.985\,\text{m}} = 87.3\,\text{Hz}$$

$$f_{2 \text{ stopped tube}} = \frac{(2n + 1)v}{4L} = \frac{(2 \times 1 + 1) \times 344\,\text{ms}^{-1}}{4 \times 0.985\,\text{m}} = 261.9\,\text{Hz}$$

In this case the first mode is at half the frequency of the pipe, that is open at both ends, and an octave below on the musical scale, which is F2. The second mode is now at three times the lowest mode, which is approximately equal to C4 on the musical scale.

1.5.9 Sound diffraction

We have all experienced the ability of sound to travel around the corners of a building or other objects. This is due to a process, known as "diffraction," in which the sound bends around objects, as shown in Figure 1.40.

FIGURE 1.39

A two-dimensional standing wave.

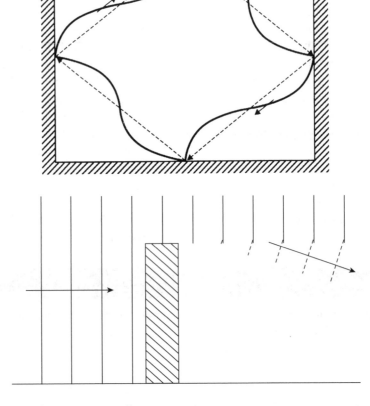

FIGURE 1.40

Diffraction around an object.

Diffraction occurs because the variations in air pressure, due to the compressions and rarefactions in the sound wave, cannot go abruptly to zero after passing the edge of an object. This is because there is interaction between the adjacent molecules that are propagating the wave. In order to allow the compressions and rarefactions to die out gracefully in the boundary between the wave and the shadow, there must be a region in which part of the propagating wave changes direction, and it is this bent part of the wave that forms the diffracted component.

The degree of diffraction depends on wavelength because it effectively takes a certain number of wavelengths for the edge of the wave to make the transition to shadow. Thus the amount of diffraction around an edge, such as a building or wall, will be greater at low and less at high frequencies. This effect is shown in Figures 1.41 and 1.42. Similar effects occur when sound has to pass through an opening, as shown in Figures 1.43 and 1.44. Here the sound wave is diffracted away from the edges of the opening.

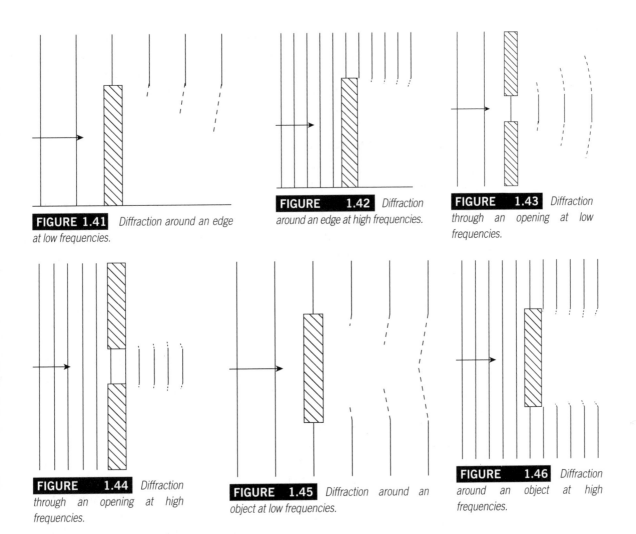

FIGURE 1.41 *Diffraction around an edge at low frequencies.*

FIGURE 1.42 *Diffraction around an edge at high frequencies.*

FIGURE 1.43 *Diffraction through an opening at low frequencies.*

FIGURE 1.44 *Diffraction through an opening at high frequencies.*

FIGURE 1.45 *Diffraction around an object at low frequencies.*

FIGURE 1.46 *Diffraction around an object at high frequencies.*

The amount of diffraction depends on the size of the opening relative to the wavelength. When the size of the wavelength is large relative to the opening the wave is diffracted strongly, and when the wavelength is small relative to the opening then the diffraction is low. The transition between the wavelength being small and large with respect to the opening occurs when the opening size is about two thirds of the wavelength ($\frac{2}{3} \lambda$).

As well as occurring through openings, diffraction also happens around solid objects; one could consider them to be anti-openings, as shown in Figures 1.45 and 1.46. Here the wave diffracts around the object and meets behind it. The effect is also a function of the size of the object with respect to the wavelength. When the size of the wavelength is large relative to the object then the wave is diffracted strongly around it and the object has little influence on the wave propagation. On the other hand, when the wavelength is small relative to the object then the diffraction is less, and the object has a significant shadow behind it; these effects are shown in Figure 1.46. The size at which an object becomes significant with respect to a wavelength is when its size is again about two thirds of a wavelength ($\frac{2}{3}\lambda$).

1.5.10 Sound scattering

Sound which is incident on an object is not just diffracted around it—some of the incident energy will be reflected, or scattered, from the side facing the incident wave, as shown in Figure 1.47. As in the case of diffraction, it is the size of the object with respect to the wavelength that determines how much, and in what way, the energy is scattered. When the object is large with respect to the wavelength then most of the sound incident on the object is scattered according to the laws of reflection and there is very little spreading or diffraction of the reflected wave, as shown in Figure 1.48.

However, when the object is small with respect to the wavelength only a small proportion of the incident energy is scattered and there is a large amount of spreading or diffraction of the reflected wave, as shown in Figure 1.47. As in the case of diffraction, the size at which the object becomes significant is when it is about two thirds of a wavelength ($\frac{2}{3}\lambda$). Thus objects smaller than this will tend to scatter the energy in all directions whereas objects bigger than this will be more directional.

In addition to the scattering effects, interference effects happen when the scattering object is about two thirds of a wavelength ($\frac{2}{3}\lambda$) in size. This is because at this frequency both reflection from, and diffraction around, the object is occurring. In this situation the waves on the two sides of the object can influence or interact with each other. This results in enhanced reflection from, or diffraction around, the object at particular wavelengths. The precise nature of these effects depends on the interaction between the front and back of the scattering object and so will be significantly affected by its shape. Thus the variation in scattering will be different for a sphere compared with rectangular plate. As the sum of energy diffracted around the object and the scattered energy must be constant, the reflection and diffraction responses will be complementary.

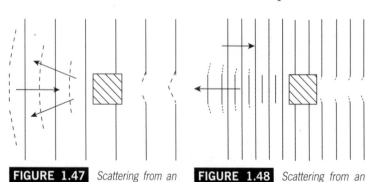

FIGURE 1.47 *Scattering from an object at low frequencies.*

FIGURE 1.48 *Scattering from an object at high frequencies.*

1.6 TIME AND FREQUENCY DOMAINS

So far we have mainly considered a sound wave to be a sinusoidal wave at a particular frequency. This is useful as it allows us to consider aspects of sound propagation in terms of the wavelength. However, most musical sounds have a waveform that is more complex than a simple sine wave; a variety of waveforms are shown in Figure 1.49. So how can we analyze real sound waveforms, and make sense of them in acoustical terms? The answer is based on the concept of superposition and a technique called Fourier analysis.

1.6.1 What is Fourier theory?

Fourier analysis is a way of building up a waveform from much simpler bits. There are many ways of building up complex waveforms from simple basic shapes, like rectangles, but Fourier analysis was developed (by a French mathematician called Jean Baptiste Joseph Fourier in 1807) to solve the problem of heat diffusion in a metal bar! His idea was very simple. He built up the more complicated shapes of the functions, or waveforms, that he was studying by using the humble sine wave. He found that by adding together sine waves with different frequencies, phases and amplitudes he could, very accurately, approximate the functions he was interested in. He found that using sine waves as a basis for approximation was a very useful thing to do. For a start different sine wave frequencies were independent, from, or; more strictly, orthogonal to, each other and so could be manipulated, or used in equations, independently of each other. That is, if you

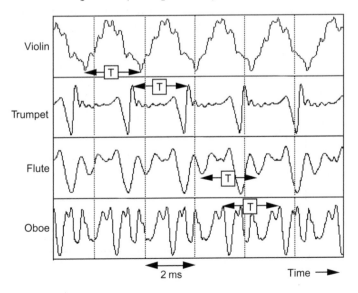

FIGURE 1.49

Waveforms from musical instruments.

solved the equations for each sine wave independently, then you had the solution for the whole waveform. As finding the solution for each sine wave was much easier, this was a great advantage. This principle—that the sum of the solutions is the same as the solution of the sum of the sine waves—is called superposition and is a powerful aspect of the Fourier approach.

Fourier further theorized that if he combined enough sine waves together he could exactly model any waveform in existence. This allowed him to develop the *Fourier transform pair*—two equations that provide a means of transforming the time domain into the frequency domain and back.

These techniques are useful because sometimes it's easier to think about some processes in one domain rather than the other. For example, it's much easier to think about filtering, resonance, and modes of vibration in the frequency domain rather than the time domain. Furthermore, the fact that the basis functions are sine waves, which have a direct physical analog, is useful when dealing with the real world. The next section gives a simple, non-mathematical introduction to Fourier analysis. More details may be found in Appendix 1 and Bracewell (1999).

1.6.2 The spectrum of periodic sound waves

Fourier theory states that any waveform can be built up by using an appropriate set of sine waves of different frequencies, amplitudes and phases. To see how this might work consider the situation shown in Figure 1.50. This shows four sine waves whose frequencies are 1F Hz, 3F Hz, 5F Hz, and 7F Hz, whose phase is zero (that is, they all start from the same value, as shown by the dotted line in Figure 1.50) and whose amplitude is inversely proportional to the frequency. This means that the 3F Hz component is

FIGURE 1.50

The effect of adding several harmonically related sine waves together.

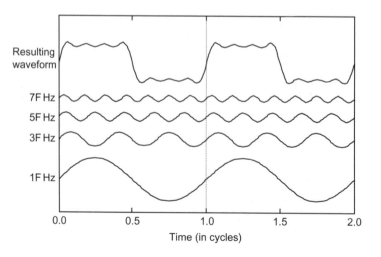

1/3 the amplitude of the component at 1F Hz, and so on. When these sine waves are added together, as shown in Figure 1.50, the result approximates a square wave, and, if more high-frequency components were added, it would become progressively closer to an ideal square wave. The higher frequency components are needed in order to provide the fast rise, and sharp corners, of the square wave.

In general, as the rise time gets faster, and/or the corners get sharper, then more high-frequency sine waves are required to represent the waveform accurately. In other words we can look at a square wave as a waveform that is formed by summing together sine waves which are odd multiples of its fundamental frequency and whose amplitudes are inversely proportional to frequency. A sine wave represents a single frequency and therefore a sine wave of a given amplitude can be plotted as a single line on a graph of amplitude versus frequency. The components of a square wave plotted in this form are shown in Figure 1.51, which clearly shows that the square wave consists of a set of progressively reducing discrete sine wave components at odd multiples of the lowest frequency. This representation is called the frequency domain representation, or spectrum, of a waveform and the waveform's amplitude versus time plot is called its time domain representation.

The individual sine wave components of the waveform are often called the partials of the waveform. If they are integer related, as in the square wave, then they can be called harmonics. The lowest frequency is called the fundamental, or first harmonic, and the higher frequency harmonics are labeled according to their frequency multiple relative to the fundamental. Thus the second harmonic is twice the frequency of the fundamental and so on. Partials on the other hand need not be harmonically related to the fundamental, and are numbered in their order of appearance with frequency.

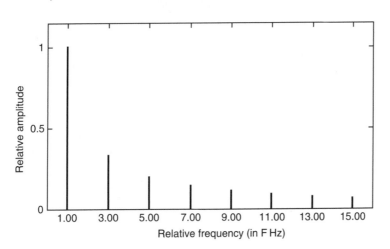

FIGURE 1.51

The frequency domain representation, or spectrum, of a square wave.

However, as we shall see later, this results in a waveform that is a periodic. So for the square wave the second partial is the third harmonic and the third partial is the fifth harmonic. Other waveforms have different frequency domain representations, because they are made up of sine waves of different amplitudes and frequencies. Some examples of other waveforms in both the time and frequency domains are shown in Chapter 3.

1.6.3 The effect of phase

The phase, which expresses the starting value of the individual sine wave components, also affects the waveshape. Figure 1.52 shows what happens to a square wave if alternate partials are subtracted rather than added, and this is equivalent to changing the phase of these components by 180°; that is, alternate frequency components start from halfway around the circle compared with the other components, as shown by the dotted line in Figure 1.52. However, although the time domain waveform is radically different the frequency domain is very similar, as the amplitudes are identical—only the phase of some of the harmonics has changed.

Interestingly, in many cases, the resulting wave is perceived as sounding the same, even though the waveform is different. This is because the ear, as we will see later, appears to be less sensitive to the phase of the individual frequency compared with the relative amplitudes. However, if the phase changes are extreme enough we can hear a difference (see Schroeder, 1975). Because of this, often only the amplitudes of the frequency components are plotted in the spectrum and, in order to handle the range of possible amplitudes and because of the way we perceive sound, the amplitudes are usually plotted as decibels. For example, Figure 4.24 in Chapter 4 shows the waveform and spectrum plotted in this fashion for middle C played on a clarinet and tenor saxophone.

FIGURE 1.52

The effect of adding harmonically related sine waves together with different phase shifts.

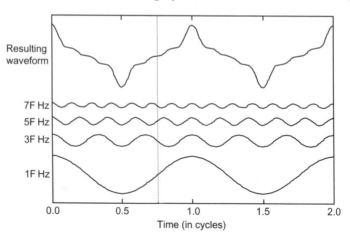

1.6.4 The spectrum of non-periodic sound waves

So far, only the spectrum of waveforms which are periodic, that is, have pitch, has been considered. However, some instruments, especially percussion, do not have pitch and hence are non-periodic, or aperiodic. How can we analyze these instruments in terms of a basic waveform, such as a sine wave, which is inherently periodic? The answer is shown in Figure 1.53. Here the square wave example discussed earlier has had four more sine waves added to it. However, these sine waves are between the harmonics of the square wave and so are unrelated to the period, but they do start off in phase with the harmonics. The effect of these other components is to start canceling out the repeat periods of the square waves, because they are not related in frequency to them. By adding more components which sit in between the harmonics, this cancellation of the repeats becomes more effective so that when in the limit, the whole space between the harmonics is filled with sine wave components of the appropriate amplitude and phase. These extra components will add constructively only at the beginning of the waveform and will interfere with successive cycles due to their different frequencies. Therefore, in this case, only one square wave will exist.

Thus the main difference between the spectrum of periodic and aperiodic waveforms is that periodic waveforms have discrete partials, which can be represented as lines in the spectrum with a spacing that is inversely proportional to the period of the waveform. Aperiodic waveforms by contrast will have a spectrum which is continuous and therefore does not have any discrete components. However, the envelope of the component amplitudes as a function of frequency will be the same for both periodic and aperiodic

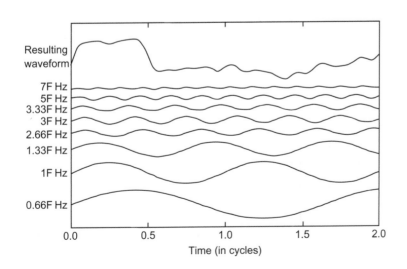

FIGURE 1.53

The effect of adding several non-harmonically related sine waves together.

waves of the same shape, as shown in Figure 1.54. Figure 3.6 in Chapter 3 shows the aperiodic waveform and spectrum of a brushed snare.

1.7 ANALYZING SPECTRA

Because the spectrum of a sound is an important part of the way we perceive it, there is often a need to look at the spectrum of a real signal. The way this is achieved is to use a bank of filters, as shown in Figure 1.55.

1.7.1 Filters and filter types

A filter is a device which separates out a portion of the frequency spectrum of a sound signal from the total; this is shown in Figure 1.56. There are four basic types of filter, which are classified in terms of their effect as a function of signal frequency. This is known as the filters' "frequency response." These basic types of filter are as follows:

- *Low-pass*: The filter only passes frequencies below a frequency known as the filter's "cut-off frequency."

- *High-pass*: The filter only passes frequencies above the cut-off frequency.

- *Band-pass*: The filter passes a range of frequencies between two cut-off frequencies. The frequency range between the cut-off frequencies is the filter's "bandwidth."

- *Band-reject*: The filter rejects a range of frequencies between two cut-off frequencies.

FIGURE 1.54

The frequency domain representation, or spectrum, of an aperiodic square wave.

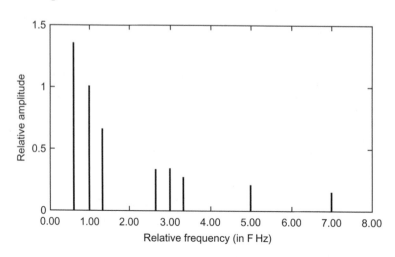

The effects of these four types of filter are shown in Figure 1.56 and one can see that a bank of band-pass filters are the most appropriate for analyzing the spectrum of a sound wave. Note that although practical filters have specified cut-off frequencies which are determined by their design, they do not cut off instantly as the frequency changes. Instead they take a finite frequency range to attenuate the signal.

The effect of a filter on a spectrum is multiplicative in that the output spectrum after filtering is the product of the filter's frequency response with the input signal's spectrum. Thus we can easily determine the effect of a given filter on a signal's spectrum. This is a useful technique which can be applied to the analysis of musical instruments, as we shall see later, by treating some of their characteristics as a form of filtering. In fact, filtering can be carried out using mechanical, acoustical and electrical means, and many instruments perform some form of filtering on the sounds they generate (see Chapter 4).

1.7.2 Filter time responses

There is a problem, however, with filtering a signal in order to derive the spectrum and this is the effect of the filter on the time response of the signal. Most filters have a time response due to the fact that they do not allow

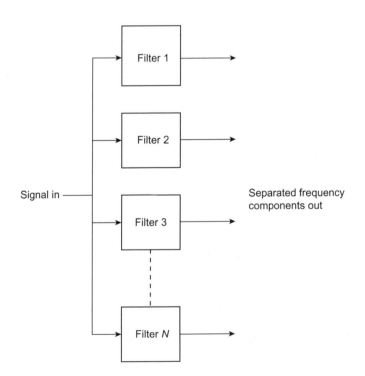

FIGURE 1.55

A filter bank for analyzing spectra.

FIGURE 1.56

The effect of different filter types.

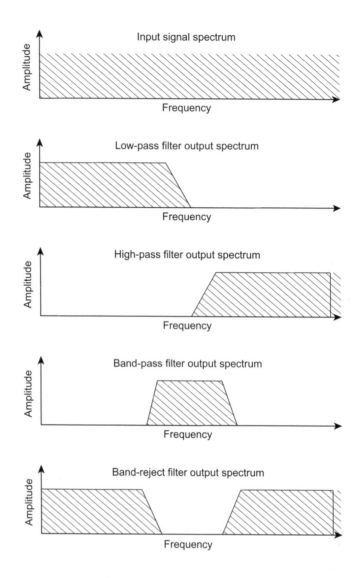

FIGURE 1.56

The effect of different filter types.

all frequencies to pass through—if they did they wouldn't be filters! Why do such filters have a time response? The answer can be obtained by reconsidering the Fourier analysis approach to analyzing a sound signal. For example, if a signal is low-pass filtered such that the maximum frequency is F_{max}, then there can be no sine waves with a frequency greater than F_{max} in the output. As a sine wave has a slope which is a function of its frequency, the maximum rate of change in the output will be determined by the frequency of F_{max}. Any faster rate of change would require higher frequency sine waves that are no longer present, as shown in Figure 1.57 which shows

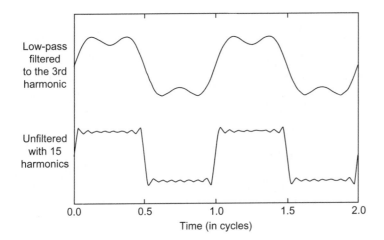

Low-pass
filtered
to the 3rd
harmonic

Unfiltered
with 15
harmonics

Time (in cycles)

The effect of low-pass filtering a square wave on the waveform.

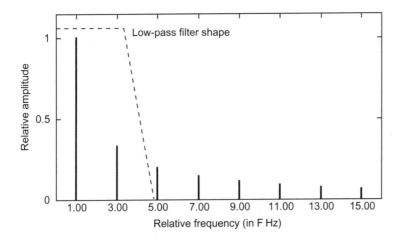

Low-pass filter shape

Relative amplitude

Relative frequency (in F Hz)

The spectral effect of low-pass filtering a square wave.

the effect of low-pass filtering a square wave such that only the first two partials are passed, as shown in Figure 1.58.

A similar argument can used for band-pass filters. In this case there is a maximum range of frequencies that can be passed around a given frequency. Although the sine waves corresponding to the band-pass frequencies will be passed, and they may be at quite a high frequency, their amplitude envelope cannot vary quickly, as shown in Figure 1.59. This is because the speed of variation of the envelope depends on the number of, and total frequency occupied by, the sine wave components in the output of the filter, as shown in Figure 1.60. As it is the amplitude variation of the output of a band-pass filter that carries the information, it too has an inherent time response which is a function of its bandwidth.

There is a special type of filter known as the "all-pass filter." This is not a piece of wire! It is a filter that changes the phase of the signal's frequency components without affecting their amplitude. Because the phases have been changed there is also an associated time response.

FIGURE 1.59

The effect of band-pass filtering on a square wave envelope.

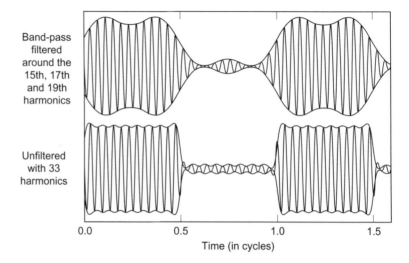

FIGURE 1.60

The spectral effect of band-pass filtering a square wave envelope.

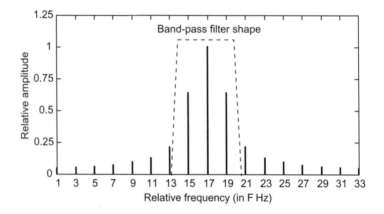

Thus all filters have a time response which is a function of the frequency range or bandwidth of the signals that they pass. Note that this is an inherent problem that cannot be solved by technology. The time response of a filter is inversely proportional to the bandwidth so a narrowband filter has a slow rise and fall time whereas a wide-band filter has faster rise and fall times. In practice this effect means that if the frequency resolution of the spectral analysis is good, implying the use of narrowband analysis filters, then the time response is poor and there is significant smearing of the signal. On the other hand if the time response is good then the frequency resolution will be poor, because the filter bandwidths will have to be wider.

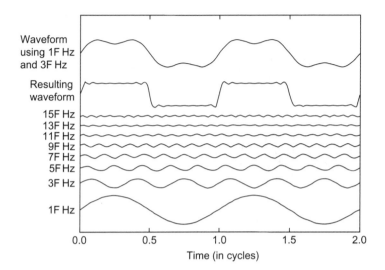

FIGURE 1.61

The effect of an increasing number of harmonics on a square wave.

1.7.3 Time responses of acoustic systems

This argument can be reversed to show that when the output of an acoustic system, such as a musical instrument, changes slowly then its spectrum occupies a narrow bandwidth, whereas when it changes quickly then it must occupy a larger bandwidth. This can be due to an increase in the number of harmonics present in the spectrum, as would be the case if the waveform becomes sharper or more spiky.

Figure 1.61 shows the effect of an increasing number of harmonics on a square wave. Or, as in the band-pass case described earlier, it would be due to an increase in the bandwidths occupied by the individual partials, or harmonics, of the sound. This would be the case if the envelope of the sound changed more rapidly. Figures 1.62 and 1.63 show this effect, which in this case compares two similar systems—one with a slow and one with a fast rate of amplitude decay, the latter being due to a higher loss of energy from the system. The figure clearly shows that in the system which decays more rapidly there are more harmonics and there is a higher bandwidth due to the more rapid change in the sound's envelope.

1.7.4 Time and frequency representations of sounds

Figure 1.64 is a useful way of showing both the time and frequency characteristics of a sound signal at the same time, called a spectrogram. In this representation the spectrum of a signal is plotted as a function of time, with frequency and time forming the main axes. The decibel amplitude of

The decay rate of two systems with different bandwidths.

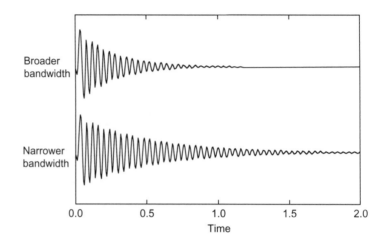

The response of two systems with different decay times.

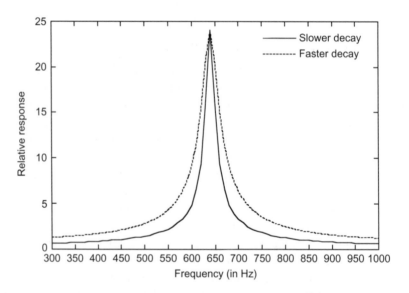

the signal is plotted as a gray scale with black representing a high amplitude and white representing a low amplitude. This representation is an excellent way of seeing the time evolution of the spectral components of a musical sound. However, it does suffer from the time smearing effects discussed earlier.

Figure 1.64 shows both narrow-band and broad-band analysis of the note at the beginning of a harpsichord piece. In the narrow-band version, although the harmonics are clearly resolved, the start of the signal is smeared due to the slow time response of the narrow-band filters. On the

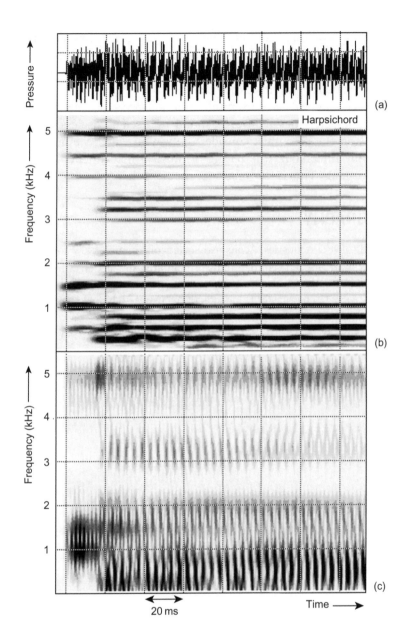

FIGURE 1.64

Acoustic pressure waveform (a) narrow-band (40 Hz analysis filter) (b) and wide-band (300 Hz analysis filter) (c) spectrograms for middle C played on a harpsichord.

other hand the broad-band version has excellent time response and even shows the periodic variation in amplitude due to the pitch of the sound, shown as vertical striations in the spectrogram; however, it is unable to resolve the harmonics of the signal.

Ideally one would like a filter which has a fast time response and a narrow bandwidth, but this is impossible. However, as we shall see later, the human hearing system is designed to provide a compromise which gives both good frequency and time resolution.

In this chapter we have examined many aspects of sound waveforms and their characteristics. However, sound by itself is useless unless it is heard and therefore we must consider the way in which we hear sound in order to fully understand the nature of musical instruments and musical signal processing. This is the subject of the next chapter.

BIBLIOGRAPHY

Beranek, L.L., 1986. Acoustics. Acoustical Society of America, New York.

Bracewell, R.N., 1999. The Fourier Transform and Its Applications, third edn. McGraw-Hill.

Everest, F.A., 1994. The Master Handbook of Acoustics. Tab Books, New York.

Kuttruff, H., 2000. Room Acoustics. E&FN Spon, London.

Schroeder, M.R., 1975. Models of Hearing. Proc. IEEE 63, 1332–1350.

Introduction to Hearing

Psychoacoustics is the study of how humans perceive sound. To begin our exploration of psychoacoustics it is first necessary to become familiar with the basic anatomy of the human hearing system to facilitate understanding of:

- the effect the normal hearing system has on sounds entering the ear;
- the origin of fundamental psychoacoustic findings relevant to the perception of music;
- how listening to very loud sounds can cause hearing damage; and
- some of the listening problems faced by the hearing impaired.

This chapter introduces the main anatomical structures of the human hearing system which form the path along which incoming music signals travel up to the point where the signal is carried by nerve fibers from the ear(s) to the brain. It also introduces the concept of "critical bands", which is the single most important psychoacoustic principle for an understanding of the perception of music and other sounds in terms of pitch, loudness and timbre.

It should be noted that many of the psychoacoustic effects have been observed experimentally, mainly as a result of playing sounds that are carefully controlled in terms of their acoustic nature to panels of listeners from whom responses are monitored. These responses are often the result of comparing two sounds and indicating, for example, which is louder or higher in pitch or "brighter." Many of the results from such experiments cannot as yet be described in terms of either where anatomically or by what physical means they occur. Psychoacoustics is a developing field of research. However, the results from such experiments give a firm foundation for understanding the nature of human perception of musical sounds, and knowledge of minimum changes that are perceived provide useful guideline bounds for those exploring the subtleties of sound synthesis.

2.1 THE ANATOMY OF THE HEARING SYSTEM

The anatomy of the human hearing system is illustrated in Figure 2.1. It consists of three sections:

- the outer ear,
- the middle ear, and
- the inner ear.

The anatomical structure of each of these is discussed below, along with the effect that each has on the incoming acoustic signal.

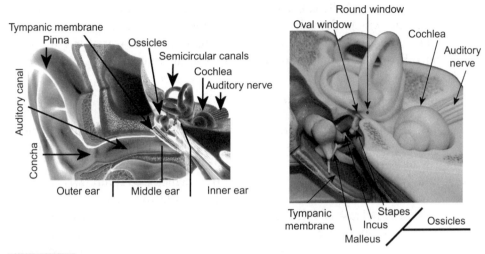

FIGURE 2.1 *The main structures of the human ear showing an overall view of the outer, middle and inner ears (left) and a detailed view of the middle and inner ear (right).*

2.1.1 Outer ear function

The outer ear (see Figure 2.1) consists of the external flap of tissue known as the "pinna" with its many grooves, ridges and depressions. The depression at the entrance to the auditory canal is known as the "concha." The auditory canal is approximately 25–35 mm long from the concha to the "tympanic membrane," more commonly known as the "eardrum." The outer ear has an acoustic effect on sounds entering the ear in that it helps us both to locate sound sources and it enhances some frequencies with respect to others.

Sound localization is helped mainly by the acoustic effect of the pinna and the concha. The concha acts as an acoustic resonant cavity. The combined acoustic effects of the pinna and concha are particularly useful for determining whether a sound source is in front or behind, and to a lesser extent whether it is above or below.

The acoustic effect of the outer ear as a whole serves to modify the frequency response of incoming sounds due to resonance effects, primarily of the auditory canal whose main resonance frequency is in the region around 4 kHz.

The tympanic membrane is a light, thin, highly elastic structure which forms the boundary between the outer and middle ears. It consists of three layers: the outside layer which is a continuation of the skin lining of the auditory canal, the inside layer which is continuous with the mucous lining of the middle ear, and the layer in between these which is a fibrous structure which gives the tympanic membrane its strength and elasticity. The tympanic membrane converts acoustic pressure variations from the outside world into mechanical vibrations in the middle ear.

2.1.2 Middle ear function

The mechanical movements of the tympanic membrane are transmitted through three small bones known as "ossicles," comprising the "malleus," "incus" and "stapes"—more commonly known as the "hammer," "anvil" and "stirrup"—to the oval window of the cochlea (see Figure 2.1). The oval window forms the boundary between the middle and inner ears.

The malleus is fixed to the middle fibrous layer of the tympanic membrane in such a way that when the membrane is at rest, it is pulled inwards. Thus the tympanic membrane when viewed down the auditory canal from outside appears concave and conical in shape. One end of the stapes, the stapes footplate, is attached to the oval window of the cochlea. The malleus and incus are joined quite firmly such that at normal intensity levels they act as a single unit, rotating together as the tympanic membrane vibrates to move the stapes via a ball and socket joint in a piston-like manner. Thus acoustic vibrations are transmitted via the tympanic membrane and ossicles as mechanical movements to the cochlea of the inner ear.

The function of the middle ear is twofold: (1) to transmit the movements of the tympanic membrane to the fluid which fills the cochlea without significant loss of energy, and (2) to protect the hearing system to some extent from the effects of loud sounds, whether from external sources or the individual concerned.

In order to achieve efficient transfer of energy from the tympanic membrane to the oval window, the effective pressure acting on the oval window is arranged by mechanical means to be greater than that acting on the tympanic membrane. This is to overcome the higher resistance to movement of the cochlear fluid compared with that of air at the input to the ear. Resistance to movement can be thought of as "impedance" to movement and the impedance of fluid to movement is high compared with that of air. The ossicles act as a mechanical "impedance converter" or "impedance transformer" and this is achieved essentially by two means:

- the lever effect of the malleus and incus; and
- the area difference between the tympanic membrane and the stapes footplate.

The lever effect of the malleus and incus arises as a direct result of the difference in their lengths. Figure 2.2 shows this effect. The force at the stapes footplate relates to the force at the tympanic membrane by the ratio of the lengths of the malleus and incus as follows:

$$F1 \times L1 = F2 \times L2$$

where $F1$ = force at tympanic membrane
$F2$ = force at stapes footplate
$L1$ = length of malleus
and $L2$ = length of incus

Therefore:

$$F2 = F1 \times \frac{L1}{L2} \qquad (2.1)$$

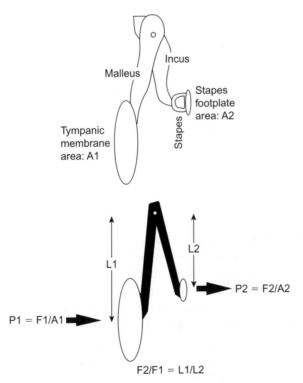

FIGURE 2.2   *The function of the ossicles of the middle ear.*

The area difference has a direct effect on the pressure applied at the stapes footplate compared with the incoming pressure at the tympanic membrane since pressure is expressed as force per unit area as follows:

$$\text{Pressure} = \frac{\text{Force}}{\text{Area}} \qquad (2.2)$$

The areas of the tympanic membrane and the stapes footplate in humans are represented in Figure 2.2 as A1 and A2 respectively. The pressure at the tympanic membrane (P1) and the pressure at the stapes footplate (P2) can therefore be expressed as follows:

$$P1 = \frac{F1}{A1}$$

$$P2 = \frac{F2}{A2}$$

The forces can therefore be expressed in terms of pressures:

$$F1 = (P1 \times A1) \qquad (2.3)$$

$$F2 = (P2 \times A2) \qquad (2.4)$$

Substituting Equations 2.3 and 2.4 into Equation 2.1 gives:

$$(P2 \times A2) = (P1 \times A1) \times \frac{L1}{L2}$$

Therefore:

$$\frac{P2}{P1} = \frac{A1 \times L1}{A2 \times L2} \qquad (2.5)$$

Pickles (1982) describes a third aspect of the middle ear which appears relevant to the impedance conversion process. This relates to a buckling motion of the tympanic membrane itself as it moves, resulting in a twofold increase in the force applied to the malleus.

In humans, the area of the tympanic membrane (A1) is approximately 13 times larger than the area of the stapes footplate (A2), and the malleus is approximately 1.3 times the length of the incus. The buckling effect of the tympanic membrane provides a force increase by a factor of 2. Thus the pressure at the stapes footplate (P2) is about $(13 \times 1.3 \times 2 = 33.8)$ times larger than the pressure at the tympanic membrane (P1).

EXAMPLE 2.1

Express the pressure ratio between the stapes footplate and the tympanic membrane in decibels.

The pressure ratio is 33.8:1. Equation 1.20 is used to convert from pressure ratio to decibels:

$$dB(SPL) = 20\log_{10}\frac{P2}{P1}$$

Substituting 33.8 as the pressure ratio gives:

$$20\log_{10}[33.8] = 30.6\,dB$$

The second function of the middle ear is to provide some protection for the hearing system from the effects of loud sounds, whether from external sources or the individual concerned. This occurs as a result of the action of two muscles in the middle ear: the tensor tympani and the stapedius muscle. These muscles contract automatically in response to sounds with levels greater than approximately 75 dB(SPL) and they have the effect of increasing the impedance of the middle ear by stiffening the ossicular chain. This reduces the efficiency with which vibrations are transmitted from the tympanic membrane to the inner ear and thus protects the inner ear to some extent from loud sounds. Approximately 12–14 dB of attenuation is provided by this protection mechanism, but this is for frequencies below 1 kHz only. The names of these muscles derive from where they connect with the ossicular chain: the tensor tympani is attached to the "handle" of the malleus, near the tympanic membranes, and the stapedius muscle attached to the stapes.

This effect is known as the "acoustic reflex." It takes some 60–120 ms for the muscles to contract in response to a loud sound. In the case of a loud impulsive sound such as the firing of a large gun, it has been suggested that the acoustic reflex is too slow to protect the hearing system. In gunnery situations, a sound loud enough to trigger the acoustic reflex, but not so loud as to damage the hearing systems, is often played at least 120 ms before the gun is fired.

2.1.3 Inner ear function

The inner ear consists of the snail-like structure known as the "cochlea." The function of the cochlea is to convert mechanical vibrations into nerve firings to be processed eventually by the brain. Mechanical vibrations reach the cochlea at the oval window via the stapes footplate of the middle ear.

The cochlea consists of a tube coiled into a spiral with approximately 2.75 turns—see Figure 2.3(a). The end with the oval and round windows is the "base" and the other end is the "apex"—see Figure 2.3(b). Figure 2.3(c) illustrates the effect of slicing through the spiral vertically, and it can be seen in (d) that the tube is divided into three sections by Reissner's membrane and the basilar membrane. The outer channels—the scala vestibuli (V) and scala tympani (T)—are filled with an incompressible fluid known as "perilymph," and the inner channel is the scala media (M). The scala vestibuli terminates at the oval window and the scala tympani at the round window. An idealized unrolled cochlea is shown in Figure 2.3(b). There is a small hole at the apex known as the "helicotrema" through which the perilymph fluid can flow.

Input acoustic vibrations result in a piston-like movement of the stapes footplate at the oval window, which moves the perilymph fluid within the cochlea. The membrane covering the round window moves to compensate for oval window movements since the perilymph fluid is essentially incompressible. Inward movements of the stapes footplate at the oval window

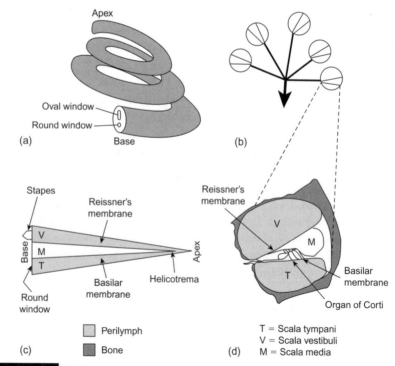

FIGURE 2.3 *(a) The spiral nature of the cochlea. (b) The cochlea "unrolled." (c) Vertical cross-section through the cochlea. (d) Detailed view of the cochlear tube.*

cause the round window to move outwards, and outward movements of the stapes footplate cause the round window to move inwards. These movements cause traveling waves to be set up in the scala vestibuli which displace both Reissner's membrane and the basilar membrane.

The basilar membrane is responsible for carrying out a frequency analysis of input sounds. In shape, the basilar membrane is both narrow and thin at the base end of the cochlea, becoming both wider and thicker along its length to the apex, as illustrated in Figure 2.4. The upper part of Figure 2.4 shows the idealized shape of the basilar membrane where it sits along the unrolled cochlea—compare with Figure 2.3(b), which illustrates that the width and depth of the basilar membrane are narrowest at the base and they increase towards the apex. The basilar membrane vibrates in response to stimulation by signals in the audio frequency range.

Small structures respond better to higher frequencies than do large structures (compare, for example, the sizes of a violin and a double bass or the strings at the treble and bass ends of a piano). The basilar membrane therefore responds best to high frequencies where it is narrow and thin (at the base) and to low frequencies where it is wide and thick (at the apex). Since its thickness and width change gradually along its length, input pure tones at different frequency will produce a maximum basilar membrane movement at different positions or "places" along its length.

This is illustrated in Figure 2.5 for a section of the length of the membrane. This is the basis of the "place" analysis of sound by the hearing

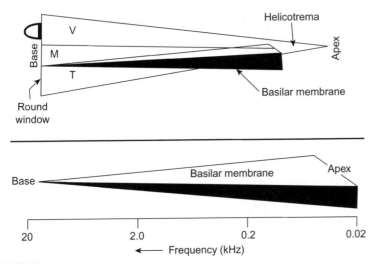

FIGURE 2.4 *Idealized shape of basilar membrane as it lies in the unrolled cochlea (upper), and the basilar membrane response with frequency (lower).*

system. The extent, or "envelope," of basilar membrane movement is plotted against frequency in an idealized manner for five input pure tones of different frequencies. If the input sound were a complex tone consisting of many components, the overall basilar membrane response is effectively the sum of the responses for each individual component. The basilar membrane is stimulated from the base end (see Figure 2.3) which responds best to high frequencies, and it is important to note that its envelope of movement for a pure tone (or individual component of a complex sound) is not symmetrical, but that it tails off less rapidly towards high frequencies than towards low frequencies. This point will be taken up again in Chapter 5.

The movement of the basilar membrane for input sine waves at different frequencies has been observed by a number of researchers following the pioneering work of von Békésy (1960). They have confirmed that the point of maximum displacement along the basilar membrane changes as the frequency of the input is altered. It has also been shown that the linear distance measured from the apex to the point of maximum basilar membrane displacement is directly proportional to the logarithm of the input frequency. The frequency axis in Figure 2.5 is therefore logarithmic. It is illustrated in the figure as being "back-to-front" (i.e., with increasing frequency changing from right to left, low frequency at the apex and high at the base) to maintain the left to right sense of flow of the input acoustic signal and to reinforce understanding of the anatomical nature of the inner ear.

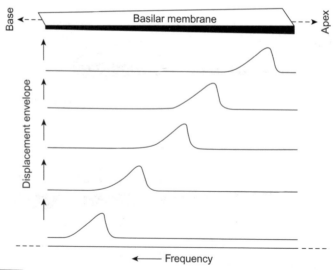

FIGURE 2.5 *Idealized envelope of basilar membrane movement to sounds at five different frequencies.*

The section of the inner ear which is responsible for the analysis of low-frequency sounds is the end farthest away from the oval window, coiled into the center of the cochlear spiral.

In order that the movements of the basilar membrane can be transmitted to the brain for further processing, they have to be converted into nerve firings. This is the function of the organ of Corti, which consists of a number of hair cells that trigger nerve firings when they are bent. These hair cells are distributed along the basilar membrane and they are bent when it is displaced by input sounds. The nerves from the hair cells form a spiral bundle known as the "auditory nerve." The auditory nerve leaves the cochlea as indicated in Figure 2.1.

2.2 CRITICAL BANDS

Section 2.1 describes how the inner ear carries out a frequency analysis of sound due to the mechanical properties of the basilar membrane and how this provides the basis behind the "place" theory of hearing. The next important aspect of the place theory to consider is how well the hearing system can discriminate between individual frequency components of an input sound. This will provide the basis for understanding the resolution of the hearing system and it will underpin discussions relating to the psychoacoustics of how we hear music, speech and other sounds.

Each component of an input sound will give rise to a displacement of the basilar membrane at a particular place, as illustrated in Figure 2.5. The displacement due to each individual component is spread to some extent on either side of the peak. Whether or not two components that are of similar amplitude and close together in frequency can be discriminated depends on the extent to which the basilar membrane displacements, due to each of the two components, are clearly separated.

Consider track 1 on the accompanying CD. Suppose two pure tones, or sine waves, with amplitudes A_1 and A_2 and frequencies F_1 and F_2 respectively are sounded together. If F_1 is fixed and F_2 is changed slowly from being equal to or in unison with F_1 either upwards or downwards in frequency, the following is generally heard (see Figure 2.6). When F_1 is equal to F_2, a single note is heard. As soon as F_2 is moved higher or lower than F_1 a sound with clearly undulating amplitude variations known as "beats" is heard. The frequency of the beats is equal to $(F_2 - F_1)$, or $(F_1 - F_2)$ if F_1 is greater than F_2, and the amplitude varies between $(A_1 + A_2)$ and $(A_1 - A_2)$, or $(A_1 + A_2)$ and $(A_2 - A_1)$ if A_2 is greater than A_1. Note that when the amplitudes are equal $(A_1 = A_2)$ the amplitude of the beats varies between $(2 \times A_1)$ and 0.

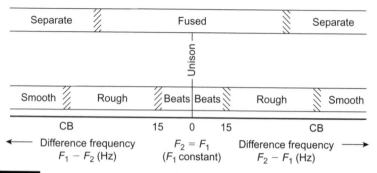

FIGURE 2.6   *An illustration of the perceptual changes which occur when a pure tone fixed at frequency F_1 is heard combined with a pure tone of variable frequency F_2.*

For the majority of listeners beats are usually heard when the frequency difference between the tones is less than about 12.5 Hz, and the sensation of beats generally gives way to one of a "fused" tone which sounds "rough" when the frequency difference is increased above 15 Hz. As the frequency difference is increased further there is a point where the fused tone gives way to two separate tones but still with the sensation of roughness, and a further increase in frequency difference is needed for the rough sensation to become smooth. The smooth separate sensation persists while the two tones remain within the frequency range of the listener's hearing.

The changes from fused to separate and from beats to rough to smooth are shown hashed in Figure 2.6 to indicate that there is no exact frequency difference at which these changes in perception occur for every listener. However, the approximate frequencies and order in which they occur is common to all listeners, and, in common with most psychoacoustic effects, average values are quoted which are based on measurements made for a large number of listeners.

The point where the two tones are heard as being separate as opposed to fused when the frequency difference is increased can be thought of as the point where two peak displacements on the basilar membrane begin to emerge from a single maximum displacement on the membrane. However, at this point the underlying motion of the membrane, which gives rise to the two peaks, causes them to interfere with each other giving the rough sensation, and it is only when the rough sensation becomes smooth that the separation of the places on the membrane is sufficient to fully resolve the two tones. The frequency difference between the pure tones at the point where a listener's perception changes from rough and separate to smooth and separate is known as the "critical bandwidth," and it is therefore marked CB in the figure. A more formal definition is given by Scharf (1970): "The critical bandwidth is that bandwidth at which subjective responses rather abruptly change."

In order to make use of the notion of critical bandwidth practically, an equation relating the effective critical bandwidth to the filter center frequency was proposed by Glasberg and Moore (1990). They define a filter with an ideal rectangular frequency response curve which passes the same power as the auditory filter in question, which is known as the "equivalent rectangular bandwidth" or "ERB." The ERB is a direct measurement of the critical bandwidth, and the Glasberg and Moore equation which allows the calculation of the ERB for any filter center frequency is as follows:

$$\text{ERB} = \{24.7 \times [(4.37 \times f_c) + 1]\} \text{ Hz} \qquad (2.6)$$

where f_c = the filter center frequency in kHz
and ERB = the equivalent rectangular bandwidth in Hz
  Equation valid for ($100\,\text{Hz} < f_c < 10\,000\,\text{Hz}$)

This relationship is plotted in Figure 2.7 and lines representing where the bandwidth is equivalent to 1, 2, 3, 4, and 5 semitones (or a semitone, whole tone, minor third, major third and perfect fourth respectively) are also plotted for comparison purposes. A third octave filter is often used in the studio as an approximation to the critical bandwidth; this is shown in the figure as the 4 semitone line (there are 12 semitones per octave, so a third of an octave is 4 semitones). A keyboard is shown on the filter center frequency axis for convenience, with middle C marked with a spot.

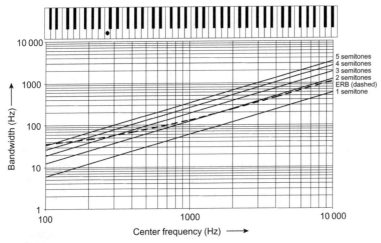

FIGURE 2.7 *The variation of equivalent rectangular bandwidth (ERB) with filter center frequency and lines indicating where the bandwidth would be equivalent to 1, 2, 3, 4 and 5 semitones. (Middle C is marked with a spot on the keyboard.)*

EXAMPLE 2.2

Calculate the critical bandwidth at 200 Hz and 2000 Hz to three significant figures.

Using Equation 2.6 and substituting 200 Hz and 2000 Hz for f_c (noting that f_c should be expressed in kHz in this equation as 0.2 kHz and 2 kHz respectively) gives the critical bandwidth (ERB) as:

$$\text{ERB at 200 Hz} = \{24.7 \times [(4.37 \times 0.2) + 1]\} = 46.3 \text{ Hz}$$

$$\text{ERB at 2000 Hz} = \{24.7 \times [(4.37 \times 2) + 1]\} = 241 \text{ Hz}$$

The change in critical bandwidth with frequency can be demonstrated if the fixed frequency F_1 in Figure 2.6 is altered to a new value and the new position of CB is found. In practice, critical bandwidth is usually measured by an effect known as "masking" (see Chapter 5) in which the "rather abrupt change" is more clearly perceived by listeners.

The response characteristic of an individual filter is illustrated in the bottom curve in Figure 2.8, the vertical axis of which is marked "filter response" (notice that increasing frequency is plotted from right to left in this figure in keeping with Figure 2.5 relating to basilar membrane displacement). The other curves in the figure are idealized envelopes of basilar membrane displacement for pure tone inputs spaced by f Hz, where f is the distance between each vertical line as marked. The filter center frequency F_c Hz is indicated with an unbroken vertical line, which also represents the place on the basilar membrane corresponding to a frequency F_c Hz. The filter response curve is plotted by observing the basilar membrane displacement at the place corresponding to F_c Hz for each input pure tone and plotting this as the filter response at the frequency of the pure tone. This results in the response curve shape illustrated as follows.

As the input pure tone is raised to F_c Hz, the membrane displacement gradually increases with the less steep side of the displacement curve. As the frequency is increased above F_c Hz, the membrane displacement falls rapidly with the steeper side of the displacement curve. This results in the filter response curve as shown, which is an exact mirror image about F_c Hz of the basilar membrane displacement curve.

Figure 2.9(a) shows the filter response curve plotted with increasing frequency and plotted more conventionally from left to right in order to facilitate discussion of the psychoacoustic relevance of its asymmetric shape in Chapter 5.

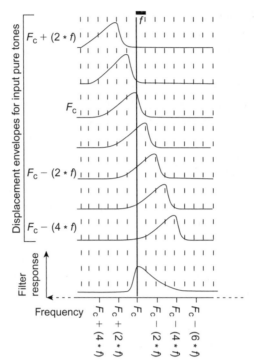

FIGURE 2.8 *Derivation of response of an auditory filter with center frequency F_c Hz based on idealized envelope of basilar membrane movement to pure tones with frequencies local to the center frequency of the filter.*

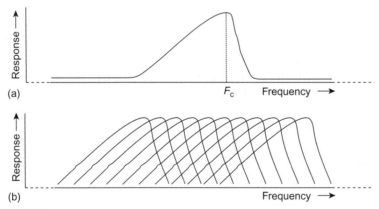

FIGURE 2.9 *(a) Idealized response of an auditory filter with center frequency F_c Hz with increasing frequency plotted in the conventional direction (left to right). (b) Idealized bank of band-pass filters model the frequency analysis capability of the basilar membrane.*

The action of the basilar membrane can be thought of as being equivalent to a large number of overlapping band-pass filters, or a "bank" of band-pass filters, each responding to a particular band of frequencies (see Chapter 1). Based on the idealized filter response curve shape in Figure 2.9(a), an illustration of the nature of this bank of filters is given in Figure 2.9(b). Each filter has an asymmetric shape to its response with a steeper roll-off on the high-frequency side than on the low-frequency side; the bandwidth of a particular filter is given by the critical bandwidth (see Figure 2.7) for any particular center frequency. It is not possible to be particularly exact with regard to the extent to which the filters overlap. A common practical compromise, for example, in studio third octave graphic equalizer filter banks, is to overlap adjacent filters at the $-3\,dB$ points on their response curves.

In terms of the perception of two pure tones illustrated in Figure 2.6, the "critical bandwidth" can be thought of as the bandwidth of the band-pass filter in the bank of filters, the center frequencies of which are exactly halfway between the frequencies of the two tones. This ignores the asymmetry of the basilar membrane response (see Figure 2.5) and the consequent asymmetry in the individual filter response curve—see Figure 2.9(a)—but it provides a good working approximation for calculations. Such a filter (and others close to it in center frequency) would capture both tones while they are perceived as "beats," "rough fused" or "rough separate," and at the point where rough changes to smooth, the two tones are too far apart to be both captured by this *or any other* filter. At this point there is no single filter which captures both tones, but there are filters which capture each of the tones individually and they are therefore resolved and the two tones are perceived as being "separate and smooth."

A musical sound can be described by the frequency components which make it up, and an understanding of the application of the critical band mechanism in human hearing in terms of the analysis of the components of musical sounds gives the basis for the study of psychoacoustics. The resolution with which the hearing system can analyze the individual components or sine waves in a sound is important for understanding psychoacoustic discussions relating to, for example, how we perceive:

- melody
- harmony
- chords
- tuning
- intonation
- musical dynamics
- the sounds of different instruments

- blend
- ensemble
- interactions between sounds produced simultaneously by different instruments.

2.3 FREQUENCY AND PRESSURE SENSITIVITY RANGES

The human hearing system is usually quoted as having an average frequency range of 20–20 000 Hz, but there can, however, be quite marked differences between individuals. This frequency range changes as part of the human aging process, particularly in terms of the upper limit which tends to reduce. Healthy young children may have a full hearing frequency range up to 20 000 Hz, but, by the age of 20, the upper limit may have dropped to 16 000 Hz. From the age of 20, it continues to reduce gradually. This is usually known as "presbyacusis," or less commonly as "presbycusis," and is a function of the normal aging process.

This reduction in the upper frequency limit of the hearing range is accompanied by a decline in hearing sensitivity at all frequencies with age, the decline being less for low frequencies than for high as shown in Figure 2.10 (consider track 2 on the accompanying CD). The figure also shows that this natural loss of hearing sensitivity and loss of upper frequencies is more marked for men than for women. Hearing losses can also be induced by other factors such as prolonged exposure to loud sounds (see Section 2.5), particularly with some of the high sound levels now readily available from electronic amplification systems, whether reproduced via loudspeakers or particularly via headphones.

The ear's sensitivity to sounds of different frequencies varies over a vast sound pressure level range. On average, the minimum sound pressure

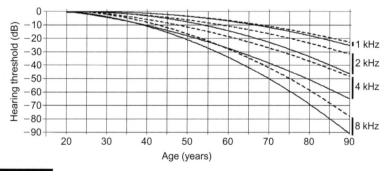

FIGURE 2.10   *The average effect of aging on the hearing sensitivity of women (dotted lines) and men (solid lines) usually known as "presbyacusis," or less commonly as "presbycusis."*

variation which can be detected by the human hearing system around 4 kHz is approximately 10 micropascals $(10 \, \mu Pa)$, or 10^{-5} Pa. The maximum average sound pressure level which is heard rather than perceived as being painful is 64 Pa. The ratio between the loudest and softest is therefore:

$$\frac{\text{Threshold of pain}}{\text{Threshold of hearing}} = \frac{64}{10^{-5}} = 6\,400\,000 = 6.4 \times 10^6$$

This is a very wide range variation in terms of the numbers involved, and it is not a convenient one to work with. Therefore sound pressure level (SPL) is represented in decibels relative to $20 \, \mu Pa$ (see Chapter 1), as dB(SPL) as follows:

$$dB(SPL) = 20 \log \left(\frac{p_{\text{actual}}}{p_{\text{ref}}} \right)$$

where p_{actual} = the actual pressure level (in Pa)
and p_{ref} = the reference pressure level $(20 \, \mu Pa)$

EXAMPLE 2.3

Calculate the threshold of hearing and threshold of pain in dB(SPL).
The threshold of hearing at 1 kHz is, in fact, p_{ref} which in dB(SPL) equals:

$$20 \log \left(\frac{p_{\text{ref}}}{p_{\text{ref}}} \right) = 20 \log \left(\frac{2 \times 10^{-5}}{2 \times 10^{-5}} \right) = 20 \log(1)$$
$$= 20 \times 0 = 0 \text{ dB(SPL)}$$

and the threshold of pain is 64 Pa which in dB(SPL) equals:

$$20 \log \left(\frac{p_{\text{actual}}}{p_{\text{ref}}} \right) = 20 \log \left(\frac{64}{2 \times 10^{-5}} \right) = 20 \log(6.4 \times 10^6)$$
$$= 20 \times 6.5 = 130 \text{ dB(SPL)}$$

Use of the dB(SPL) scale results in a more convenient range of values (0 to 130) to consider, since values in the range of about 0 to 100 are common in everyday dealings. Also, it is a more appropriate basis for expressing acoustic amplitude values, changes in which are primarily perceived as variations in loudness since loudness perception turns out to be essentially logarithmic in nature.

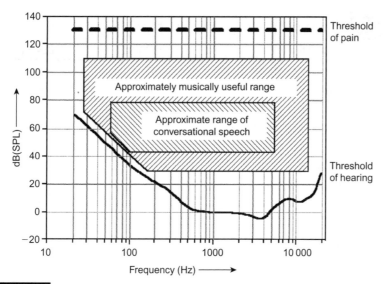

FIGURE 2.11 *The general shape of the average human threshold of hearing and the threshold of pain with approximate conversational speech and musically useful ranges.*

The threshold of hearing varies with frequency. The ear is far more sensitive in the middle of its frequency range than at the high and low extremes. The lower curve in Figure 2.11 is the general shape of the average threshold of hearing curve for sinusoidal stimuli between 20 Hz and 20 kHz. The upper curve in the figure is the general shape of the threshold of pain, which also varies with frequency but not to such a great extent. It can be seen from the figure that the full 130 dB(SPL) range, or "dynamic range," between the threshold of hearing and the threshold of pain exists at approximately 4 kHz, but that the dynamic range available at lower and higher frequencies is considerably less. For reference, the sound level and frequency range for both average normal conversational speech and music are shown in Figure 2.11, while Table 2.1 shows approximate sound levels of everyday sounds for reference.

2.4 LOUDNESS PERCEPTION

Although the perceived loudness of an acoustic sound is related to its amplitude, there is not a simple one-to-one functional relationship. As a psychoacoustic effect it is affected by both the context and nature of the sound. It is also difficult to measure because it is dependent on the interpretation by listeners of what they hear. It is neither ethically appropriate

Table 2.1	Typical sound levels in the environment	
Example sound/situation	**dB(SPL)**	**Description**
Long range gunfire at gunner's ear	140	
Threshold of pain	130	Ouch!
Jet take-off at approximately 100 m	120	
Peak levels on a night club dance floor	110	
Loud shout at 1 m	100	Very noisy
Heavy truck at about 10 m	90	
Heavy car traffic at about 10 m	80	
Car interior	70	Noisy
Normal conversation at 1 m	60	
Office noise level	50	
Living room in a quiet area	40	Quiet
Bedroom at night time	30	
Empty concert hall	20	
Gentle breeze through leaves	10	Just audible
Threshold of hearing for a child	0	

nor technologically possible to put a probe in the brain to ascertain the loudness of a sound.

In Chapter 1 the concepts of the sound pressure level and sound intensity level were introduced. These were shown to be approximately equivalent in the case of free space propagation in which no interference effects were present. The ear is a pressure sensitive organ that divides the audio spectrum into a set of overlapping frequency bands whose bandwidth increases with frequency. These are both objective descriptions of the amplitude and the function of the ear. However, they tell us nothing about the perception of loudness in relation to the objective measures of sound amplitude level. Consideration of such issues will allow us to understand some of the effects that occur when one listens to musical sound sources.

The pressure amplitude of a sound wave does not directly relate to its perceived loudness. In fact it is possible for a sound wave with a larger pressure amplitude to sound quieter than a sound wave with a lower pressure amplitude. How can this be so? The answer is that the sounds are at different frequencies and the sensitivity of our hearing varies as the frequency varies. Figure 2.12 shows the equal loudness contours for the human ear. These contours, originally measured by Fletcher and Munson (1933) and by others since, represent the relationship between the measured sound pressure level and the perceived loudness of the sound. The curves show how loud a sound must be in terms of the measured sound pressure level to be perceived as being of the same loudness as a 1 kHz tone of a given level. There are two major features to take note of, which are discussed below.

The first is that there are some humps and bumps in the contours above 1 kHz. These are due to the resonances of the outer ear. Within the outer ear there is a tube about 25 mm long with one open and one closed end. This will have a first resonance at about 3.4 kHz and, due to its non-uniform shape, a second resonance at approximately 13 kHz, as shown in the figure. The effect of these resonances is to enhance the sensitivity of the ear around the resonant frequencies. Note that because this enhancement is due to an acoustic effect in the outer ear it is independent of signal level.

The second effect is an amplitude dependence of sensitivity which is due to the way the ear transduces and interprets the sound and, as a result, the frequency response is a function of amplitude. This effect is particularly noticeable at low frequencies but there is also an effect at higher frequencies. The net result of these effects is that the sensitivity of the ear is a function of both frequency and amplitude. In other words the frequency response of

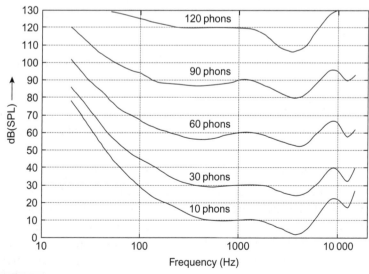

FIGURE 2.12 *Equal loudness contours for the human ear.*

the ear is not flat and is also dependent on sound level. Therefore two tones of equal sound pressure level will rarely sound equally loud. For example, a sound at a level which is just audible at 20 Hz would sound much louder if it was at 4 kHz. Tones of different frequencies therefore have to be at different sound pressure levels to sound equally loud and their relative loudness will also be a function of their absolute sound pressure levels.

The loudness of sine wave signals, as a function of frequency and sound pressure levels, is given by the "phon" scale. The phon scale is a subjective scale of loudness based on the judgments of listeners to match the loudness of tones to reference tones at 1 kHz. The curve for N phones intersects 1 kHz at N dB(SPL) by definition, and it can be seen that the relative shape of the phon curves flattens out at higher sound levels, as shown in Figure 2.12. The relative loudness of different frequencies is not preserved, and therefore the perceived frequency balance of sound varies as the listening level is altered. This is an effect that we have all heard when the volume of a recording is turned down and the bass and treble components appear suppressed relative to the midrange frequencies, and the sound becomes "duller" and "thinner." Ideally we should listen to reproduced sound at the level at which it was originally recorded. However, in most cases this would be antisocial, especially as much rock material is mixed at levels in excess of 100 dB(SPL)!

In the early 1970s hi-fi manufacturers provided a "loudness" button which put in a bass and treble boost in order to flatten the Fletcher–Munson curves, and so provide a simple compensation for the reduction in hearing sensitivity at low levels. The action of this control was wrong in two important respects:

- Firstly, it directly used the equal loudness contours to perform the compensation, rather than the difference between the curves at two different absolute sound pressure levels, which would be more accurate. The latter approach has been used in professional products to allow nightclubs to achieve the equivalent effect of a louder replay level.

- Secondly, the curves are a measure of the equal loudness for *sine waves* at a similar level. Real music on the other hand consists of many different frequencies at many different amplitudes and does not directly follow these curves as its level changes. We shall see later how we can analyze the loudness of complex sounds. In fact because the response of the ear is dependent on both absolute sound pressure level and frequency it cannot be compensated for simply by using treble and bass boost.

2.4.1 Measuring loudness

These effects make it difficult to design a meter which will give a reading which truly relates to the perceived loudness of a sound, so an instrument which gives an approximate result is usually used. This is achieved by using the sound pressure level but frequency weighting it to compensate for the variation of sensitivity of the ear as a function of frequency. Clearly the optimum compensation will depend on the absolute value of the sound pressure level being measured and so some form of compromise is necessary.

Figure 2.13 shows two frequency weightings which are commonly used to perform this compensation—termed "A" and "C" weightings. The "A" weighting is most appropriate for low amplitude sounds as it broadly compensates for the low-level sensitivity versus frequency curve of the ear. The "C" weighting on the other hand is more suited to sound at higher absolute sound pressure levels and because of this is more sensitive to low-frequency components than the "A" weighting. The sound levels measured using the "A" weighting are often given in the unit dBA, and levels using the "C" weighting in dBA. Despite the fact that it is most appropriate for low sound levels, and is a reasonably good approximation there, the "A" weighting is now recommended for any sound level in order to provide a measure of consistency between measurements.

The frequency weighting is not the only factor which must be considered when using a sound level meter. In order to obtain an estimate of the sound pressure level it is necessary to average over at least one cycle, and

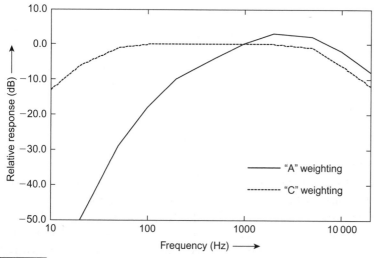

FIGURE 2.13 The frequency response of "A" and "C" weightings.

preferably more, of the sound waveform. Thus most sound level meters have slow and fast time response settings. The slow time response gives an estimate of the average sound level whereas the fast response tracks more rapid variations in the sound pressure level.

Sometimes it is important to be able to calculate an estimate of the equivalent sound level experienced over a period of time. This is especially important when measuring people's noise exposure in order to see if they might suffer noise-induced hearing loss. This cannot be done using the standard fast or slow time responses on a sound level meter; instead a special form of measurement known as the "L_{eq}" (pronounced L E Q) measurement is used.

This measure integrates the instantaneous squared pressure over some time interval, such as 15 min or 8 h, and then takes the square root of the result. This provides an estimate of the root mean square level of the signal over the time period of the measurement and so gives the equivalent sound level for the time period. That is, the output of the L_{eq} measurement is the constant sound pressure level, which is equivalent to the varying sound level over the measurement period. The L_{eq} measurement also provides a means of estimating the total energy in the signal by squaring its output. A series of L_{eq} measurements over short times can also be easily combined to provide a longer time L_{eq} measurement by simply squaring the individual results, adding them together, and then taking the square root of the result, as shown in Equation 2.7.

$$L_{eq(total)} = \sqrt{L^2_{eq1} + L^2_{eq2} + \cdots + L^2_{eqn}}$$

(2.7)

where $L_{eq(1-n)}$ = the individual short time L_{eq} measurements

This extendibility makes the L_{eq} measurement a powerful method of noise monitoring. As with a conventional instrument, the "A" or "C" weightings can be applied.

2.4.2 Loudness of simple sounds

In Figure 2.10 the two limits of loudness are illustrated: the threshold of hearing and the threshold of pain. As we have already seen, the sensitivity of the ear varies with frequency and therefore so does the threshold of hearing, as shown in Figure 2.14. The peak sensitivities shown in this figure are equivalent to a sound pressure amplitude in the sound wave of $10\,\mu Pa$ or about $-6\,dB(SPL)$. Note that this is for monaural listening to a sound presented at the front of the listener. For sounds presented on the listening

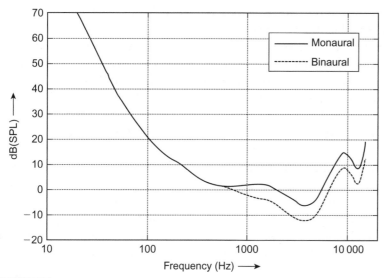

FIGURE 2.14   *The threshold of hearing as a function of frequency.*

side of the head there is a rise in peak sensitivity of about 6 dB due to the increase in pressure caused by reflection from the head. There is also some evidence that the effect of hearing with two ears is to increase the sensitivity by between 3 and 6 dB.

At 4 kHz, which is about the frequency of the sensitivity peak, the pressure amplitude variations caused by the Brownian motion of air molecules, at room temperature and over a critical bandwidth, correspond to a sound pressure level of about −23 dB. Thus the human hearing system is close to the theoretical physical limits of sensitivity. In other words there would be little point in being much more sensitive to sound, as all we would hear would be a "hiss" due to the thermal agitation of the air! Many studio and concert hall designers now try to design the building such that environmental noise levels are lower than the threshold of hearing, and so are inaudible.

The second limit is the just noticeable change in amplitude. This is strongly dependent on the nature of the signal, its frequency, and its amplitude. For broad-band noise the just noticeable difference in amplitude is 0.5 to 1 dB when the sound level lies between 20 and 100 dB(SPL) relative to a threshold of 0 dB(SPL). Below 20 dB(SPL) the ear is less sensitive to changes in sound level. For pure sine waves, however, the sensitivity to change is markedly different and is a strong function of both amplitude and frequency. For example, at 1 kHz the just noticeable amplitude change varies from 3 dB at 10 dB(SPL) to 0.3 dB at 80 dB(SPL).

This variation occurs at other frequencies as well but in general the just noticeable difference at other frequencies is greater than the values for 1–4 kHz. These different effects make it difficult to judge exactly what difference in amplitude would be noticeable as it is clearly dependent on the precise nature of the sound being listened to. There is some evidence that once more than a few harmonics are present the just noticeable difference is closer to the broad-band case, of 0.5–1 dB, rather than the pure tone case. As a general rule of thumb the just noticeable difference in sound level is about 1 dB.

The mapping of sound pressure change to loudness variation for larger changes is also dependent on the nature of the sound signal. However, for broad-band noise, or sounds with several harmonics, it is generally accepted that a change of about 10 dB in SPL corresponds to a doubling or halving of perceived loudness. However, this scaling factor is dependent on the nature of the sound, and there is some dispute over both its value and its validity.

EXAMPLE 2.4

Calculate the increase in the number of violinists required to double the loudness of a string section, assuming all the violinists play at the same sound level.

From Chapter 1 the total level from combining several uncorrelated sources is given by:

$$P_{N\ uncorrelated} = P\sqrt{N}$$

This can be expressed in terms of the SPL as:

$$SPL_{N\ uncorrelated} = SPL_{single\ source} + 10\log_{10}(N)$$

In order to double the loudness we need an increase in SPL of 10 dB. Since 10 log(10) = 10, ten times the number of sources will raise the SPL by 10 dB.

Therefore we must increase the number of violinists in the string section by a factor of ten in order to double their volume.

As well as frequency and amplitude, duration also has an effect on the perception of loudness, as shown in Figure 2.15 for a pure tone. Here we can see that once the sound lasts more than about 200 milliseconds then its perceived level does not change. However, when the tone is shorter than this the perceived amplitude reduces. The perceived amplitude is inversely proportional to the length of the tone burst. This means that when we listen to sounds which vary in amplitude the loudness level is not perceived significantly by short amplitude peaks, but more by the sound level averaged over 200 milliseconds.

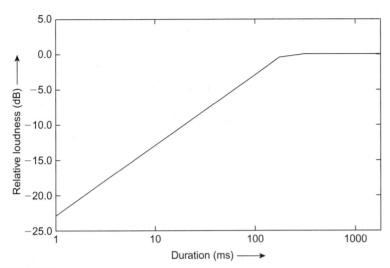

FIGURE 2.15 *The effect of tone duration on loudness.*

2.4.3 Loudness of complex sounds

Unlike tones, real sounds occupy more than one frequency. We have already seen that the ear separates sound into frequency bands based on critical bands. The brain seems to treat sounds within a critical band differently to those outside its frequency range and there are consequential effects on the perception of loudness.

The first effect is that the ear seems to lump all the energy within a critical band together and treat it as one item of sound. So when all the sound energy is concentrated within a critical band the loudness is proportional to the total intensity of the sound within the critical band. That is:

$$\text{Loudness} \quad \propto \quad P_1^2 + P_2^2 + \cdots + P_n^2 \tag{2.8}$$

where P_{1-n} = the pressures of the n individual frequency components

As the ear is sensitive to sound pressures, the sound intensity is proportional to the square of the sound pressures, as discussed in Chapter 1. Because the acoustic intensity of the sound is also proportional to the sum of the squared pressure, the loudness of a sound within a critical band is independent of the number of frequency components so long as their total acoustic intensity is constant. When the frequency components of the sound extend beyond a critical band, an additional effect occurs due to the presence of components in other critical bands. In this case more than one critical band is contributing to the perception of loudness and the brain appears to add the individual critical band responses together. The effect is to increase the perceived loudness of the sound even though the total acoustic intensity is unchanged.

Figure 2.16 shows a plot of the subjective loudness perception of a sound at a constant intensity level as a function of the sound's bandwidth, which illustrates this effect. In the cochlea the critical bands are determined by the place at which the peak of the standing wave occurs; therefore all energy within a critical band will be integrated as one overall effect at that point on the basilar membrane and transduced into nerve impulses as a unit. On the other hand, energy which extends beyond a critical band will cause other nerves to fire and it is these extra nerve firings which give rise to an increase in loudness.

The interpretation of complex sounds which cover the whole frequency range is further complicated by psychological effects, in that a listener will attend to particular parts of the sound, such as the soloist, or conversation, and ignore or be less aware of other sounds, and will tend to base their perception of loudness on what they have attended to.

Duration also has an effect on the perception of the loudness of complex tones in a similar fashion to that of pure tones. As is the case for pure tones, complex tones have an amplitude which is independent of duration once the sound is longer than about 200 milliseconds, and is inversely proportionate to duration when the duration is less than this.

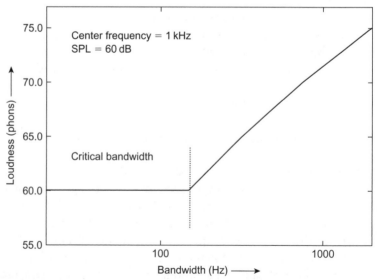

FIGURE 2.16 *The effect of tone bandwidth on loudness.*

2.5 NOISE-INDUCED HEARING LOSS

The ear is a sensitive and accurate organ of sound transduction and analysis. However, the ear can be damaged by exposure to excessive levels of sound or noise. This damage can manifest itself in two major forms:

- *A loss of hearing sensitivity*: The effect of noise exposure causes the efficiency of the transduction of sound into nerve impulses to reduce. This is due to damage to the hair cells in each of the organs of Corti. Note this is different from the threshold shift due to the acoustic reflex, which occurs over a much shorter time period and is a form of built-in hearing protection. This loss of sensitivity manifests itself as a shift in the threshold of hearing of someone who has been exposed to excessive noise, as shown in Figure 2.17. This shift in the threshold can be temporary, for short times of exposures, but ultimately it becomes permanent as the hair cells are permanently flattened as a result of the damage, due to long-term exposure, which does not allow them time to recover.

- *A loss of hearing acuity*: This is a more subtle effect, but in many ways is more severe than the first effect. We have seen that a crucial part of our ability to hear and analyze sounds is our ability to separate out the sounds into distinct frequency bands, called critical bands. These bands are very narrow. Their narrowness is due to an

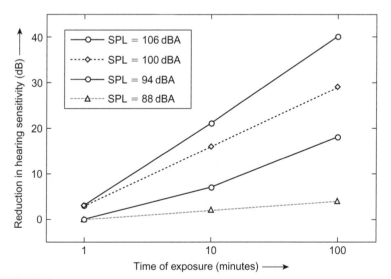

FIGURE 2.17 *The effect of noise exposure on hearing sensitivity (data from Tempest, 1985).*

active mechanism of positive feedback in the cochlea which enhances the standing wave effects mentioned earlier. This enhancement mechanism is very easily damaged; it appears to be more sensitive to excessive noise than the main transduction system. The effect of the damage though is not just to reduce the threshold but also to increase the bandwidth of our acoustic filters, as shown in idealized form in Figure 2.18. This has two main effects: firstly, our ability to separate out the different components of the sound is impaired, and this will reduce our ability to understand speech or separate out desired sound from competing noise. Interestingly it may well make musical sounds that were consonant more dissonant because of the presence of more than one frequency harmonic in a critical band; this will be discussed in Chapter 3. The second effect is a reduction in the hearing sensitivity, also shown in Figure 2.18, because the enhancement mechanism also increases the amplitude sensitivity of the ear. This effect is more insidious because the effect is less easy to measure and perceive; it manifests itself as a difficulty in interpreting sounds rather than a mere reduction in their perceived level.

Another related effect due to damage to the hair cells is noise-induced tinnitus. Tinnitus is the name given to a condition in which the cochlea spontaneously generates noise, which can be tonal or random noises, or a mixture of the two. In noise-induced tinnitus exposure to loud noise triggers this, and, as well as being disturbing, there is some evidence that people who

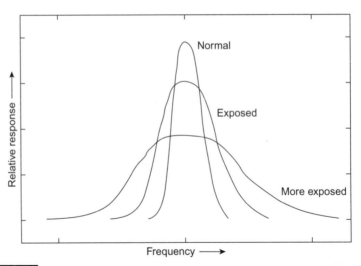

FIGURE 2.18 *Idealized form of the effect of noise exposure on hearing bandwidth.*

suffer from this complaint may be more sensitive to noise-induced hearing damage.

Because the damage is caused by excessive noise exposure, it is more likely at the frequencies at which the acoustic level at the ear is enhanced. The ear is most sensitive at the first resonance of the ear canal, or about 4 kHz, and this is the frequency at which most hearing damage first shows up. Hearing damage in this region is usually referred to as an "audiometric notch" because of its distinctive shape on an audiogram (once the results have been plotted following a hearing test); see Figure 2.19. (Hearing testing using audiograms and the dBHL scale are described in detail in Section 7.2.) This distinctive pattern is evidence that the hearing loss measured is due to noise exposure rather than some other condition, such as the inevitable high-frequency loss due to aging.

How much noise exposure is acceptable? There is some evidence that the levels of noise generated by our normal noisy Western society have some long-term effects because measurements on the hearing of other cultures show a much lower threshold of hearing at a given age compared with that of Westerners. However, this may be due to other factors as well, for example the level of pollution. But strong evidence exists demonstrating that exposure to noises with amplitudes of greater than 90 dBA can cause permanent hearing damage. This fact is recognized, for example, by UK legislation, which requires that the noise exposure of workers be less than

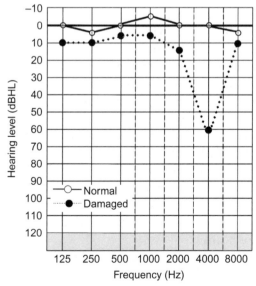

FIGURE 2.19   *Example audiograms of normal and damaged (notched) hearing.*

this limit, known as the "second action limit." This level has been reduced since April 2006 under European legislation to 85 dBA. If the work environment has a noise level greater than the second action limit, then employers are obliged to provide hearing protection for employees of a sufficient standard to bring the noise level at the ear below this figure. There is also a "first action level," which is 5 dB below the second action level, and if employees are subjected to this level (now 80 dBA in Europe), then employees can request hearing protection which must be made available.

Please be aware that the regulations vary by country. Readers should check their local regulations for noise exposure in practice.

2.5.1 Integrated noise dose

However, in many musical situations the noise level is greater than 90 dBA for short periods. For example, the audience at a concert may well experience brief peaks above this, especially at particular instants in works such as Elgar's *Dreams of Gerontius*, or Orff's *Carmina Burana*. Also, in many practical industrial and social situations the noise level may be louder than the second action level of 85 dBA, in Europe, for only part of the time. How can we relate intermittent periods of noise exposure to continuous noise exposure? For example, how damaging is a short exposure to a sound of 96 dBA? The answer is to use a similar technique to that used in assessing the effect of radiation exposure, that is, "integrated dose."

The integrated noise dose is defined as the equivalent level of the sound over a fixed period of time, which is currently 8 hours. In other words the noise exposure can be greater than the second action level provided that it is for an appropriately shorter time, which results in a noise dose that is less than that which would result from being exposed to noise at the second action level for 8 hours. The measure used is the L_{eq} mentioned earlier and the maximum dose is $85 \, \text{dB}L_{eq}$ over 8 hours. This means that one can be exposed to 88 dBA for 4 hours, 91 dBA for 2 hours, and so on.

Figure 2.20 shows how the time of exposure varies with the sound level on linear and logarithmic timescales for the second action level in Europe. It can be seen that exposure to extreme sound levels, greater than 100 dBA, can only be tolerated for a very short period of time, less than half an hour. There is also a limit to how far this concept can be taken because very loud sounds can rupture the eardrum causing instant, and sometimes permanent, loss of hearing.

This approach to measuring the noise dose takes no account of the spectrum of the sound which is causing the noise exposure, because to do so would be difficult in practice. However, it is obvious that the effect of a

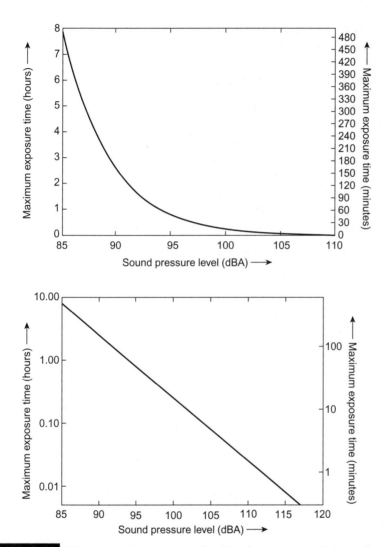

FIGURE 2.20 *Maximum exposure time as a function of sound level plotted on a linear scale (upper) and a logarithmic scale (lower).*

pure tone at 85 dBA on the ear is going to be different from the same level spread over the full frequency range. In the former situation there will be a large amount of energy concentrated at a particular point on the basilar membrane and this is likely to be more damaging than the second case in which the energy will be spread out over the full length of the membrane. Note that the specification for noise dose uses "A" weighting for the measurement which, although it is more appropriate for low rather than high sound levels, weights the sensitive 4 kHz region more strongly.

2.5.2 Protecting your hearing

Hearing loss is insidious and permanent, and by the time it is measurable it is too late. Therefore in order to protect hearing sensitivity and acuity one must be proactive. The first strategy is to avoid exposure to excess noises. Although 85 dB(SPL) is taken as a damage threshold if the noise exposure causes ringing in the ears, especially if the ringing lasts longer than the length of exposure, it may be that damage may be occurring even if the sound level is less than 85 dB(SPL).

There are a few situations where potential damage is more likely:

- The first is when listening to recorded music through headphones, as even small ones are capable of producing damaging sound levels.

- The second is when one is playing music, with either acoustic or electric instruments, as these are also capable of producing damaging sound levels, especially in small rooms with a "live" acoustic; see Chapter 6.

In both cases the levels are under your control and so can be reduced. However, there is an effect called the acoustic reflex (see Section 2.1.2), which reduces the sensitivity of your hearing when loud sounds occur. This effect, combined with the effects of temporary threshold shifts, can result in a sound level increase spiral where there is a tendency to increase the sound level "to hear it better," which results in further dulling, etc. The only real solution is to avoid the loud sounds in the first place. However, if this situation does occur then a rest away from the excessive noise will allow some sensitivity to return.

There are sound sources over which one has no control, such as bands, discos, nightclubs, and power tools. In these situations it is a good idea either to limit the noise dose or, better still, use some hearing protection. For example, one can keep a reasonable distance away from the speakers at a concert or disco. It takes a few days, or even weeks in the case of hearing acuity, to recover from a large noise dose so one should avoid going to a loud concert, or nightclub, every day of the week!

The authors regularly use small "in-ear" hearing protectors when they know they are going to be exposed to high sound levels, and many professional sound engineers also do the same. These have the advantage of being unobtrusive and reduce the sound level by a modest, but useful, amount (15–20 dB) while still allowing conversation to take place at the speech levels required to compete with the noise! These devices are also available with a "flat" attenuation characteristic with frequency and so do not alter the sound balance too much, and cost less than a CD recording. For very loud sounds, such as those emitted by power tools, a more extreme form of hearing protection may be required, such as headphone style ear defenders.

Your hearing is essential, and irreplaceable, for the enjoyment of music, for communicating, and for socializing with other people. Now and in the future, it is worth taking care of.

2.6 PERCEPTION OF SOUND SOURCE DIRECTION

How do we perceive the direction that a sound arrives from? The answer is that we make use of our two ears, but how? Because our two ears are separated by our head, this has an acoustic effect which is a function of the direction of the sound. There are two effects of the separation of our ears on the sound wave: firstly, the sounds arrive at different times and, secondly, they have different intensities. These two effects are quite different so let us consider them in turn.

2.6.1 Interaural time difference (ITD)

Consider the model of the head, shown in Figure 2.21, which shows the ears relative to different sound directions in the horizontal plane. Because

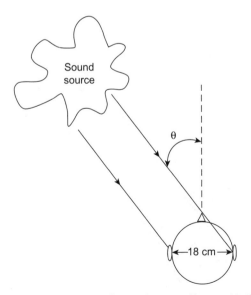

FIGURE 2.21   *The effect of the direction of a sound source with respect to the head.*

the ears are separated by about 18 cm there will be a time difference between the sound arriving at the ear nearest the source and the one further away. So when the sound is off to the left the left ear will receive the sound first, and when it is off to the right the right ear will hear it first. If the sound is directly in front, or behind, or anywhere on the median plane, the sound will arrive at both ears simultaneously. The time difference between the two ears will depend on the difference in the distances that the two sounds have to travel. A simplistic view might just allow for the fact that the ears are separated by a distance d and therefore calculate the effect of angle on the relative time difference by considering only the extra length introduced due to the angle of incidence, as shown in Figure 2.22. This assumption will give the following equation for the time difference due to sound angle:

$$\Delta t = \frac{d \ \sin(\theta)}{c}$$

where Δt = the time difference between the ears (in s)

d = the distance between the ears (in m)

θ = the angle of arrival of the sound from the median (in radians)

and c = the speed of sound (in ms^{-1})

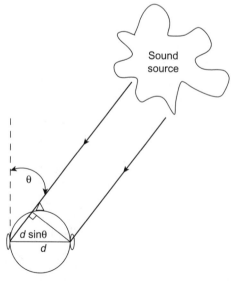

FIGURE 2.22 *A simple model for the interaural time difference.*

Unfortunately this equation is wrong. It underestimates the delay between the ears because it ignores the fact that the sound must travel around the head in order to get to them. This adds an additional delay to the sound, as shown in Figure 2.23. This additional delay can be calculated, provided one assumes that the head is spherical, by recognizing that the distance traveled around the head for a given angle of incidence is given by:

$$\Delta d = r\theta$$

where Δd = the extra path round the head at a given angle of
    incidence (in m)
    and r = half the distance between the ears (in m)

This equation can be used in conjunction with the extra path length due to the angle of incidence, which is now a function of r, as shown in Figure 2.24, to give a more accurate equation for the ITD as:

$$ITD = \frac{r(\theta + \sin(\theta))}{c} \tag{2.9}$$

Using this equation we can find that the maximum ITD, which occurs at 90° or ($\pi/2$ radians), is:

$$ITD_{max} = \frac{0.09 \text{ m} \times (\pi/2 + \sin(\pi/2))}{344 \text{ ms}^{-1}} = 6.73 \times 10^{-4} \text{ s } (673 \text{ μs})$$

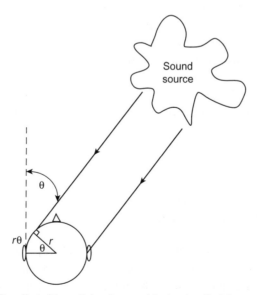

FIGURE 2.23 *The effect of the path length around the head on the interaural time difference.*

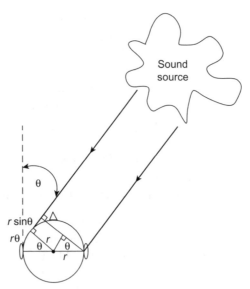

FIGURE 2.24 *A better model for the interaural time difference.*

This is a very small delay but a variation from this to zero determines the direction of sounds at low frequencies. Figure 2.25 shows how this delay varies as a function of angle, where positive delay corresponds to a source at the right of the median plane and negative delay corresponds to a source on the left. Note that there is no difference in the delay between front and back positions at the same angle. This means that we must use different mechanisms and strategies to differentiate between front and back sounds. There is also a frequency limit to the way in which sound direction can be resolved by the ear in this way. This is due to the fact that the ear appears to use the phase shift in the wave caused by the interaural time difference to resolve the direction; that is, the ear measures the phase shift given by:

$$\Phi_{ITD} = 2\pi fr(\theta + \sin(\theta))$$

where Φ_{ITD} = the phase difference between the ears (in radians)
and f = the frequency (in Hz)

When this phase shift is greater than π radians (180°) there will be an unresolvable ambiguity in the direction because there are two possible angles—one to the left and one to the right—that could cause such a phase shift.

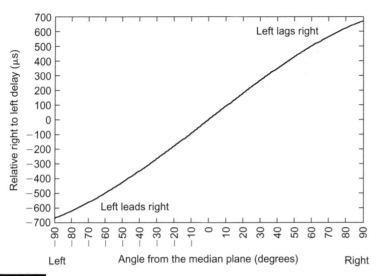

FIGURE 2.25   *The interaural time difference (ITD) as a function of angle.*

This sets a maximum frequency, at a particular angle, for this method of sound localization, which is given by:

$$f_{max}(\theta) = \frac{1}{2 \times 0.09 \text{ m} \times (\theta + \sin(\theta))}$$

which for an angle of 90° is:

$$f_{max}(\theta = \pi/2) = \frac{1}{2 \times 0.09 \text{ m} \times (\pi/2 + \sin(\pi/2))} = 743 \text{ Hz}$$

Thus for sounds at 90° the maximum frequency that can have its direction determined by phase is 743 Hz. However, the ambiguous frequency limit would be higher at smaller angles.

2.6.2 Interaural intensity difference (IID)

The other cue that is used to detect the direction of the sound is the differing levels of intensity that result at each ear due to the shading effect of the head. This effect is shown in Figure 2.26 which shows that the levels at each ear are equal when the sound source is on the median plane but that the level at one ear progressively reduces, and increases at the other, as the source moves away from the median plane. The level reduces in the ear that is furthest away from the source. The effect of the shading of the head is harder to calculate

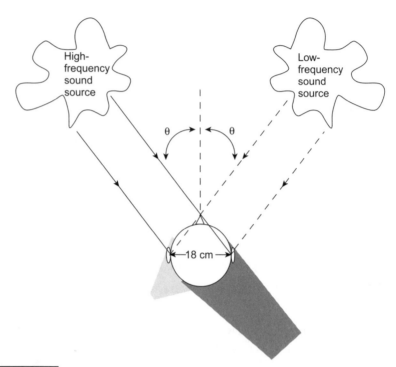

FIGURE 2.26 *The effect of the head on the interaural intensity difference.*

but experiments seem to indicate that the intensity ratio between the two ears varies sinusoidally in a frequency dependent fashion, from 0 dB up to 20 dB, depending on the sound direction, as shown in Figure 2.27.

However, as we saw in Chapter 1, an object is not significant as a scatterer or shader of sound until its size is about two thirds of a wavelength $(\frac{1}{2}\lambda)$, although it will be starting to scatter an octave below that frequency. This means that there will be a minimum frequency below which the effect of intensity is less useful for localization, which will correspond to when the head is about one third of a wavelength in size $(\frac{1}{3}\lambda)$. For a head the diameter of which is 18 cm, this corresponds to a minimum frequency of:

$$f_{\min(\theta=\pi/2)} = \frac{1}{3}\left(\frac{c}{d}\right) = \frac{1}{3} \times \left(\frac{344 \text{ ms}^{-1}}{0.18 \text{ m}}\right) = 637 \text{ Hz}$$

Thus the interaural intensity difference is a cue for direction at high frequencies whereas the interaural time difference is a cue for direction at low frequencies. Note that the crossover between the two techniques starts at about 700 Hz and would be complete at about four times this frequency

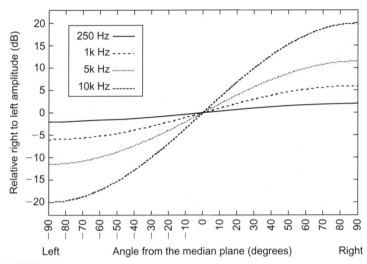

FIGURE 2.27 *The interaural intensity difference (IID) as a function of angle and frequency (data from Gulick, 1971).*

at 2.8 kHz. In between these two frequencies the ability of our ears to resolve direction is not as good as at other frequencies.

2.6.3 Pinnae and head movement effects

The above models of directional hearing do not explain how we can resolve front to back ambiguities or the elevation of the source. There are in fact two ways which are used by the human being to perform these tasks.

The first is to use the effect of our ears on the sounds we receive to resolve the angle and direction of the sound. This is due to the fact that sounds striking the pinnae are reflected into the ear canal by the complex set of ridges that exist on the ear. These pinnae reflections will be delayed, by a very small but significant amount, and so will form comb filter interference effects on the sound the ear receives. The delay that a sound wave experiences will be a function of its direction of arrival, in all three dimensions, and we can use these cues to help resolve the ambiguities in direction that are not resolved by the main directional hearing mechanism. The delays are very small and so these effects occur at high audio frequencies, typically above 5 kHz.

The effect is also person specific, as we all have differently shaped ears and learn these cues as we grow up. Thus we get confused for a while when we change our acoustic head shape radically, for example by cutting very long hair short. We also find that if we hear sound recorded through other people's ears we may have a different ability to localize the sound, because the interference patterns are not the same as those for our ears. In fact, sometimes this localization capability is worse than when using our own ears and sometimes it is better.

The second, and powerful, means of resolving directional ambiguities is to move our heads. When we hear a sound that we wish to attend to, or whose direction we wish to resolve, we turn our head towards the sound and may even attempt to place it in front of us in the normal direction, where all the delays and intensities will be the same. The act of moving our head will change the direction of the sound arrival and this change of direction will depend on the sound source position relative to us. Thus a sound from the rear will move in a different direction compared with a sound in front of or above the listener. This movement cue is one of the reasons that we perceive the sound from headphones as being "in the head." Because the sound source tracks our head movement it cannot be outside and hence must be in the head. There is also an effect due to the fact that the headphones also do not model the effect of the head. Experiments with headphone listening which correctly model the head and keep the source direction constant as the head moves give a much more convincing illusion.

2.6.4 ITD and IID trading

Because both intensity and delay cues are used for the perception of sound source direction one might expect the mechanisms to be in similar areas of the brain and linked together. If this were the case one might also reasonably expect that there was some overlap in the way the cues were interpreted such that intensity might be confused with delay and vice versa in the brain. This allows for the possibility that the effect of one cue, for example delay, could be canceled out by the other, for example intensity. This effect does in fact happen and is known as "interaural time difference versus interaural intensity difference trading." In effect, within limits, an interaural time delay can be compensated for by an appropriate interaural intensity difference, as shown in Figure 2.28, which has several interesting features.

Firstly, as expected, time delay versus intensity trading is only effective over the range of delay times which correspond to the maximum interaural time delay of 673 µs. Beyond this amount of delay, small intensity differences will not alter the perceived direction of the image. Instead the sound will appear to come from the source which arrives first. This effect occurs between 673 µs and 30 ms. However, if the delayed sound's amplitude is more than 12 dB greater than the first arrival then we will perceive the direction of the sound to be towards the delayed sound. After 30 ms the delayed signal is perceived as an echo and so the listener will be able to differentiate

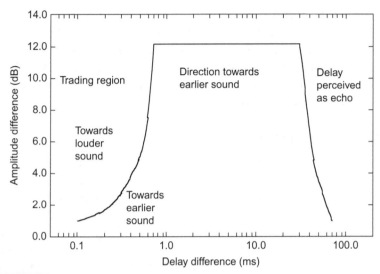

FIGURE 2.28 *Delay versus intensity trading (data from Madsen, 1990).*

between the delayed and undelayed sound. The implications of these results are twofold; firstly, it should be possible to provide directional information purely through either only delay cues or only intensity cues. Secondly, once a sound is delayed by greater than about 700 μs the ear attends to the sound that arrives first almost irrespective of the relative levels of the incoming sounds, although clearly if the earlier arriving sound is significantly lower in amplitude, compared with the delayed sound, then the effect will disappear.

2.6.5 The Haas effect

The second of the ITD and IID trading effects is also known as the "Haas," or "precedence," effect—named after the experimenter who quantified this behavior of our ears. The effect can be summarized as follows:

- The ear will attend to the direction of the sound that arrives first and will not attend to the reflections provided they arrive within 30 ms of the first sound.

- The reflections arriving before 30 ms are fused into the perception of the first arrival. However, if they arrive after 30 ms they will be perceived as echoes.

These results have important implications for studios, concert halls and sound reinforcement systems. In essence, it is important to ensure that the first reflections arrive at the audience earlier than 30 ms to avoid them being perceived as echoes. In fact it seems that our preference is for a delay gap of less than 20 ms if the sound of the hall is to be classed as "intimate." In sound reinforcement systems the output of the speakers will often be delayed with respect to their acoustic sound but, because of this effect, we perceive the sound as coming from the acoustic source, unless the level of sound from the speakers is very high.

2.6.6 Stereophonic listening

Because of the way we perceive directional sound it is possible to fool the ear into perceiving a directional sound through just two loudspeakers or a pair of headphones in stereo listening. This can be achieved in basically three ways: two using loudspeakers and one using headphones. The first two ways are based on the concept of providing only one of the two major directional cues in the hearing system; that is, using either intensity or delay cues and relying on the effect of the ear's time–intensity trading mechanisms to fill in the gaps. The two systems are as follows:

- *Delay stereo*: This system is shown in Figure 2.29 and consists of two omni-directional microphones spaced a reasonable distance apart and away from the performers. Because of the distance of the microphones a change in performer position does not alter the sound intensity much, but does alter the delay. So the two channels when

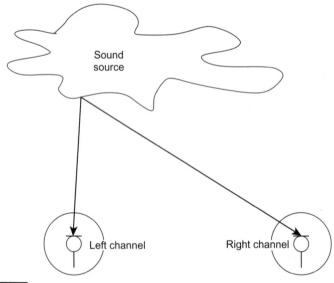

FIGURE 2.29 *Delay stereo recording.*

presented over loudspeakers contain predominantly directional cues based on delay to the listener.

- *Intensity stereo*: This system is shown in Figure 2.30 and consists of two directional microphones placed together and pointing at the left and right extent of the performers' positions. Because the microphones are closely spaced, a change in performer position does not alter the delay between the two sounds. However, because the microphones are directional the intensity received by the two microphones does vary. So the two channels when presented over loudspeakers contain predominantly directional cues based on intensity to the listener. Intensity stereo is the method that is mostly used in pop music production, as the pan-pots on a mixing desk, which determine the position of a track in the stereo image, vary the relative intensities of the two channels, as shown in Figure 2.31.

These two methods differ primarily in the method used to record the original performance and are independent of the listening arrangement, so which method is used is determined by the producer or engineer on the recording. It is also possible to mix the two cues by using different types of microphone arrangement—for example, slightly spaced directional microphones—and these can give stereo based on both cues. Unfortunately they also provide spurious cues, which confuse the ear, and getting the balance

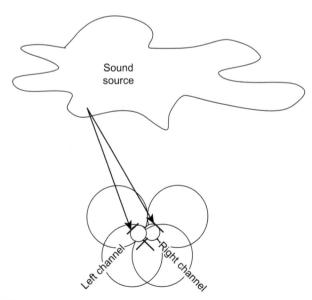

FIGURE 2.30 *Intensity stereo recording.*

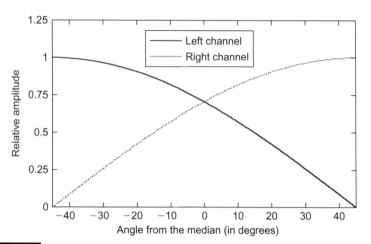

FIGURE 2.31 *The effect of the "pan-pots" in a mixing desk on intensity of the two channels.*

between the spurious and wanted cues, and so providing a good directional illusion, is difficult.

- *Binaural stereo*: The third major way of providing a directional illusion is to use binaural stereo techniques. This system is shown in Figure 2.32 and consists of two omni-directional microphones placed on a head—real or more usually artificial—and presenting the result over headphones. The distance of the microphones is identical to the ear spacing, and they are placed on an object which shades the sound in the same way as a human head, and, possibly, torso. This means that a change in performer position provides both intensity and delay cues to the listener: the results can be very effective. However, they must be presented over headphones because any cross-ear coupling of the two channels, as would happen with loudspeaker reproduction, would cause spurious cues and so destroy the illusion. Note that this effect happens in reverse when listening to loudspeaker stereo over headphones, because the cross coupling that normally exists in a loudspeaker presentation no longer exists. This is another reason why the sound is always "in the head" when listening via conventional headphones.

The main compromise in stereo sound reproduction is the presence of spurious direction cues in the listening environment because the loudspeakers and environment will all contribute cues about their position in the room, which have nothing to do with the original recording. More information about directional hearing and stereophonic listening can be found in Blauert (1997) and Rumsey (2001).

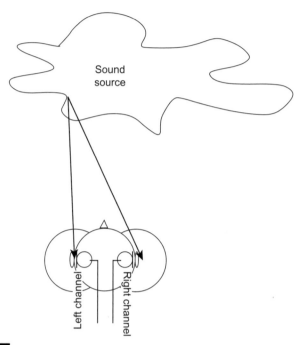

FIGURE 2.32 *Binaural stereo recording.*

REFERENCES

Blauert, J., 1997. Spatial Hearing: The Psychophysics of Human Sound Location, revised edn. MIT Press, Cambridge.

Fletcher, H., Munson, W., 1933. Loudness, its measurement and calculation. J. Acoust. Soc. Am. 5, 82–108.

Glasberg, B.R., Moore, B.C.J., 1990. Derivation of auditory filter shapes from notched-noise data. Hear. Res. 47, 103–138.

Gulick, L.W., 1971. Hearing: Physiology and Psychophysics. Oxford University Press, Oxford, pp. 188–189.

Madsen, E.R., 1990. In: The Science of Sound. Rossing, T.D. (ed.), second edn. Addison Wesley, p. 500.

Pickles, J.O., 1982. An Introduction to the Physiology of Hearing. Academic Press, London.

Rumsey, F., 2001. Spatial Audio. Focal Press, Oxford.

Scharf, B., 1970. Critical bands. In: Tobias, J.V. (ed.), Foundations of Modern Auditory Theory, Vol. 1. Academic Press, London, pp. 159–202.

Tempest, W., 1985. The Noise Handbook, (ed.) Academic press.

von Békésy, G., 1960. Experiments in Hearing. McGraw-Hill, New York.

Notes and Harmony

3.1 MUSICAL NOTES

Music of all cultures is based on the use of instruments (including the human voice) which produce notes of different pitches. The particular set of pitches used by any culture may be unique but the psychoacoustic basis on which pitch is perceived is basic to all human listeners. This chapter explores the acoustics of musical notes which are perceived as having a pitch, and the psychoacoustics of pitch perception. It then considers the acoustics and psychoacoustics of different tuning systems that have been used in Western music.

The representation of musical pitch can be confusing because a number of different notation systems are in use. In this book the system which uses A4 to represent the A above middle C has been adopted. The number changes between the notes B and C, and capital letters are always used for the names of the notes. Thus middle C is C4, the B immediately below it is B3, etc. The bottom note on an 88-note piano keyboard is therefore A0 since it is the fourth A below middle C, and the top note on an 88-note piano keyboard is C8. (This notation system is shown for reference against a keyboard later in the chapter in Figure 3.21.)

3.1.1 Musical notes and their fundamental frequency

When we listen to a note played on a musical instrument and we perceive it as having a clear unambiguous musical pitch, this is because that instrument produces an acoustic pressure wave which repeats regularly. For example, consider the acoustic pressure waveforms recorded by means of a microphone and shown in Figure 3.1 for A4 played on four orchestral instruments: violin, trumpet, flute and oboe. Notice that in each case, the waveshape repeats regularly, or the waveform is "periodic" (see Chapter 1). Each section that repeats is known as a "cycle" and the time for which each cycle lasts is known as the "fundamental period" or "period" of the waveform. The number of cycles which occur in one second gives the fundamental frequency of the note in hertz (or Hz). The fundamental frequency is often notated as "f_0", pronounced "F zero" or "F nought", a practice which will be used throughout the rest of this book. Thus f_0 of any waveform can be found from its period as:

$$(f_0 \text{ in Hz}) = \frac{1}{(\text{period in seconds})} \qquad (3.1)$$

and the period from a known f_0 as:

$$(\text{period in seconds}) = \frac{1}{f_0 \text{ in Hz}} \qquad (3.2)$$

EXAMPLE 3.1

Find the period of the note G5, and the note an instrument is playing if its measured period is 5.41 ms.

Figure 3.21 gives the f_0 of G5 as 784.0 Hz; therefore its period from Equation 3.2 is:

$$\text{Period of G5 in seconds} = \frac{1}{784.0} = 1.276 \times 10^{-3} \text{ or } 1.276 \text{ms}$$

The f_0 of a note whose measured period is 5.405 ms can be found using Equation 3.1 as:

$$f_0 \text{ in Hz} = \frac{1}{5.41 \times 10^{-3}} = 184.8 \text{ Hz}$$

The note whose f_0 is nearest to 184.8 Hz (from Figure 3.21) is F#3.

For the violin note shown in Figure 3.1, the f_0 equivalent to any cycle can be found by measuring the period of that cycle from the waveform plot from which the f_0 can be calculated. The period is measured from any point in one cycle to the point in the next (or last) cycle where it repeats, for example a positive peak, a negative peak or a point where it crosses the zero amplitude line. The distance marked "T" for the violen in the figure shows where the period

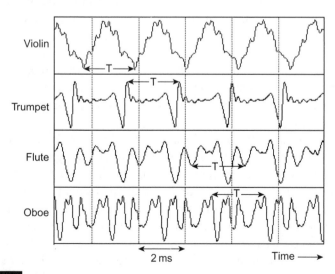

FIGURE 3.1 *Acoustic pressure waveform of A4 (440 Hz) played on a violin, trumpet, flute and oboe. (Note: T indicates one cycle of the waveform.)*

could be measured between negative peaks, and this measurement was made in the laboratory to give the period as 2.27 ms. Using Equation 3.1:

$$f_0 = \frac{1}{2.27 \text{ ms}} = \frac{1}{(2.27 \times 10^{-3})\text{s}} = 440.5 \text{ Hz}$$

This is close to 440 Hz, which is the tuning reference f_0 for A4 (see Figure 3.21). Variation in tuning accuracy, intonation or, for example, vibrato if the note were not played on an open string, will mean that the f_0 measured for any particular individual cycle is not likely to be exactly equivalent to one of the reference f_0 values in Figure 3.21. An average f_0 measurement over a number of individual periods might be taken in practice.

3.1.2 Musical notes and their harmonics

Figure 3.1 also shows the acoustic pressure waveforms produced by other instruments when A4 is played. While the periods and therefore the f_0 values of these notes are similar, their waveform shapes are very different. The perceived pitch of each of these notes will be A4 and the distinctive sound of each of these instruments is related to the differences in the detailed shape of their acoustic pressure waveforms, which is how listeners recognize the difference between, for example, a violin, a clarinet and an oboe. This is because acoustic pressure variations produced by a musical instrument that impinge on the listener's tympanic membrane are responsible for the pattern of vibration set up on the basilar membrane of that ear. It is this pattern of vibration that is then analyzed in terms of the frequency components of which they are comprised (see Chapter 2). If the pattern of vibration on the basilar membrane varies when comparing different sounds, for example from a violin and a clarinet, then the sounds are perceived as having a different "timbre" (see Chapter 5) whether or not they have the same pitch.

Every instrument therefore has an underlying set of partials in its spectrum (see Chapter 1) from which we are able to recognize it from other instruments. These can be thought of as the frequency component "recipe" underlying the particular sound of that instrument. Figure 3.1 shows the acoustic pressure waveform for different notes played on four orchestral instruments and Figure 3.2 shows the amplitude–frequency spectrum for each. Notice that the shape of the waveform for each of the notes is different and so is the recipe of frequency components. Each of these notes would be perceived as being the note A4 but as having different timbres. The frequency components of notes produced by any pitched instrument,

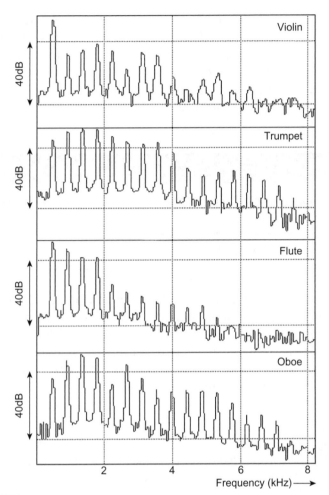

FIGURE 3.2 *Spectra of waveforms shown in Figure 3.1 for A4 (f₀ = 440 Hz) played on a violin, trumpet, flute and oboe.*

such as a violin, oboe, clarinet, trumpet, etc., are harmonics, or integer (1, 2, 3, 4, 5, etc.) multiples of f_0 (see Chapter 1). Thus the only possible frequency components for the acoustic pressure waveform of the violin note shown in Figure 3.1 whose f_0 is 440.5 Hz are: 440.5 Hz (1 × 440.5 Hz); 881.0 Hz (2 × 440.5 Hz); 1321.5 Hz (3 × 440.5 Hz); 1762 Hz (4 × 440.5 Hz); 2202.5 Hz (5 × 440.5 Hz); etc. Figure 3.2 shows that these are the only frequencies at which peaks appear in each spectrum (see Chapter 1). These harmonics are generally referred to by their "harmonic number," which is the integer by which f_0 is multiplied to calculate the frequency of the particular component of interest.

An earlier term still used by many authors for referring to the components of a periodic waveform is "overtones." The first overtone refers to the first frequency component that is "over" or above f_0, which is the second harmonic. The second overtone is the third harmonic, and so on. Table 3.1 summarizes the relationship between f_0, overtones and harmonics for integer multipliers from 1 to 10.

EXAMPLE 3.2

Find the fourth harmonic of a note whose f_0 is 101 Hz, and the sixth overtone of a note whose f_0 is 120 Hz.

The fourth harmonic has a frequency that is ($4\ f_0$), which is (4×101) Hz = 404 Hz.

The sixth overtone has a frequency that is ($7\ f_0$), which is (7×120) Hz = 840 Hz.

There is no theoretical upper limit to the number of harmonics which could be present in the output from any particular instrument, although for many instruments there are acoustic limits imposed by the structure of the instrument itself. An upper limit can be set though, in terms of the number of harmonics which could be present based on the upper frequency limit of the hearing system, for which a practical limit might be 16 000 Hz (see

Table 3.1   The relationship between overtone series, harmonic series and fundamental frequency for the first 10 components of a period waveform

Integer (N)	Overtone series $((N - 1) \times f_0)$ when $N > 1$	Harmonic series $(N \times f_0)$	Component frequency (Hz)
1	fundamental frequency (f_0)	1st harmonic	$1\ f_0$
2	1st overtone	2nd harmonic	$2\ f_0$
3	2nd overtone	3rd harmonic	$3\ f_0$
4	3rd overtone	4th harmonic	$4\ f_0$
5	4th overtone	5th harmonic	$5\ f_0$
6	5th overtone	6th harmonic	$6\ f_0$
7	6th overtone	7th harmonic	$7\ f_0$
8	7th overtone	8th harmonic	$8\ f_0$
9	8th overtone	9th harmonic	$9\ f_0$
10	9th overtone	10th harmonic	$10\ f_0$

Chapter 2). Thus an instrument playing the A above middle C, which has an f_0 of 440 Hz, could theoretically contain 36 ($=16000/440$) harmonics within the human hearing range. If this instrument played a note an octave higher, f_0 is doubled to 880 Hz, and the output could now theoretically contain 18 ($=16\,000/880$) harmonics. This is an increasingly important consideration since although there is often an upper frequency limit to an acoustic instrument which is well within the practical upper frequency range of human hearing, it is quite possible with electronic synthesizers to produce sounds with harmonics which extend beyond this upper frequency limit.

3.1.3 Musical intervals between harmonics

Acoustically, a note perceived to have a distinct pitch contains frequency components that are integer multiples of f_0 usually known as "harmonics." Each harmonic is a sine wave and since the hearing system analyzes sounds in terms of their frequency components it turns out to be highly instructive, in terms of understanding how to analyze and synthesize periodic sounds, as well as being central to the development of Western musical harmony, to consider the musical relationship between the individual harmonics themselves. The frequency ratios of the harmonic series are known (see Table 3.1) and their equivalent musical intervals, frequency ratios and staff notation in the key of C are shown in Figure 3.3 for the first 10 harmonics. The musical intervals (apart from the octave) are only approximated on a modern keyboard due to the tuning system used, as discussed in Section 3.3.

The musical intervals of adjacent harmonics in the natural harmonic series starting with the fundamental or first harmonic, illustrated on a

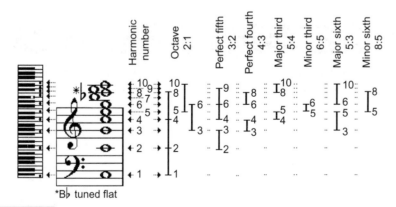

*B♭ tuned flat

FIGURE 3.3 *Frequency ratios and common musical intervals between the first 10 harmonics of the natural harmonic series of C3 against a musical stave and keyboard.*

musical stave and as notes on a keyboard in Figure 3.3, are: octave (2:1), perfect fifth (3:2), perfect fourth (4:3), major third (5:4), minor third (6:5), flat minor third (7:6), sharp major second (8:7), a major whole tone (9:8), and a minor whole tone (10:9). The frequency ratios for intervals between non-adjacent harmonics in the series can also be inferred from the figure. For example, the musical interval between the fourth harmonic and the fundamental is two octaves and the frequency ratio is 4:1, equivalent to a doubling for each octave. Similarly the frequency ratio for three octaves is 8:1, and for a twelfth (octave and a fifth) is 3:1.

Ratios for other commonly used musical intervals can be found from the ones just mentioned (musical intervals which occur within an octave are illustrated in Figure 3.15). To demonstrate this for a known result, the frequency ratio for a perfect fourth (4:3) can be found from that for a perfect fifth (3:2) since together they make one octave (2:1): C to G (perfect fifth) and G to C (perfect fourth). The perfect fifth has a frequency ratio 3:2 and the octave a ratio of 2:1. Bearing in mind that musical intervals are ratios in terms of their frequency relationships and that any mathematical manipulation must therefore be carried out by means of division and multiplication, the ratio for a perfect fourth is that for an octave divided by that for a perfect fifth, or up one octave and down a fifth:

$$\text{Frequency ratio for a perfect fourth } = \frac{2}{1} \div \frac{3}{2} = \frac{2}{1} \times \frac{2}{3} = \frac{4}{3}$$

Two other common intervals are the major sixth and minor sixth, and their frequency ratios can be found from those for the minor third and major third respectively since in each case they combine to make one octave.

EXAMPLE 3.3

Find the frequency ratio for a major and a minor sixth given the frequency ratios for an octave (2:1), a minor third (6:5) and a major third (5:4).

A major sixth and a minor third together span one octave. Therefore:

$$\text{Frequency ratio for a major sixth } = \frac{2}{1} \div \frac{6}{5} = \frac{2}{1} \times \frac{5}{6} = \frac{10}{6} = \frac{5}{3}$$

A minor sixth and a major third together span one octave. Therefore:

$$\text{Frequency ratio for a minor sixth } = \frac{2}{1} \div \frac{5}{4} = \frac{2}{1} \times \frac{4}{5} = \frac{8}{5}$$

These ratios can also be inferred from knowledge of the musical intervals and the harmonic series. Figure 3.3 shows that the major sixth is the interval between the fifth and third harmonics—in this example these are G4 and E5—and therefore their frequency ratio is 5:3. Similarly the interval of a minor sixth is the interval between the fifth and eighth harmonics, in this case E5 and C6; therefore the frequency ratio for the minor sixth is 8:5. Knowledge of the notes of the harmonic series is both musically and acoustically useful and is something that all brass players and organists who understand mutation stops (see Section 5.4) are particularly aware of.

Figure 3.4 shows the positions of the first 10 harmonics of A3 ($f_0 = 220.0$ Hz), plotted on a linear and a logarithmic axis. Notice that the distance between the harmonics is equal on the linear plot and therefore the harmonics becomes progressively closer together as frequency increases on the logarithmic axis. While the logarithmic plot might appear more complex than the linear plot at first glance in terms of the distribution of the harmonics themselves, particularly given that nature often appears to make use of the most efficient process, notice that when different notes are plotted, in this case E4 ($f_0 = 329.6$ Hz) and A4 ($f_0 = 440.0$ Hz), the patterning of the harmonics remains constant on the logarithmic scale but they are spaced differently on the linear scale. This is an important aspect of timbre perception which will be explored further in Chapter 5.

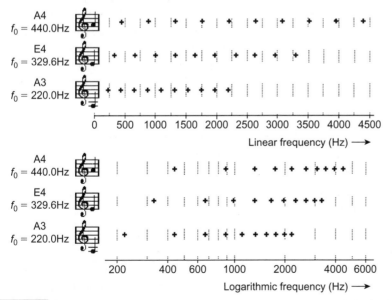

FIGURE 3.4   *The positions of the first 10 harmonics of A3 ($f_0 = 220$ Hz), E4 ($f_0 = 330$ Hz), and A4 ($f_0 = 440$ Hz) on linear (upper) and logarithmic (lower) axes.*

Bearing in mind that the hearing system carries out a frequency analysis due to the place analysis which is based on a logarithmic distribution of position with frequency on the basilar membrane, the logarithmic plot most closely represents the perceptual weighting given to the harmonics of a note played on a pitched instrument.

The use of a logarithmic representation of frequency in perception has the effect of giving equal weight to the frequencies of components analyzed by the hearing system that are in the same ratio. Figure 3.5 shows a number of musical intervals plotted on a logarithmic scale and in each case they continue up to around the upper useful frequency limit of the hearing system. In this case they are all related to A1 ($f_0 = 55\,Hz$) for convenience. Such a plot could be produced relative to an f_0 value for any note and it is important to notice that the intervals themselves would remain a constant distance on a given logarithmic scale. This can be readily verified with a ruler, for example by measuring the distance equivalent to an octave from 100 Hz (between 100 and 200 Hz, 200 and 400 Hz, 400 and 800 Hz, etc.) on the x axis of Figure 3.5 and comparing this with the distance between any of the points on the octave plot. The distance anywhere on a given logarithmic axis that is equivalent to a particular ratio such as 2:1, 3:2, 4:3, etc. will be the same no matter where on the axis it is measured.

A musical interval ruler could be made which is calibrated in musical intervals to enable the frequencies of notes separated by particular intervals to be readily found on a logarithmic axis. Such a calibration must, however, be carried out with respect to the length of the ratios of interest: octave (2:1), perfect fifth (3:2), major sixth (5:3), etc. If the distance equivalent to a perfect fifth is added to the distance equivalent to a perfect fourth, the distance for one octave will be obtained since a fifth plus a fourth equals one

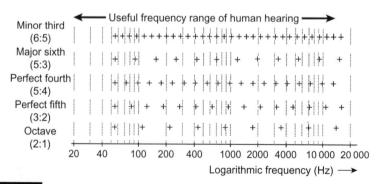

FIGURE 3.5 *Octaves, perfect fifths, perfect fourths, major sixths and minor thirds plotted on a logarithmic scale relative to A1 ($f_0 = 55\,Hz$).*

octave. Similarly, if the distance equivalent to a major sixth is added to that for a minor third, the distance for one octave will again be obtained since a major sixth plus a minor third equals one octave (see Example 3.3).

A doubling (or halving) of a value anywhere on a logarithmic frequency scale is equivalent perceptually to a raising (or lowering) by a musical interval of one octave, and multiplying by $\frac{3}{2}$ (or by $\frac{2}{3}$) is equivalent perceptually to a raising (or lowering) by a musical interval of a perfect fifth, and so on. We perceive a given musical interval (octave, perfect fifth, perfect fourth, major third, etc.) as being similar no matter where in the frequency range it occurs. For example, a two-note chord a major sixth apart whether played on two double basses or two flutes gives a similar perception of the musical interval. In this way, the logarithmic nature of the place analysis mechanism provides a basis for understanding the nature of our perception of musical intervals and of musical pitch.

By way of contrast and to complete the story, sounds which have no definite musical pitch (but a pitch, nevertheless—see below) associated with them, such as the "ss" in *sea* (Figure 3.9 later in the chapter), have an acoustic pressure waveform that does not repeat regularly, and is often random in its variation with time and is therefore not periodic. Such a waveform is referred to as being "aperiodic" (see Chapter 1). The spectrum of such sounds contains frequency components that are not related as integer multiples of some frequency, and there are no harmonic components. The spectrum will often contain all frequencies, in which case it is known as a "continuous" spectrum. An example of an acoustic pressure waveform and spectrum for a non-periodic sound is illustrated in Figure 3.6 for a snare drum being brushed.

3.2 HEARING PITCH

The perception of pitch is basic to the hearing of tonal music. Familiarity with current theories of pitch perception as well as other aspects of psychoacoustics enables a well founded understanding of musically important matters such as tuning, intonation, perfect pitch, vibrato, electronic synthesis of new sounds, and pitch paradoxes (see Chapter 5).

Pitch relates to the perceived position of a sound on a scale from low to high and its formal definition by the American National Standards Institute (1960) is couched in these terms as: "pitch is that attribute of auditory sensation in terms of which sounds may be ordered on a scale extending from low to high." The measurement of pitch is therefore "subjective" because it requires a human listener (the "subject") to make a perceptual judgment.

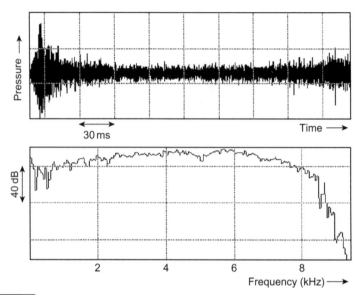

FIGURE 3.6   *Acoustic pressure waveform (upper) and spectrum (lower) for a snare drum being brushed.*

This is in contrast to the measurement in the laboratory of, for example, the fundamental frequency (f_0) of a note, which is an "objective" measurement.

In general, sounds which have a periodic acoustic pressure variation with time are perceived as having a pitch associated with them, and sounds whose acoustic pressure waveform is non-periodic are perceived as having no pitch. The relationship between the waveforms and spectra of pitched and non-pitched sounds is summarized in Table 3.2 and examples of each have been discussed in relation to Figures 3.2 and 3.6. The terms "time domain" and "frequency domain" are widely used when considering time (waveform) and frequency (spectral) representations of signals.

Table 3.2	The nature of the waveforms and spectra for pitched and non-pitched sounds	
	Pitched	**Non-pitched**
Waveform	Periodic	Non-periodic
(time domain)	*regular repetitions*	*no regular repetitions*
Spectrum	Line	Continuous
(frequency domain)	*harmonic components*	*no harmonic components*

The pitch of a note varies as its f_0 is changed: the greater the f_0 the higher the pitch and vice versa. Although the measurement of pitch and f_0 are subjective and objective and measured on a scale of high/low and Hz respectively, a measurement of pitch can be given in Hz. This is achieved by asking a listener to compare the sound of interest by switching between it and a sine wave with a variable frequency. The listener would adjust the frequency of the sine wave until the pitches of the two sounds are perceived as being equal, at which point the pitch of the sound of interest is equal to the frequency of the sine wave in Hz.

Two basic theories of pitch perception have been proposed to explain how the human hearing system is able to locate and track changes in the f_0 of an input sound: the "place" theory and the "temporal" theory. These are described below along with their limitations in terms of explaining observed pitch perception effects.

3.2.1 Place theory of pitch perception

The place theory of pitch perception relates directly to the frequency analysis carried out by the basilar membrane in which different frequency components of the input sound stimulate different positions, or places, on the membrane. Neural firing of the hair cells occurs at each of these places, indicating to higher centers of neural processing and the brain which frequency components are present in the input sound. For sounds in which all the harmonics are present, the following are possibilities for finding the value of f_0 based on a place analysis of the components of the input sound and allowing for the possibility of some "higher processing" of the component frequencies at higher centers of neural processing and/or the brain.

■ *Method 1*: Locate the f_0 component itself.

■ *Method 2*: Find the minimum frequency difference between adjacent harmonics. The frequency difference between the $(n + 1)$th and the (n)th harmonic, which are adjacent by definition if all harmonics are present, is:

$$((n + 1)f_0) - (n\,f_0) = (n\,f_0) + (1\,f_0) - (n\,f_0) = f_0$$
where $n = 1, 2, 3, 4,...$

■ *Method 3*: Find the highest common factor (the highest number that will divide into all the frequencies present giving an integer result) of the components present. Table 3.3 illustrates this for a sound

Table 3.3	Processing method to find the highest common factor of the frequencies of the first 10 harmonics of a sound whose $f_0 = 100\,Hz$ (calculations to four significant figures)

Place analysis					Higher processing				
$n \times f_0$ $(\div 1)$ (Hz)	÷2 (Hz)	÷3 (Hz)	÷4 (Hz)	÷5 (Hz)	÷6 (Hz)	÷7 (Hz)	÷8 (Hz)	÷9 (Hz)	÷10 (Hz)
100	50.00	33.33	25.00	20.00	16.67	14.29	12.50	11.11	10.00
200	100.0	66.67	50.00	40.00	33.33	28.57	25.00	22.22	20.00
300	150.0	100.0	75.00	60.00	50.00	42.86	37.30	33.33	30.00
400	200.0	133.3	100.0	80.00	66.67	57.14	50.00	44.44	40.00
500	250.0	166.7	125.0	100.0	83.33	71.43	62.50	55.56	50.00
600	300.0	200.0	150.0	120.0	100.0	85.71	75.00	66.67	60.00
700	350.0	233.3	175.0	140.0	116.7	100.0	87.50	77.78	70.00
800	400.0	266.7	200.0	160.0	133.3	114.3	100.0	88.89	80.00
900	450.0	300.0	225.0	180.0	150.0	128.6	112.5	100.0	90.00
1000	500.0	333.3	250.0	200.0	166.7	142.9	125.0	111.1	100.0

consisting of the first 10 harmonics whose f_0 is 100 Hz, by dividing each frequency by integers, in this case up to 10, and looking for the largest number in the results which exists for every frequency. The frequencies of the harmonics are given in the left-hand column (the result of a place analysis), and each of the other columns shows the result of dividing the frequency of each component by integers ($m = 2$ to 10). The highest common factor is the highest value appearing in all rows of the table, including the frequencies of the components themselves ($f_0 \div 1$) or ($m = 1$), and is 100 Hz, which would be perceived as the pitch.

■ In addition, it is of interest to notice that every value which appears in the row relating to the f_0, in this case 100 Hz, will appear in each of the other rows if the table were extended far enough to the right. This is the case because by definition, 100 divides into each harmonic frequency to give an integer result (n) and all values appearing in the 100 Hz row are found by integer (m) division of 100 Hz; therefore all values in the 100 Hz row can be gained by division of harmonic frequencies by ($m \times n$), which must itself be an integer. These are f_0 values (50 Hz, 33 Hz, 25 Hz, 20 Hz, etc.) whose harmonic series also contain all the given components, and they are known as "sub-harmonics." This is why it is the *highest* common factor which is used.

One of the earliest versions of the place theory suggests that the pitch of a sound corresponds to the place stimulated by the lowest frequency component in the sound which is f_0 (Method 1 above). The assumption underlying this is that f_0 is always present in sounds and the theory was encapsulated by Ohm in his second or "acoustical" law[1]: "a pitch corresponding to a certain frequency can only be heard if the acoustic wave contains power at that frequency."

This theory came under close scrutiny when it became possible to carry out experiments in which sounds could be synthesized with known spectra. Schouten (1940) demonstrated that the pitch of a pulse wave remained the same when the fundamental component was removed, thus demonstrating: (i) that f_0 did not have to be present for pitch perception, and (ii) that the lowest component present is not the basis for pitch perception because the pitch does not jump up by one octave (since the second harmonic is now the lowest component after f_0 has been removed). This experiment has become known as "the phenomenon of the missing fundamental," and suggests that Method 1 cannot account for human pitch perception.

Method 2 seems to provide an attractive possibility since the place theory gives the positions of the harmonics, whether or not f_0 is present, and it should provide a basis for pitch perception provided some adjacent harmonics are present. For most musical sounds, adjacent harmonics are indeed present. However, researchers are always looking for ways of testing psychoacoustic theories, in this case pitch perception, by creating sounds for which the perceived pitch cannot be explained by current theories. Such sounds are often generated electronically to provide accurate control over their frequency components and temporal development.

Figure 3.7 shows an idealized spectrum of a sound which contains just odd harmonics ($1\,f_0$, $3\,f_0$, $5\,f_0$, ...) and shows that measurement of the frequency distance between adjacent harmonics would give f_0, $2\,f_0$, $2\,f_0$, $2\,f_0$, etc. The minimum spacing between the harmonics is f_0, which gives a possible basis for pitch perception. However, if the f_0 component were removed (imagine removing the dotted f_0 component in Figure 3.7), the perceived pitch would not change. Now, however, the spacings between adjacent harmonics is $3\,f_0$, $2\,f_0$, $2\,f_0$, $2\,f_0$, etc. and the minimum spacing is $2\,f_0$, but the pitch does not jump up by an octave.

The third method will give an appropriate f_0 for: (i) sounds with missing f_0 components (see Table 3.3 and ignore the f_0 row), (ii) sounds with odd harmonic components only (see Table 3.3 and ignore the rows for the even harmonics), and (iii) sounds with odd harmonic components only with a missing f_0 component (see Table 3.3 and ignore the rows for f_0 and the even harmonics).

---

[1]His first law being basic to electrical work: voltage = current × resistance.

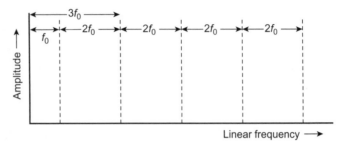

FIGURE 3.7 *An idealized spectrum for a sound with odd harmonics only to show the spacing between adjacent harmonics when the fundamental frequency component (shown dashed) is present or absent.*

In each case, the highest common factor of the components is f_0. This method also provides a basis for understanding how a pitch is perceived for non-harmonic sounds, such as bells or chime bars, whose components are not exact harmonics (integer multipliers) of the resulting f_0.

As an example of such a non-harmonic sound, Schouten in one of his later experiments produced sounds whose component frequencies were 1040 Hz, 1240 Hz, and 1440 Hz and found that the perceived pitch was approximately 207 Hz (consider track 4C on the accompanying CD). The f_0 for these components, based on the minimum spacing between the components (Method 2), is 200 Hz. Table 3.4 shows the result of applying Method 3 (searching for the highest common factor of these three components) up to an integer divisor of 10. Schouten's proposal can be interpreted in terms of this table by looking for the closest set of values in the table that would be consistent with the three components being true harmonics and taking their average to give an estimate of f_0. In this case, taking 1040 Hz as the fifth "harmonic," 1240 Hz as the sixth "harmonic" and 1440 Hz as the

Table 3.4	Illustration of how finding the highest common factor of the frequencies of the three components—1040 Hz, 1240 Hz and 1440 Hz—gives a basis for explaining a perceived pitch of approximately 207 Hz (calculations to four significant figures)								
Component frequency (Hz)	÷2 (Hz)	÷3 (Hz)	÷4 (Hz)	÷5 (Hz)	÷6 (Hz)	÷7 (Hz)	÷8 (Hz)	÷9 (Hz)	÷10 (Hz)
1040	520.0	346.7	260.0	208.0	173.3	148.6	130.0	115.6	104.0
1240	620.0	413.3	310.0	248.0	206.7	177.1	155.0	137.8	124.0
1440	720.0	480.0	360.0	288.0	240.0	205.7	180.0	160.0	144.0

seventh "harmonic" gives 208 Hz, 207 Hz and 206 Hz respectively. The average of these values is 207 Hz, and Schouten referred to the pitch perceived in such a situation as the "residue pitch" or "pitch of the residue." It is also sometimes referred to as "virtual pitch."

By way of a coda to this discussion, it is interesting to note that these components 1040 Hz, 1240 Hz and 1440 Hz do, in fact, have a true f_0 of 40 Hz of which they are the 26th, 31st and 36th harmonics, which would appear if the table were continued well over to the right. However, the auditory system appears to find an f_0 for which the components present are adjacent harmonics.

3.2.2 Problems with the place theory

The place theory provides a basis for understanding how f_0 could be found from a frequency analysis of components. However, there are a number of problems with the place theory because it does not explain:

- the fine degree of accuracy observed in human pitch perception;

- pitch perception of sounds whose frequency components are not resolved by the place mechanism;

- the pitch perceived for some sounds which have continuous (non-harmonic) spectra; or

- pitch perception for sounds with an f_0 less than 50 Hz.

Each will be considered in turn.

Psychoacoustically, the ability to discriminate between sounds that are nearly the same except for a change in one aspect (f_0, intensity, duration, etc.) is measured as a "difference limen" (DL), or "just noticeable difference" (JND). JND is preferred in this book. The JND for human pitch perception is shown graphically in Figure 3.8 along with the critical bandwidth curve. This JND graph is based on an experiment by Zwicker *et al.* (1957) in which sinusoidal stimuli were used (fixed waveshape) and the sound intensity level and sound duration remained constant. It turns out that the JND is approximately one thirtieth of the critical bandwidth across the hearing range. Musically, this is equivalent to approximately one twelfth of a semitone. Thus the JND in pitch is much smaller than the resolution of the analysis filters (critical bandwidth).

The place mechanism will resolve a given harmonic of an input sound provided that the critical bandwidth of the filter concerned is sufficiently narrow to exclude adjacent harmonics. It turns out that, no matter what the f_0 of the sound is, only the first five to seven harmonics are resolved by

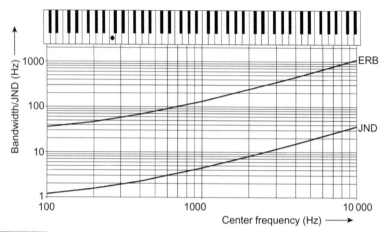

FIGURE 3.8 *Just noticeable difference (JND) for pitch perception and the equivalent rectangular bandwidth.*

the place analysis mechanism. This can be illustrated with an example as follows with reference to Table 3.5.

Consider a sound consisting of all harmonics (f_0, 2 f_0, 3 f_0, 4 f_0, 5 f_0, etc.) whose f_0 is 110 Hz. The frequencies of the first 10 harmonics are given in the left-hand column of the table. The next column shows the critical bandwidth of a filter centered on each of these harmonics by calculation using Equation 2.6. The critical bandwidth increases with filter center frequency (see Figure 3.8), and the frequency analysis action of the basilar membrane is equivalent to a bank of filters. Harmonics will cease to be resolved by the place mechanism when the critical bandwidth exceeds the frequency spacing between adjacent harmonics, which is f_0 when all adjacent harmonics are present.

In the table, it can be seen that the critical bandwidth is greater than f_0 for the filter centered at 770 Hz (the seventh harmonic), but this filter will resolve the seventh harmonic since it is centered over it and its bandwidth extends half above and half below 770 Hz. In order to establish when harmonics are not resolved, consider the filters centered midway between adjacent harmonic positions (their center frequencies and critical bandwidths are shown in Table 3.5).

The filter centered between the seventh and eighth harmonics has a critical bandwidth of 113.7 Hz which exceeds f_0 (110 Hz in this example) and therefore the seventh and eighth harmonics will not be resolved by this filter. Due to the continuous nature of the wave traveling along the basilar membrane, no harmonics will be resolved in this example above the sixth,

| Table 3.5 | Illustration of resolution of place mechanism for an input consisting of the first 10 harmonics and an f_0 of 110 Hz (calculations to four significant figures). Key: CB = critical bandwidth, CF = center frequency |

Harmonic frequency (Hz)	CB of local filter (Hz)	CF of mid harmonic filter (Hz)	CB of mid harmonic filter (Hz)	Resolved?
110	36.57	165.0	42.51	Yes
220	48.45	275.0	54.38	Yes
330	60.32	385.0	66.26	Yes
440	72.19	495.0	78.13	Yes
550	84.07	605.0	90.00	Yes
660	95.94	715.0	101.8	Yes
770	107.8	825.0	113.7	No
880	119.6	935.0	125.6	No
990	131.5	1045	137.5	No
1100	143.4	1155	149.3	No

since there will be areas on the membrane responding to at least adjacent pairs of harmonics everywhere above the place where the sixth harmonic stimulates it. Appendix 2 shows a method for finding the filter center frequency whose critical bandwidth is equal to a given f_0 by solving Equation 2.6 mathematically.

EXAMPLE 3.4

Confirm the result illustrated in Table 3.5 that the sixth harmonic will be resolved but the seventh harmonic will not be resolved for an f_0 of 110 Hz.
Using Equation A1.2 in Appendix 1:

$$f_c(\text{kHz}) = \left| \frac{\left(\frac{\text{ERB}}{24.7}\right) - 1}{4.37} \right|$$

Find the center frequency (f_c) for which the critical bandwidth (ERB) equals 110.0 Hz by substituting 110 Hz for ERB. (Bear in mind that the center frequency is in kHz in this equation.)

$$f_c(\text{kHz}) = \left| \frac{\left[\left(\dfrac{110}{24.7}\right) - 1\right]}{4.37} \right| = \left[\frac{4.453 - 1}{4.37}\right] = \left[\frac{3.453}{4.37}\right] = 0.79\,\text{kHz or } 790\,\text{Hz}$$

The center frequency of the filter whose critical bandwidth is 110.0 Hz is 790 Hz. All filters above 790 Hz will have bandwidths that are greater than its bandwidth of 110 Hz, because we know that the critical bandwidth increases with center frequency (see Figure 3.8). As it lies below the center frequency of the filter midway between the seventh (770 Hz) and eighth (880 Hz) harmonics, those harmonics will *not* be resolved, since the filter that does lie between them will have a bandwidth that is greater than 110 Hz. Therefore harmonics up to the sixth will be resolved and harmonics from the seventh upwards will not be resolved.

Observation of the relationship between the critical bandwidth and center frequency plotted in Figure 3.8 allows the general conclusion that no harmonic above about the fifth to seventh is resolved for any f_0 to be approximately validated as follows. The center frequency for which the critical band exceeds the f_0 of the sound of interest is found from the graph and no harmonic above this center frequency will be resolved. To find the center frequency, plot a horizontal line for the f_0 of interest on the y axis, and find the frequency on the x axis where the line intersects the critical band curve. Only harmonics below this frequency will be resolved and those above will not. It is worth trying this exercise for a few f_0 values to reinforce the general conclusion about resolution of harmonics, since this is vital to the understanding of other aspects of psychoacoustics as well as pitch perception.

There are sounds which have non-harmonic spectra for which a pitch is perceived; these are exceptions to the second part of the general statement given earlier that "sounds whose acoustic pressure waveform is non-periodic are perceived as having no pitch." For example, listen to examples of the "ss" in *sea* and the "sh" in *shell* (produce these yourself or ask someone else to) in terms of which one has the higher pitch. Most listeners will readily respond that "ss" has a higher pitch than "sh." The spectrum of both sounds is continuous and an example for each is shown in Figure 3.9. Notice that the energy is biased more towards lower frequencies for the "sh" with a peak around 2.5 kHz, compared with the "ss" where the energy has a peak at about 5 kHz.

This "center of gravity" of the spectral energy of a sound is thought to convey a sense of higher or lower pitch for such sounds which are noise

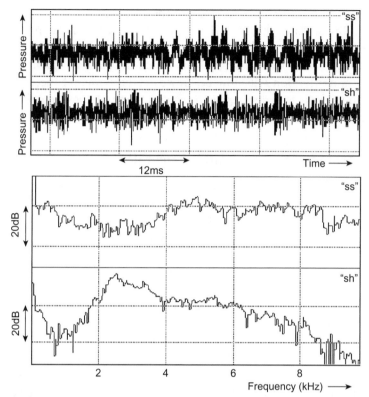

FIGURE 3.9 *Waveforms and spectra for "ss" as in sea and "sh" as in shoe.*

based, but the pitch sensation is far weaker than for that perceived for periodic sounds. This pitch phenomenon is, however, important in music when considering the perception of the non-periodic sounds produced, for example, by some groups of instruments in the percussion section (consider track 5 on the accompanying CD), but the majority of instruments on which notes are played in musical performances produce periodic acoustic pressure variations.

The final identified problem is that the pitch perceived for sounds with components only below 50 Hz cannot be explained by the place theory, because the pattern of vibration on the basilar membrane does not appear to change in that region. Sounds of this type are rather unusual, but not impossible to create by electronic synthesis. Since the typical lowest audible frequency for humans is 20 Hz, a sound with an f_0 of 20 Hz would have harmonics at 40 Hz, 60 Hz, etc., and only the first two fall within this region where no change is observed in basilar membrane response. Harmonics falling above 50 Hz will be analyzed by the place mechanism in the usual manner. Sinusoids in the 20–50 Hz range are perceived as having different pitches and the place mechanism cannot explain this.

These are some of the key problems which the place mechanism cannot explain, and attention will now be drawn to the temporal theory of pitch perception which was developed to explain some of these problems with the place theory.

3.2.3 Temporal theory of pitch perception

The temporal theory of pitch perception is based on the fact that the waveform of a sound with a strong musical pitch repeats or is periodic (see Table 3.2). An example is shown in Figure 3.1 for A4 played on four instruments. The f_0 for a periodic sound can be found from a measurement of the period of a cycle of the waveform using Equation 3.1.

The temporal theory of pitch perception relies on the timing of neural firings generated in the organ of Corti (see Figure 2.3) which occur in response to vibrations of the basilar membrane. The place theory is based on the fact that the basilar membrane is stimulated at different places along its length according to the frequency components in the input sound. The key to the temporal theory is the detailed nature of the actual waveform exciting the different places along the length of the basilar membrane. This can be modeled using a bank of electronic band-pass filters whose center frequencies and bandwidths vary according to the critical bandwidth of the human hearing system as illustrated, for example, in Figure 3.8.

Figure 3.10 shows the output waveforms from such a bank of electronic filters, implemented using transputers by Howard *et al.* (1995), with critical bandwidths based on the ERB equation (Equation 2.6) for C4 played on a violin. The nominal f_0 for C4 is 261.6 Hz (see Figure 3.21 later in the chapter). The output waveform from the filter with a center frequency just above 200 Hz, the lowest center frequency represented in the figure, is a sine wave at f_0. This is because the f_0 component is resolved by the analyzing filter, and an individual harmonic of a complex periodic waveform is a sine wave (see Chapter 1).

The place theory suggests (see calculation associated with Table 3.5) that the first six harmonics will be resolved by the basilar membrane. It can be seen in the example note shown in Figure 3.10 that the second (around 520 Hz), third (around 780 Hz), fourth (around 1040 Hz) and fifth (around 1300 Hz) harmonics are resolved and their waveforms are sinusoidal. Some amplitude variation is apparent on these sine waves, particularly on the fourth and fifth, indicating the dynamic nature of the acoustic pressure output from

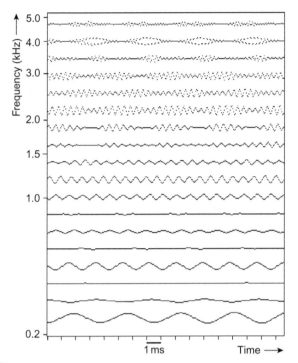

FIGURE 3.10   *Output from a transputer-based model of human hearing to illustrate the nature of basilar membrane vibration at different places along its length for C4 played on a violin.*

a musical instrument. The sixth harmonic (around 1560 Hz) has greater amplitude variation, but the individual cycles are clear.

Output waveforms for filter center frequencies above the sixth harmonic in this example are not sinusoidal because these harmonics are not resolved individually. At least two harmonics are combined in the outputs from filters which are not sinusoidal in Figure 3.10. When two components close in frequency are combined, they produce a "beat" waveform whose amplitude rises and falls regularly if the components are harmonics of some fundamental. The period of the beat is equal to the difference between the frequencies of the two components. Therefore if the components are adjacent harmonics, then the beat frequency is equal to their f_0 and the period of the beat waveform is $(1/f_0)$. This can be observed in the figure by comparing the beat period for filter outputs above 1.5 kHz with the period of the output sinewave at f_0. Thus the period of output waveforms for filters with center frequencies higher than the sixth harmonic will be at $(1/f_0)$ for an input consisting of adjacent harmonics.

The periods of all the output waveforms which stimulate the neural firing in the organ of Corti form the basis of the temporal theory of pitch perception. There are nerve fibers available to fire at all places along the basilar membrane, and they do so in such a manner that a given nerve fiber may only fire at one phase or instant in each cycle of the stimulating waveform, a process known as "phase locking." Although the nerve firing is phase locked to one instant in each cycle of the stimulating waveform, it has been observed that no single nerve fiber is able to fire continuously at frequencies above approximately 300 Hz. It turns out that the nerve does not necessarily fire in every cycle and that the cycle in which it fires tends to be random, which according to Pickles (1982) may be "perhaps as little as once every hundred cycles on average."

However, due to phase locking, the time between firings for any particular nerve will always be an integer (1, 2, 3, 4, ...) multiple of periods of the stimulating waveform and there are a number of nerves involved at each place. A "volley firing" principle has also been suggested by Wever (1949) in which groups of nerves work together, each firing in different cycles to enable frequencies higher than 300 Hz to be coded. A full discussion of this area is beyond the scope of this book, and the interested reader is encouraged to consult, for example, Pickles (1982), Moore (1982, 1986) and Roederer (1975). What follows relies on the principle of phase locking.

The minimum time between firings (1 period of the stimulating waveform) at different places along the basilar membrane can be inferred from Figure 3.10 for the violin playing C4, since it will be equivalent to the period of the output waveform from the analysis filter. For places which respond to frequencies below about the sixth harmonic, the minimum time between firings is at the period of the harmonic itself, and, for places above, the minimum time between firings is the period of the input waveform itself (i.e., $1/f_0$).

The possible instants of nerve firing are illustrated in Figure 3.11. This figure enables the benefit to be illustrated that results from the fact that nerves fire phase locked to the stimulating waveform but not necessarily during *every* cycle. The figure shows an idealized unrolled basilar membrane with the places corresponding to where maximum stimulation would occur for input components at multiples of f_0 up to the sixteenth harmonic, for *any* f_0 of input sound. The assumption on which the figure is based is that harmonics up to and including the seventh are analyzed separately. The main part of the figure shows the possible instants where nerves could fire based on phase locking and the fact that nerves may not fire every cycle; the lengths of the vertical lines illustrate the proportion of firings which might occur at that position, on the basis that more firings are likely with reduced times between them. These approximate to the idea

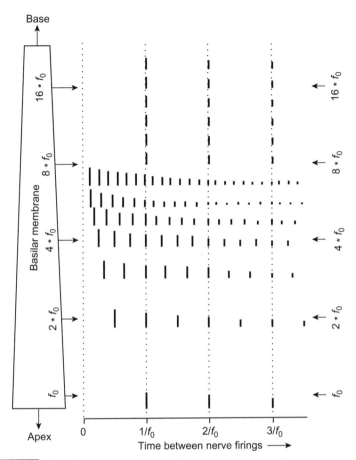

FIGURE 3.11   *The possible instants for nerve firing across the places on the basilar membrane for the first 16 harmonics of an input sound.*

of a histogram of firings being built up, sometimes referred to as an "inter-spike interval" histogram, where a "spike" is a single nerve firing.

Thus at the place on the basilar membrane stimulated by the f_0 component, possible times between nerve firing are: $(1/f_0)$, $(2/f_0)$ and $(3/f_0)$ in this figure as shown, with fewer firings at the higher intervals. For the place stimulated by the second harmonic, possible firing times are: $[1/(2\,f_0)]$, $[2/(2\,f_0)]$ or $(1/f_0)$, $[3/(2\,f_0)]$, $[4/(2\,f_0)]$ or $(2/f_0)$, and so on. This is the case for each place stimulated by a harmonic of f_0 up to the seventh. For places corresponding to higher frequencies than $(7\,f_0)$, the stimulating waveform is beat-like and its fundamental period is $(1/f_0)$, and therefore the possible firing times are: $(1/f_0)$, $(2/f_0)$ and $(3/f_0)$ in this figure as shown.

Visually it can be seen in Figure 3.11 that if the entries in all these inter-spike interval histograms were added together vertically (i.e., for each firing time interval), then the maximum entry would occur for the period of f_0. This is reinforced when it is remembered that all places higher than those shown in the figure would exhibit outputs similar to those shown above the eighth harmonic. Notice how all the places where harmonics are resolved have an entry in their histograms at the fundamental period as a direct result of the fact that nerves may not fire in every cycle. This is the basis on which the temporal theory of pitch perception is thought to function.

3.2.4 Problems with the temporal theory

The temporal theory gives a basis for understanding how the fundamental period could be found from an analysis of the nerve firing times from all places across the basilar membrane. However, not all observed pitch perception abilities can be explained by the temporal theory alone, the most important being the pitch perceived for sounds whose f_0 is greater than 5 kHz. This cannot be explained by the temporal theory because phase locking breaks down above 5 kHz. Any ability to perceive the pitches of sounds with f_0 greater than 5 kHz is therefore thought to be due to the place theory alone.

Given that the upper frequency limit of human hearing is at best 20 kHz for youngsters, with a more practical upper limit being 16 kHz for those over 20 years of age, a sound with an f_0 greater than 5 kHz is only going to provide the hearing system with two harmonics (f_0 and $2 f_0$) for analysis. In practice it has been established that human pitch perception for sounds whose f_0 is greater than 5 kHz is rather poor, with many musicians finding it difficult to judge accurately musical intervals in this frequency range. Moore (1982) notes that this ties in well with f_0 for the upper note of the piccolo being approximately 4.5 kHz. On large organs, some stops can have pipes whose f_0 exceeds 8 kHz, but these are provided to be used in conjunction with other stops (see Section 5.4).

3.2.5 Contemporary theory of pitch perception

Psychoacoustic research has tended historically to consider human pitch perception with reference to the place or the temporal theory, and it is clear that neither theory alone can account for all observed pitch perception abilities. In reality, place analysis occurs giving rise to nerve firings from each place on the basilar membrane that is stimulated. Thus nerve centers and the parts of the brain concerned with auditory processing are provided not only with an indication of the place where basilar membrane stimulation occurs (frequency analysis) but also with information about the nature of that stimulation (temporal analysis). Therefore neither theory is likely to explain human pitch perception completely, since the output from either the place or temporal analysis makes use of the other in communicating itself on the auditory nerve.

Figure 3.12 shows a model for pitch perception of complex tones based on that of Moore (1982) which encapsulates the benefits described for both theories. The acoustic pressure wave is modified by the frequency response of the outer and middle ears (see Chapter 2), and analyzed by the place mechanism which is equivalent to a filter bank analysis. Neural firings occur stimulated by the detailed vibration of the membrane at places equivalent to frequency components of the input sound based on phase locking but not always once per cycle—the latter is illustrated on the right-hand side of the figure. The fact that firing is occurring from particular places provides the basis for the place theory of pitch perception. The intervals between neural firings (spikes) are analyzed and the results are combined to allow common intervals to be found which will tend to be at the fundamental period and its multiples, but predominantly at $(1/f_0)$. This is the

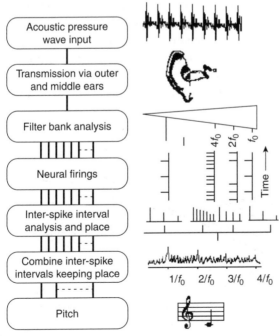

FIGURE 3.12 *A model for human pitch perception based on Moore (1982).*

basis of the temporal theory of pitch perception. The pitch of the sound is based on the results.

3.2.6 Secondary aspects of pitch perception

The perceived pitch of a sound is primarily affected by changes in f_0, which is why the pitch of a note is usually directly related to its f_0, for example by stating that A4 has an f_0 of 440 Hz as a standard pitch reference. The estimation of f_0 forms the basis of both the place and temporal theories of pitch perception. A change in pitch of a particular musical interval manifests itself if the f_0 values of the notes concerned are in the appropriate frequency ratio to give the primary acoustic (objective) basis for the perceived (subjective) pitch of the notes and hence the musical interval. Changes in pitch are also, however, perceived by modifying the intensity or duration of a sound while keeping f_0 constant. These are by far secondary pitch change effects compared with the result of varying f_0, and they are often very subtle.

These secondary pitch effects are summarized as follows. If the intensity of a sine wave is varied between 40 dBSPL and 90 dBSPL while keeping its f_0 constant, a change in pitch is perceived for all f_0 values other than those around 1–2 kHz. For f_0 values greater than 2 kHz the pitch becomes sharper as the intensity is raised, and for f_0 values below 1 kHz the pitch becomes flatter as the intensity is raised. This effect is illustrated in Figure 3.13, and the JND for pitch is shown with reference to the pitch at 60 dBSPL to enable the frequencies and intensities of sine waves for which the effect might be perceived to be inferred. This effect is for sine waves which are rarely encountered in music, although electronic synthesizers have made them widely available.

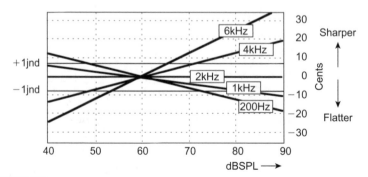

FIGURE 3.13 *The pitch shifts perceived when the intensity of a sine wave with a constant fundamental frequency is varied (after Rossing, 2001).*

With complex tones the effect is less well defined; Rossing (2001) suggests around 17 cents (0.17 of a semitone) for an intensity change between 65 dBSPL and 95 dBSPL. Rossing gives two suggestions as to where this effect could have musical consequences: (i) he cites Parkin (1974) to note that this pitch shift phenomenon is apparent when listening in a highly reverberant building to the release of a final loud organ chord which appears to sharpen as the sound level diminishes, and (ii) he suggests that the pitch shift observed for sounds with varying rates of waveform amplitude change, while f_0 is kept constant, should be "taken into account when dealing with percussion instruments."

The effect that the duration of a sound has on the perception of the pitch of a note is not a simple one, but it is summarized graphically in Figure 3.14 in terms of the minimum number of cycles required at a given f_0 for a definite distinct pitch to be perceived. Shorter sounds may be perceived as being pitched rather than non-pitched, but the accuracy with which listeners can make such a judgment worsens as the duration of the sound drops below that shown in the figure.

By way of a coda to this section on the perception of pitch, a phenomenon known as "repetition pitch" is briefly introduced, particularly now that electronic synthesis and studio techniques make it relatively straightforward to reproduce (consider track 6 on the accompanying CD). Repetition pitch is perceived (by most but not all listeners) if a non-periodic noise-based signal, for example the sound of a waterfall, the consonants in *see*, *shoe*, *fee*, or a noise generator, is added to a delayed version of itself and played to listeners. When the delay is altered a change in pitch is perceived.

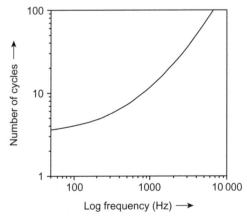

FIGURE 3.14 *The effect of duration on pitch in terms of the number of cycles needed for a definite distinct pitch to be perceived for a given fundamental frequency (data from Rossing, 2001).*

The pitch is equivalent to a sound whose f_0 is equal to (1/delay), and the effect works for delays between approximately 1 ms and 10 ms depending on the listener, giving an equivalent f_0 range for the effect of 100 to 1000 Hz. With modern electronic equipment it is quite possible to play tunes using this effect!

3.3 HEARING NOTES

The music of different cultures can vary considerably in many aspects including, for example, pitch, rhythm, instrumentation, available dynamic range, and the basic melodic and harmonic usage in the music. Musical taste is always evolving with time; what one composer is experimenting with may well become part of the established tradition a number of years later. The perception of chords and the development of different tuning systems are discussed in this section from a psychoacoustic perspective to complement the acoustic discussion earlier in this chapter in consideration of the development of melody and harmony in Western music.

3.3.1 Harmonics and the development of Western harmony

Hearing harmony is basic to music appreciation, and in its basic form harmony is sustained by means of chords. A chord consists of at least two notes sounding together and it can be described in terms of the musical intervals between the individual notes which make it up.

A basis for understanding the psychoacoustics of a chord is given by considering the perception of any two notes sounding together. The full set of commonly considered two-note intervals and their names are shown in Figure 3.15 relative to middle C. Each of the augmented and diminished intervals sounds the same as another interval shown if played on a modern keyboard, for example the augmented unison and minor second, the

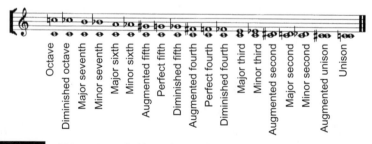

FIGURE 3.15 *All two-note musical intervals occurring up to an octave related to C4.*

augmented fourth and diminished fifth, the augmented fifth and minor sixth, and the major seventh and diminished octave, but they are notated differently on the stave and, depending on the tuning system in use, these "enharmonics" would sound different also.

The development of harmony in Western music can be viewed in terms of the decreasing musical interval size between adjacent members of the natural harmonic series as the harmonic number is increased. Figure 3.3 shows the musical intervals between the first 10 harmonics of the natural harmonic series. The musical interval between adjacent harmonics must reduce as the harmonic number is increased since it is determined in terms of the f_0 of the notes concerned by the ratio of the harmonic numbers themselves (e.g. 2:1 > 3:2 > 4:3 > 5:4 > 6:5, etc.).

The earliest polyphonic Western music, known as "organum," made use of the octave, the perfect fifth, and its inversion, the perfect fourth. These are the intervals between the 1st and 2nd, the 2nd and 3rd, and the 3rd and 4th members of the natural harmonic series respectively (see Figure 3.3). Later, the major and minor third began to be accepted, the intervals between the 4th and 5th, and the 5th and 6th natural harmonics, with their inversions, the minor and major sixth respectively which are the intervals between the 5th and 8th, and the 3rd and 5th harmonics respectively. The major triad, consisting of a major third and a minor third, and the minor triad, a minor third and a major third, became the building block of Western tonal harmony. The interval of the minor seventh started to be incorporated, and its inversion the major second, the intervals between the 4th and 7th, and the 7th and 8th harmonics respectively. Twentieth century composers have explored music composed using major and minor whole tones (the intervals between the 8th and 9th, and between the 9th and 10th harmonics respectively), semitones (adjacent harmonics above the 11th are spaced by intervals close to semitones) and microtones or intervals of less than a semitone (adjacent harmonics above the 16th are spaced by microtones.)

3.3.2 Consonance and dissonance

The development of Western harmony follows a pattern where the intervals central to musical development have been gradually ascending the natural harmonic series. These changes have occurred partly as a function of increasing acceptance of intervals which are deemed to be musically "consonant," or pleasing to listen to, as opposed to "dissonant," or unpleasant to the listener. The psychoacoustic basis behind consonance and dissonance relates to critical bandwidth, which provides a means for determining the degree of consonance (or dissonance) of musical intervals.

Figure 2.6 illustrates the perceived effect of two sine waves heard together when the difference between their frequencies was increased from 0 to above one critical bandwidth. Listeners perceive a change from "rough" to "smooth" when the frequency difference crosses the critical bandwidth. In addition, a change occurs between "rough fused" to "rough separate" as the frequency difference is increased within the critical bandwidth. Figure 3.16 shows the result of an experiment by Plomp and Levelt (1965) to determine to what extent two sine waves played together sound consonant or dissonant as their frequency difference is altered. Listeners with no musical training were asked to indicate the consonance, or pleasantness, of two sine waves played together. (Musicians were not used in the experiment since they would have preconceived ideas about musical intervals which are consonant.) The result is the continuous pattern of response shown in the figure, with no particular musical interval being prominent in its degree of perceived consonance. Intervals greater than a minor third were judged to be consonant for all frequency ratios. The following can be concluded:

- When the frequencies are equal (unison) the tones are judged to be "perfectly consonant."

- When their frequency difference is greater than one critical bandwidth, they are judged consonant.

- For frequency differences of between 5 and 50% of the critical bandwidth the interval is dissonant.

- Maximum dissonance occurs when the frequency difference is a quarter of a critical bandwidth.

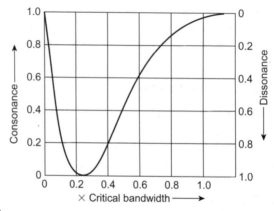

FIGURE 3.16   *The perceived consonance and dissonance of two pure tones (after Plomp and Levelt, 1965, reproduced with permission).*

Few musical instruments ever produce a sinusoidal acoustic waveform, and the results relating consonance and dissonance to pure tones can be extended to the perception of musical intervals heard when instruments which produce complex periodic tones play together. For each note of the chord, each harmonic that would be resolved by the hearing system if the note were played alone, that is all harmonics up to about the seventh, contributes to the overall perception of consonance or dissonance depending on its frequency proximity to a harmonic of another note in the chord. This contribution can be assessed based on the conclusions from Figure 3.16. The overall consonance (dissonance) of a chord is based on the total consonance (dissonance) contribution from each of these harmonics.

3.3.3 Hearing musical intervals

Musical intervals can be ordered by decreasing consonance on this psychoacoustic basis. To determine the degree of consonance of a musical interval consisting of two complex tones, each with all harmonics present, the frequencies up to the frequency of the seventh harmonic of the lower notes are found. Then the critical bandwidth at each frequency midway between harmonics of each note that are closest in frequency is found to establish whether or not they are within 5 to 50% of a critical bandwidth and therefore adding a dissonance contribution to the overall perception when the two notes are played together. If the harmonic of the upper note is midway between harmonics of the lower note, the test is carried out with the higher frequency pair since the critical bandwidth will be larger and the positions of table entries indicate this. (This exercise is similar to that carried out using the entries in Table 3.5.)

For example, Table 3.6 shows this calculation for two notes whose f_0 values are a perfect fifth apart (f_0 frequency ratio is 3:2), the lower note having an f_0 of 220 Hz. The frequency difference between each harmonic of each note and its closest neighbor harmonic in the other note is calculated (the higher of the two is used in the case of a tied distance) to give the entries in column 3, the frequency midway between these harmonic pairs is found (column 4), and the critical bandwidth for these mid-frequencies is calculated (column 5). The contribution to dissonance of each of the harmonic pairs is given in the right-hand column as follows:

(i) If they are in unison (equal frequencies) they are "perfectly consonant," shown as "C" (note that their frequency difference is less than 5% of the critical bandwidth).

(ii) If their frequency difference is greater than the critical bandwidth of the frequency midway between them (i.e., the entry in column 3 is greater than that in column 5) they are "consonant," shown as "c."

(iii) If their frequency difference is less than half the critical bandwidth of the frequency midway between them (i.e., the entry in column 3 is less than that in column 6) they are "highly dissonant," shown as "D."

(iv) If their frequency difference is less than the critical bandwidth of the frequency midway between them but greater than half that critical bandwidth (i.e., the entry in column 3 is less than that in column 5 and greater than that in column 6) they are "dissonant," shown as "d."

The contribution to dissonance depends on where the musical interval occurs between adjacent harmonics in the natural harmonic series. The higher up the series it occurs, the greater the dissonant contribution made by harmonics of the two notes concerned. The case of a two-note unison is trivial in that all harmonics are in unison with each other and all contribute as "C." For the octave, all harmonics of the upper note are in unison with harmonics of the lower note contributions as "C." Tables 3.6–3.10 show the contribution to dissonance and consonance for the intervals perfect fifth (3:2), perfect fourth (4:3), major third (5:4), minor third (6:5) and major whole tone (9:8) respectively. The dissonance of the chord in each case is related to the entries in the final column which indicate increased

Table 3.6	The degree of consonance and dissonance of a two-note chord in which all harmonics are present for both notes a perfect fifth apart, the f_0 for the lower note being 220 Hz

Perfect fifth (3:2); f_0 of lower note = 220 Hz						
First seven harmonics of lower note (Hz)	Harmonics of upper note (Hz)	Frequency difference (Hz)	Mid-frequency (Hz)	Mid-frequency critical band-width (Hz)	Mid-frequency half-critical bandwidth (Hz)	Consonant consonant dissonant Dissonant (C, c, d, D)
220						
440	330	110	385	66.3	33	c
660	660	0	Unison	–	–	C
880	990	110	1045	138	69	d
1100						
1320	1320	0	Unison	–	–	C
1540	1650	110	1595	197	99	d

Table 3.7 The degree of consonance and dissonance of a two-note chord in which all harmonics are present for both notes a perfect fourth apart, the f_0 for the lower note being 220 Hz

Perfect fourth (4:3); f_0 of lower note = 220 Hz

First seven harmonics of lower note (Hz)	Harmonics of upper note (Hz)	Frequency difference (Hz)	Mid-frequency (Hz)	Mid-frequency critical band-width (Hz)	Mid-frequency half-critical bandwidth (Hz)	Consonant consonant dissonant Dissonant (C, c, d, D)
220	293	73.0	330	60	30	c
440						
660	586	73.3	623	92	46	d
880	879	Unison	–	–	–	C
1100	1172	73.3	1170	151	76	D
1320						
1540	1465	73.3	1500	187	93	D

Table 3.8 The degree of consonance and dissonance of a two-note chord in which all harmonics are present for both notes a major third apart, the f_0 for the lower note being 220 Hz

Major third (5:4); f_0 of lower note = 220 Hz

First seven harmonics of lower note (Hz)	Harmonics of upper note (Hz)	Frequency difference (Hz)	Mid-frequency (Hz)	Mid-frequency critical band-width (Hz)	Mid-frequency half-critical bandwidth (Hz)	Consonant consonant dissonant Dissonant (C, c, d, D)
220	275	55.0	248	52	26	c
440						
660	550	110	605	90	45	c
880	825	55.0	853	117	58	D
1100	1100	Unison	–	–	–	C
1320	1375	55.0	1350	170	85	D
1540	1650	110	1600	197	99	d

dissonance in the order C, c, d and D; it can be seen that the dissonance increases as the harmonic number increases and the musical interval decreases.

The harmonics which are in unison with each other can be predicted from the harmonic number. For example, in the case of the perfect fourth

Table 3.9 The degree of consonance and dissonance of a two-note chord in which all harmonics are present for both notes a minor third apart, the f_0 for the lower note being 220 Hz

Minor third (6:5); f_0 of lower note = 220 Hz

First seven harmonics of lower note (Hz)	Harmonics of upper note (Hz)	Frequency difference (Hz)	Mid-frequency (Hz)	Mid-frequency critical band-width (Hz)	Mid-frequency half-critical bandwidth (Hz)	Consonant consonant dissonant Dissonant (C, c, d, D)
220	264.0	44.0	242	51	26	d
440	528.0	88.0	484	77	39	d
660						
880	792.0	82.0	833	115	58	d
1100	1056.0	44.0	1080	141	71	D
1320	1320.0	Unison	–	–	–	C
1540	1584.0	44.0	1560	193	97	D

Table 3.10 The degree of consonance and dissonance of a two-note chord in which all harmonics are present for both notes a major whole tone apart, the f_0 for the lower note being 220 Hz

Major whole tone (9:8); f_0 of lower note = 220 Hz

First seven harmonics of lower note (Hz)	Harmonics of upper note (Hz)	Frequency difference (Hz)	Mid-frequency (Hz)	Mid-frequency critical band-width (Hz)	Mid-frequency half-critical bandwidth (Hz)	Consonant consonant dissonant Dissonant (C, c, d, D)
220	247.5	27.5	234	50	25	d
440	495.0	55.0	477	76	38	d
660	742.5	82.5	701	100	50	d
880						
1100	990.0	110	1050	138	69	d
1320	1237.5	82.5	1280	163	82	d
1540	1485.0	55.0	1510	188	94	D

the fourth harmonic of the lower note is in unison with the third of the upper note because their f_0 values are in the ratio (4:3). For the major whole tone (9:8), the unison will occur between harmonics (the eighth of the upper note and the ninth of the lower) which are not resolved by the auditory system for each individual note.

As a final point, the degree of dissonance of a given musical interval will vary depending on the f_0 value of the lower note, due to the nature of the critical bandwidth with center frequency (e.g., see Figure 3.8). Tables 3.11 and 3.12 illustrate this effect for the major third where the f_0 of the lower note is one octave and two octaves below that used in Table 3.8 at 110 Hz

Table 3.11 The degree of consonance and dissonance of a two-note chord in which all harmonics are present for both notes a major third apart, the f_0 for the lower note being 110 Hz

Major third (5:4); f_0 of lower note = 110 Hz

First seven harmonics of lower note (Hz)	Harmonics of upper note (Hz)	Frequency difference (Hz)	Mid-frequency (Hz)	Mid-frequency critical band-width (Hz)	Mid-frequency half-critical bandwidth (Hz)	Consonant consonant dissonant Dissonant (C, c, d, D)
110	137.5	27.5	124	38	19	d
220						
330	275.0	55.0	303	57	29	d
440	412.5	27.5	426	71	36	D
550	550.0	Unison	–	–	–	C
660	687.5	27.5	674	97	49	D
770	825.0	55.0	798	111	56	D

Table 3.12 The degree of consonance and dissonance of a two-note chord in which all harmonics are present for both notes a major third apart, the f_0 for the lower note being 55.0 Hz

Major third (5:4); f_0 of lower note = 55 Hz

First seven harmonics of lower note (Hz)	Harmonics of upper note (Hz)	Frequency difference (Hz)	Mid-frequency (Hz)	Mid-frequency critical band-width (Hz)	Mid-frequency half-critical bandwidth (Hz)	Consonant consonant dissonant Dissonant (C, c, d, D)
55.0	68.75	13.8	61.9	31	16	D
110						
165	137.5	27.5	151	41	21	d
220	206.3	13.8	213	48	24	D
275	275.0	Unison	–	–	–	C
330	343.8	13.8	337	61	31	D
385	412.5	27.5	399	68	34	D

and 55 Hz respectively. The number of "D" entries increases in each case as the f_0 values of the two notes are lowered.

This increase in dissonance of any given interval, excluding the unison and octave which are equally consonant at any pitch on this basis, manifests itself in terms of preferred chord spacings in classical harmony. As a rule when writing four-part harmony such as SATB (soprano, alto, tenor, bass) hymns, the bass and tenor parts are usually no closer together than a fourth except when they are above the bass staff, because the result would otherwise sound "muddy" or "harsh."

Figure 3.17 shows a chord of C major in a variety of four-part spacings and inversions which illustrate this effect when the chords are played, preferably on an instrument producing a continuous steady sound for each note such as a pipe organ, instrumental group or suitable synthesizer sound. To realize the importance of this point, it is essential to *listen* to the effect. The psychoacoustics of music is, after all, about how music is perceived, not what it looks like on paper!

3.4 TUNING SYSTEMS

Musical scales are basic to most Western music. Modern keyboard instruments have 12 notes per octave with a musical interval of one semitone between adjacent notes. All common Western scales incorporate octaves whose frequency ratios are (2:1). Therefore it is only necessary to consider notes in a scale over a range of one octave, since the frequencies of notes in other octaves can be found from them. Early scales were based on one or more of the musical intervals found between members of the natural harmonic series (e.g. see Figure 3.3).

3.4.1 Pythagorean tuning

The Pythagorean scale is built up from the perfect fifth. Starting, for example, from the note C and going up in 12 steps of a perfect fifth produces

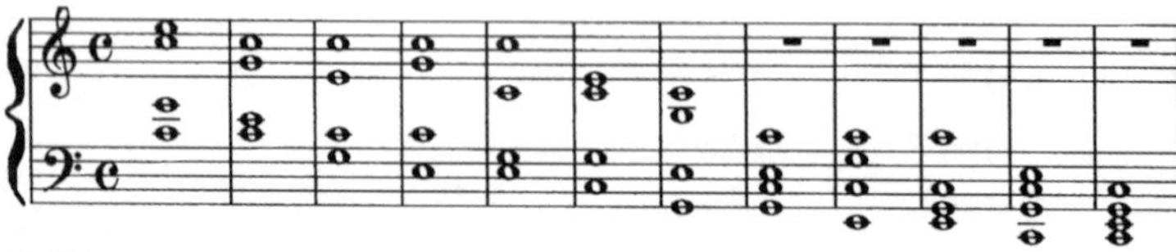

FIGURE 3.17 *Different spacings of the chord of C major. Play each chord and listen to the degree of "muddiness" or "harshness" each produces (see text).*

the "circle of fifths:" C, G, D, A, E, B, F#, C#, G#, D#, A#, E#, c. The final note after 12 steps around the circle of fifths, shown as c, has a frequency ratio to the starting note, C, of the frequency ratio of the perfect fifth (3:2) multiplied by itself 12 times, or:

$$\frac{c}{C} = \left(\frac{3}{2}\right)^{12} = 129.746$$

An interval of 12 fifths is equivalent to seven octaves, and the frequency ratio for the note (c') which is seven octaves above C is:

$$\frac{c'}{C} = 2^7 = 128.0$$

Thus 12 perfect fifths (C to c) is slightly sharp compared with seven octaves (C to c') of the so-called "Pythagorean comma" which has a frequency ratio:

$$\frac{c}{c'} = \frac{129.746}{128.0} = 1.01364$$

If the circle of fifths were established by descending by perfect fifths instead of ascending, the resulting note 12 fifths below the starting notes would be flatter than seven octaves by 1.0136433, and *every* note of the descending circle would be slightly different from the members of the ascending circle. Figure 3.18 shows this effect and the manner in which the notes can be notated. For example, notes such as D# and Eb, A# and Bb, Bbb and A are not the same and are known as "enharmonics," giving rise to the pairs of intervals such as major third and diminished fourth, and major seventh and diminished octave shown in Figure 3.15. The Pythagorean scale can be built up on the starting note C by making F and G an exact perfect fourth and perfect fifth respectively (maintaining a perfect relationship for the subdominant and dominant respectively):

$$\frac{F}{C} = \frac{4}{3}$$
$$\frac{G}{C} = \frac{3}{2}$$

The frequency ratios for the other notes of the scale are found by ascending in perfect fifths from G and, when necessary, bringing the result down to be

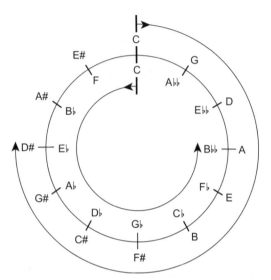

FIGURE 3.18 *The Pythagorean scale is based on the circle of fifths formed either by ascending by 12 perfect fifths (outer) or descending by 12 perfect fifths (inner).*

within an octave of the starting note. The resulting frequency ratios relative to the starting note C are:

$$\frac{D}{C} = \frac{3}{2} \times \frac{G}{C} \times \frac{1}{2} = \frac{3}{2} \times \frac{3}{2} \times \frac{1}{2} = \frac{9}{8}$$

$$\frac{A}{C} = \frac{3}{2} \times \frac{D}{C} = \frac{3}{2} \times \frac{9}{8} = \frac{27}{16}$$

$$\frac{E}{C} = \frac{3}{2} \times \frac{A}{C} \times \frac{1}{2} = \frac{3}{2} \times \frac{27}{16} \times \frac{1}{2} = \frac{81}{64}$$

$$\frac{B}{C} = \frac{3}{2} \times \frac{E}{C} = \frac{3}{2} \times \frac{81}{64} = \frac{243}{128}$$

The frequency ratios of the members of the Pythagorean major scale are shown in Figure 3.19 relative to C for convenience. The frequency ratios between adjacent notes can be calculated by dividing the frequency ratios of the upper note of the pair to C by that of the lower. For example:

$$\text{Frequency ratio between A and B} = \frac{243}{128} \div \frac{27}{16} = \frac{243}{128} \times \frac{16}{27} = \frac{9}{8}$$

$$\text{Frequency ratio between E and F} = \frac{4}{3} \div \frac{81}{64} = \frac{4}{3} \times \frac{64}{81} = \frac{256}{243}$$

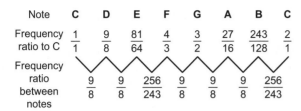

FIGURE 3.19 *Frequency ratios between the notes of a C major Pythagorean scale and the tonic (C).*

Figure 3.18 shows the frequency ratios between adjacent notes of the Pythagorean major scale. A major scale consists of the following intervals: tone, tone, semitone, tone, tone, tone, semitone, and it can be seen that:

$$\text{Frequency ratio of the Pythagorean semitone} = \frac{256}{243}$$

$$\text{Frequency ratio of the Pythagorean tone} = \frac{9}{8}$$

3.4.2 Just tuning

Another important scale is the "just diatonic" scale which is made by keeping the intervals that make up the major triads pure: the octave (2:1), the perfect fifth (3:2) and the major third (5:4) for triads on the tonic, dominant and sub-dominant. The dominant and sub-dominant keynotes are a perfect fifth above and below the key note respectively. This produces all the notes of the major scale (any of which can be harmonized using one of these three chords). Taking the note C being used as a starting reference for convenience, the major scale is built as follows. The notes E and G are a major third (5:4) and a perfect fifth (3:2) respectively above the tonic, C:

$$\frac{E}{C} = \frac{5}{4}$$

$$\frac{G}{C} = \frac{3}{2}$$

The frequency ratios of B and D are a major third (5:4) and a perfect fifth (3:2) respectively above the dominant, G, and they are related to C as:

$$\frac{B}{C} = \frac{5}{4} \times \frac{G}{C} = \frac{5}{4} \times \frac{3}{2} = \frac{15}{8}$$

$$\frac{D}{C} = \frac{3}{2} \times \frac{G}{C} \times \frac{1}{2} = \frac{3}{2} \times \frac{3}{2} \times \frac{1}{2} = \frac{9}{8}$$

(The result for the D is brought down one octave to keep it within an octave of the C.)

The frequency ratios of A and C are a major third (5:4) and a perfect fifth (3:2) respectively above the sub-dominant F. The F is therefore a perfect fourth (4:3) above the C (perfect fourth plus a perfect fifth is an octave):

$$\frac{F}{C} = \frac{4}{3}$$

$$\frac{A}{C} = \frac{5}{4} \times \frac{F}{C} = \frac{5}{4} \times \frac{4}{3} = \frac{20}{12} = \frac{5}{3}$$

> The "just diatonic" scale has both a major and a minor whole tone.

The frequency ratios of the members of the just diatonic major scale are shown in Figure 3.20 relative to C for convenience, along with the frequency ratios between adjacent notes (calculated by dividing the frequency ratio of the upper note of each pair to C by that of the lower). The figure shows that the just diatonic major scale (tone, tone, semitone, tone, tone, tone, semitone) has equal semitone intervals, but two different tone intervals, the larger of which is known as a "major whole tone" and the smaller as a "minor whole tone:"

$$\text{Frequency ratio of the just diatonic semitone} = \frac{16}{15}$$

$$\text{and frequency ratio of the just diatonic major whole tone} = \frac{9}{8}$$

$$\text{and frequency ratio of the just diatonic minor whole tone} = \frac{10}{9}$$

The two whole tone and the semitone intervals appear as members of the musical intervals between adjacent members of the natural harmonic series (see Figure 3.3), which means that the notes of the scale are as consonant with each other as possible for both melodic and harmonic musical phrases. However, the presence of two whole tone intervals means that this

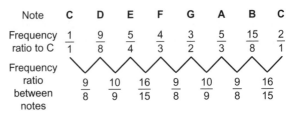

FIGURE 3.20 *Frequency ratios between the notes of a C major "just diatonic" scale and the tonic (C).*

scale can only be used in one key since each key requires its own tuning. This means, for example, that the interval between D and A is:

$$\frac{A}{D} = \frac{5}{3} \div \frac{9}{8} = \frac{5}{3} \times \frac{8}{9} = \frac{40}{27}$$

which is a musically flatter fifth than the perfect fifth (3:2).

In order to tune a musical instrument for practical purposes to enable it to be played in a number of different keys, the Pythagorean comma has to be distributed among some of the fifths in the circle of fifths such that the note reached after 12 fifths is exactly seven octaves above the starting note (see Figure 3.18). This can be achieved by flattening some of the fifths, possibly by different amounts, while leaving some perfect, or flattening all of the fifths by varying amounts, or even by additionally sharpening some and flattening others to compensate. There is therefore an infinite variety of possibilities, but none will result in just tuning in all keys. Many tuning systems were experimented with to provide tuning of thirds and fifths which were close to just tuning in some keys at the expense of other keys whose tuning could end up being so out-of-tune as to be unusable musically.

Padgham (1986) gives a fuller discussion of tuning systems. A number of keyboard instruments have been experimented with which had split black notes (in either direction) to provide access to their enharmonics, giving C# and Db, D# and Eb, F# and Gb, G# and Ab, and A# and Bb—for example, the McClure pipe organ in the Faculty of Music at the University of Edinburgh discussed by Padgham (1986)—but these have never become popular with keyboard players.

3.4.3 Equal tempered tuning

The spreading of the Pythagorean comma unequally among the fifths in the circle results in an "unequal temperament." Another possibility is to spread it evenly to give "equal temperament", which makes modulation to all keys possible where each one is equally out-of-tune with the just scale. This is the tuning system commonly found on today's keyboard instruments. All semitones are equal to one twelfth of an octave. Therefore the frequency ratio (r) for an equal tempered semitone is a number which when multiplied by itself 12 times is equal to 2, or:

> A *cent* is one hundredth of a semitone.

$$r = \sqrt[12]{2} = 1.0595$$

The equal tempered semitone is subdivided into "cents," where one cent is one hundredth of an equal tempered semitone. The frequency ratio for one cent (c) is therefore:

$$c = \sqrt[100]{r} = \sqrt[100]{1.0595} = 1.000578$$

Cents are widely used in discussions of pitch intervals and the results of psychoacoustic experiments involving pitch. Appendix 3 gives an equation for converting frequency ratios to cents and vice versa.

Music can be played in all keys when equal tempered tuning is used, as all semitones and tones have identical frequency ratios. However, no interval is in-tune in relation to the intervals between adjacent members of the natural harmonic series (see Figure 3.3); therefore none is perfectly consonant. However, intervals of the equal tempered scale can still be considered

> In today's equal tempered scale, *no* interval is in-tune with integer ratios between frequencies except the octave.

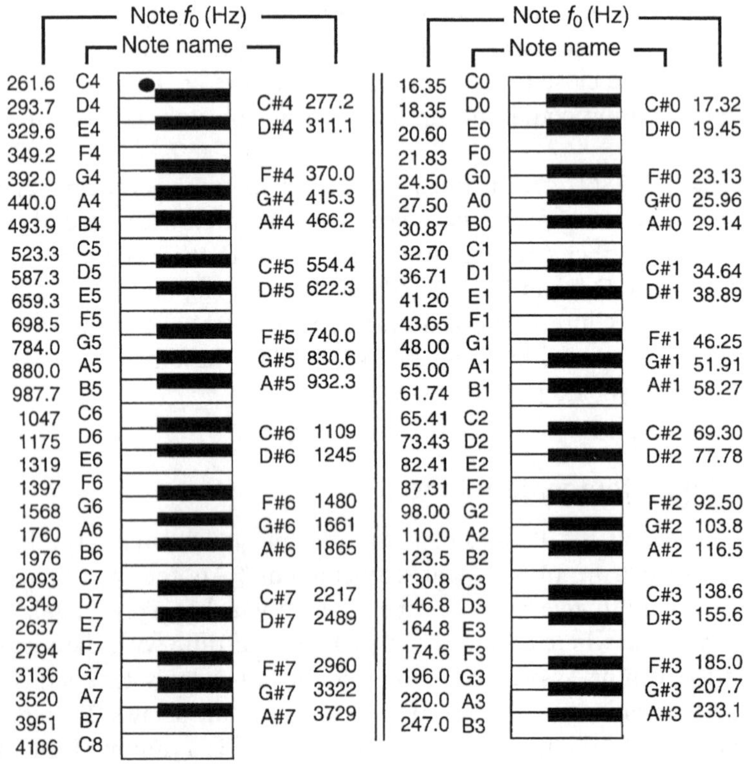

FIGURE 3.21 *Fundamental frequency values to four significant figures for eight octaves of notes, four either side of middle C, tuned in equal temperament with a tuning reference of A4 = 440 Hz. (Middle C is marked with a black spot.)*

in terms of their consonance and dissonance, because although harmonics of pairs of notes that are in unison for pure intervals (see Tables 3.6 to 3.12) are not identical in equal temperament, the difference is within the 5% critical bandwidth criterion for consonance. Beats (see Figure 2.6) will exist between some harmonics in equal tempered chords which are not present in their pure counterparts.

Figure 3.21 shows the f_0 values and the note naming convention used in this book for eight octaves, four either side of middle C, tuned in equal temperament with a tuning reference of 440 Hz for A4: the A above middle C. The equal tempered system is found on modern keyboard instruments, but there is increasing interest among performing musicians and listeners alike in the use of unequal temperament. This may involve the use of original instruments or electronic synthesizers which incorporate various tuning systems. Padgham (1986) lists approximately 100 pipe organs in Britain which are tuned to an unequal temperament in addition to the McClure organ.

REFERENCES

Howard, D.M., Hirson, A., Brookes, T., Tyrrell, A.M., 1995. Spectrography of disputed speech samples by peripheral human hearing modeling. Forensic Linguist. 2 (1), 28–38.

Moore, B.C.J., 1982. An Introduction to the Psychology of Hearing. Academic Press, London.

Padgham, C.A., 1986. The Well-tempered Organ. Positif Press, Oxford.

Parkin, P.H., 1974. Pitch change during reverberant decay. J. Sound Vib. 32, 530.

Pickles, J.O., 1982. An Introduction to the Physiology of Hearing. Academic Press, London.

Plomp, R., Levelt, W.J.M., 1965. Tonal consonance and critical bandwidth. J. Acoust. Soc. Am. 38, 548.

Roederer, J.G., 1975. Introduction to the Physics and Psychophysics of Music. Springer-Verlag, New York.

Rossing, T.D., 2001. The Science of Sound, Third edition. Addison Wesley, New York.

Schouten, J.F., 1940. The perception of pitch. Philips Tech. Rev. 5, 286.

Wever, E.G., 1949. Theory of Hearing. Wiley, New York.

Zwicker, E., Flottorp, G., Stevens, S.S., 1957. Critical bandwidth in loudness summation. J. Acoust. Soc. Am. 29, 548.

Acoustic Model for Musical Instruments

4.1 A "BLACK BOX" MODEL OF MUSICAL INSTRUMENTS

In this chapter a simple model is developed which allows the acoustics of all musical instruments to be discussed and, it is hoped, readily understood. The model is used to explain the acoustics of stringed, wind and percussion instruments as well as the singing voice. A selection of anechoic (no reverberation) recordings of a variety of acoustic instruments and the singing voice is provided on tracks 8–61 on the accompanying CD. Any acoustic instrument has two main components:

- a sound source, and
- sound modifiers.

For the purposes of our simple model, the sound source is known as the "input" and the sound modifiers are known as the "system." The result of the input passing through the system is known as the "output." Figure 4.1 shows the complete input/system/output model.

This model provides a framework within which the acoustics of musical instruments can be usefully discussed, reviewed and understood. Notice that the "output" relates to the actual output from the instrument, which is not that which the listener hears since it is modified by the acoustics of the environment in which the instrument is being played. The input/system/output model can be extended to include the acoustic effects of the environment as follows.

If we are modeling the effect of an instrument being played in a room, then the output we require is the sound heard by the listener and not the output from the instrument itself. The environment itself acts as a sound modifier and therefore it too acts as a "system" in terms of the input/system/output model. The input to the model of the environment is the output from the instrument being played. Thus the complete practical input/system/output model for an instrument being played in a room is shown in Figure 4.2. Here, the output from the instrument is equal to the input to the room.

In order to make use of the model in practice, acoustic details are required for the "input" and "system" boxes to enable the output(s) to be determined. The effects of the room are described in Chapter 6. In this chapter, the "input" and "system" characteristics for stringed, wind and percussion instruments as well as the singing voice are discussed. Such details can be calculated theoretically from first principles, or measured experimentally in which case they must be carried out in an environment which either has no effect on the acoustic recording or has a known effect which can be accounted for mathematically. An environment which has no acoustic effect

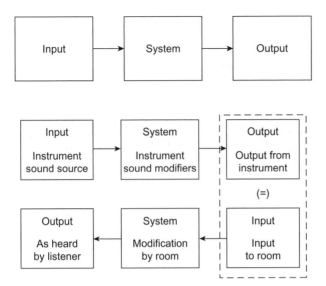

FIGURE 4.1

An input/system/output model for describing the acoustics of musical instruments.

FIGURE 4.2

The input/system/output model applied to an instrument being played in a room.

is one where there are no reflections of sound—ideally this is known as "free space." In practice, free space is achieved in a laboratory in an anechoic ("no echo") room in which all sound reaching the walls, floor and ceiling is totally absorbed by large wedges of sound-absorbing material. However, anechoic rooms are rare, and a useful practical approximation to free space for experimental purposes is outside on grass during a windless day, with the experiment being conducted at a reasonable height above the ground.

This chapter considers the acoustics of stringed, wind, and percussion instruments. In each case, the sound source and the sound modifiers are discussed. These discussions are not intended to be exhaustive since the subject is a large one. Rather they focus on one or two instruments by way of examples as to how their acoustics can be described using the sound source and sound modifier model outlined above. References are included to other textbooks in which additional information can be found for those wishing to explore a particular area more fully.

Finally, the singing voice is considered. It is often the case that budding music technologists are able to make good approximations with their voices to sounds they wish to synthesize electronically or acoustically, and a basic understanding of the acoustics of the human voice can facilitate this. As a starting point for the consideration of the acoustics of musical instruments, the playing fundamental frequency ranges of a number of orchestral instruments, as well as the organ, piano and singers, are illustrated in Figure 4.3. A nine octave keyboard is provided for reference on which middle C is marked with a black square.

"Free space" is sometimes called "free field".

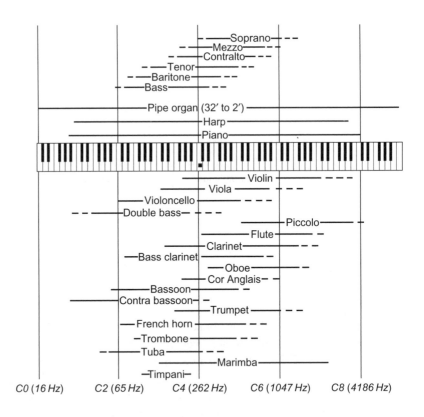

FIGURE 4.3

Playing fundamental frequency ranges of selected acoustic instruments and singers.

4.2 STRINGED INSTRUMENTS

The string family of musical instruments includes the violin, viola, violon 'cello' and double bass and all their predecessors, as well as keyboard instruments which make use of strings, such as the piano, harpsichord, clavichord and spinet. In each case, the acoustic output from the instrument can be considered in terms of an input sound source and sound modifiers as illustrated in Figure 4.1. A more detailed discussion on stringed instruments can be found in Hutchins (1975a, 1975b), Benade (1976), Rossing (2001), Hall (2001) and Fletcher and Rossing (1999). The playing fundamental frequency (f_0) ranges of the orchestral stringed instruments are shown in Figure 4.3.

All stringed instruments consist of one or more strings stretched between two points, and the f_0 produced by the string is dependent on its mass per unit length, length and tension. For any practical musical instrument, the mass per unit length of an individual string is constant, and changes are made to the tension and/or the length to enable different notes to be played. Figure 4.4 shows a string fixed at one end, supported on two single-point contact bridges, and passed over a pulley with a variable mass hanging on

Different notes are played on stringed instruments by changing either the length or tension of the string.

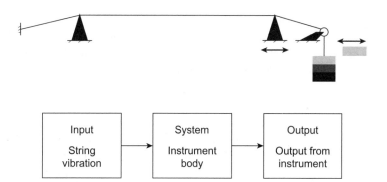

FIGURE 4.4
Idealized string whose tension and length can be varied.

FIGURE 4.5
Input/system/output model for a stringed instrument.

the other end. The variable mass enables the string tension to be altered, and the length of the string can be altered by moving the right-hand bridge. In a practical musical instrument, the tension of each string is usually altered by means of a peg with which the string can be wound, or winched in, to tune the string, and the position of one of the points of support is varied to enable different notes to be played—except in instruments such as stringed keyboard instruments where individual strings are provided to play each note.

The string is set into vibration to provide the sound source to the instrument. A vibrating string on its own is extremely quiet because little energy is imparted to the surrounding air due to the small size of a string with respect to the air particle movement it can initiate. All practical stringed instruments have a body which is set in motion by the vibrations of the string(s) of the instrument, giving a large area from which vibration can be imparted to the surrounding air. The body of the instrument is the sound modifier. It imparts its own mechanical properties onto the acoustic input provided by the vibrating string (see Figure 4.5).

There are three main methods by which energy is provided to a stringed instrument. The strings are either "plucked," "bowed" or "struck." Instruments which are usually plucked or bowed include those in the violin family; instruments whose strings are generally only plucked include the guitar, lute, and harpsichord; and the piano is an instrument whose strings are struck.

A vibrating string fixed at both ends, for example by being stretched across two bridge-like supports as illustrated in Figure 4.4, has a unique set of standing waves (see Chapter 1). Any observed instantaneous shape adopted by the string can be analyzed (and synthesized) as a combination of some or all of these standing wave modes. The first 10 modes of a string fixed at both ends are shown in Figure 4.6. In each case the mode is illustrated in terms of the extreme positions of the string between which it oscillates. Every mode of a string fixed at both ends is constrained not to move; therefore there cannot be any velocity, or displacement, at the ends themselves

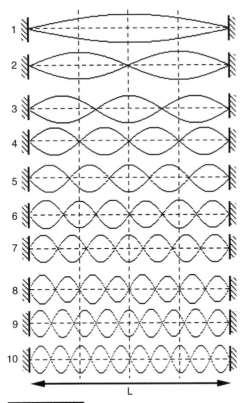

FIGURE 4.6 *The first 10 possible modes of vibration of a string of length (L) fixed at both ends.*

and so these points are known as "velocity nodes" or, more usually, "displacement nodes." Points of maximum movement are known as "velocity antinodes" or "displacement antinodes."

It can be seen in Figure 4.6 that the first mode has two displacement nodes (at the ends of the string) and one displacement antinode (in the center). The sixth mode has seven displacement nodes and six displacement antinodes. In general, a particular mode (n) of a string fixed at both ends has $(n + 1)$ displacement nodes and (n) displacement antinodes. The frequencies of the standing wave modes are related to the length of the string and the velocity of a transverse wave in a string by Equation 1.35.

4.2.1 Sound source from a plucked string

When a string is plucked, it is pulled a small distance away from its rest position and released. The nature of the sound source it provides to the body of the instrument depends in part on the position on the string at which it is plucked. This is directly related to the displacement component modes that a string can adopt. For example, if the string is plucked at the center, as indicated by the central dashed vertical line in Figure 4.6, modes which have a node at the center of the string (the 2nd, 4th, 6th, 8th, 10th, etc., or the even modes) are not excited, and those with an antinode at the center (the 1st, 3rd, 5th, 7th, 9th, etc., or the odd modes) are maximally excited. If the string is plucked at a quarter of its length from either end (as indicated by the other dashed vertical lines in the figure), modes with a node at the plucking point (the 4th, 8th, etc.) are not excited and other modes are excited to a greater or lesser degree. In general, the modes that are not excited for a plucking point a distance (d) from the closest end of a string fixed at both ends are those with a node at the plucking position. They are given by:

$$\text{Modes not excited} = m\left[\frac{L}{d}\right] \tag{4.1}$$

where m = 1, 2, 3, 4,
L = length of string
and d = distance to plucking point from closest end of the string

Thus if the plucking point is a third of the way along the string, the modes not excited are the 3rd, 6th, 9th, 12th, 15th, etc. For a component mode not to be excited at all, it should be noted that the plucking distance has to be exactly an integer fraction of the length of the string in order that it exactly coincides with nodes of that component.

This gives the sound input to the body of a stringed instrument when it is plucked. The frequencies (f_n) of the component modes of a string supported at both ends can be related to the length, tension (T) and mass per unit length (μ) of the string by substituting Equation 1.7 for the transverse wave velocity in Equation 1.35 to give:

$$f_n = \left[\frac{n}{2L}\right]\sqrt{\frac{T}{\mu}} \qquad (4.2a)$$

where $n = 1, 2, 3, 4, \ldots$
$L = $ length
$T = $ tension
and $\mu = $ mass per unit length

The frequency of the lowest mode is given by Equation 4.2a when $(n = 1)$:

$$f_1 = \left[\frac{1}{2L}\right]\sqrt{\frac{T}{\mu}} \qquad (4.2b)$$

This is the f_0 of the string which is also known as the "first harmonic" (see Table 3.1). Thus the first mode (f_1) in Equation 4.2a is the f_0 of string vibration. Equation 4.2a shows that the frequencies of the higher modes are harmonically related to f_0.

4.2.2 Sound source from a struck string

The piano is an instrument in which strings are struck to provide the sound source, and the relationship discussed in the last Section (4.2.1) concerning the modes that will be missing in the sound source is equally relevant here. There is, however, an additional effect that is particularly relevant to the sound source in the piano, and this relates to the fact that the strings of a piano are under very high tension and therefore very hard compared with those on a harpsichord or plucked orchestral stringed instrument. Strings on a piano are struck by a hammer which is "fired" at the string from which it immediately bounces back so as not to interfere with the free vibration of the string(s). When a piano string is struck by the hammer, it behaves partly like a bar because it is not completely flexible due to its considerable stiffness. This results in a slight raising in frequency of all the component modes with respect to the fundamental, an effect known as "inharmonicity," and this effect is greater for the higher modes.

Equation 4.2b assumes an ideal string; that is, a string with zero radius. Substituting Equation 4.2b into 4.2a gives the simple relationship between the frequency of any mode and that of the first mode:

$$f_n = nf_1$$

Any practical string must have a finite radius, and the effect is given in Equation 4.2c. This is the effect of inharmonicity, or the amount by which the actual frequencies of the modes vary from integer multiples of the fundamental.

$$f_n = nf_1 \left[1 + (n^2 - 1) \frac{\pi^3 r^4 E}{8TL^2} \right] \qquad (4.2c)$$

where f_n = the frequency of the nth mode
n = 1, 2, 3, 4, ...
r = string radius
E = Young's modulus (see Section 1.1.2)
T = tension
and L = length

It can be seen that inharmonicity increases as the square of the component mode (n^2) and as the fourth power of the string radius (r^4), and that it decreases with increased tension and as the square of increased length. Inharmonicity can be kept low if the strings are thin (small r), long (large L), and under high tension (high T). The effect would therefore be particularly marked for bass strings if they were simply made thicker (larger r) to give them greater mass, since the variation is to the fourth power of r. Therefore in many stringed instruments, including pianos, guitars and violins, the bass strings are wrapped with wire to increase their mass without increasing the underlying core string's radius (r). (A detailed discussion of the acoustics of pianos is given in: Benade, 1976; Askenfelt, 1990; Fletcher and Rossing, 1999.)

The notes of a piano are usually tuned to equal temperament (see Chapter 3) and octaves are then tuned by minimizing the beats between pairs of notes an octave apart. When tuning two notes an octave apart, the components which give rise to the strongest sensation of beats are the first harmonic of the upper note and the second harmonic of the lower note. These are tuned in unison to minimize the beats between the notes. This results in the f_0 of the lower note being slightly lower than half the f_0 of the higher note due to the inharmonicity between the first and second components of the lower note.

EXAMPLE 4.1

If the f_0 of a piano note is 400 Hz and inharmonicity results in the second component being stretched to 801 Hz, how many cents sharp will the note an octave above be if it is tuned for no beats between it and the octave below?

Tuning for no beats will result in the f_0 of the upper note being 801 Hz, slightly greater than an exact octave above 400 Hz which would be 800 Hz. The frequency ratio (801/800) can be converted to cents using Equation A3.4 in Appendix 3:

$$\text{Number of cents} = 3986.3137 \, \log_{10}\left[\frac{801}{800}\right] = 2.16 \text{ cents}$$

Inharmonicity on a piano increases as the strings become shorter and therefore the octave stretching effect increases with note pitch. The stretching effect is usually related to middle C and it becomes greater the further away the note of interest is in pitch. Figure 4.7 illustrates the effect in terms of the average deviation from equal tempered tuning across the keyboard of a small piano. Thus high notes and low notes on the piano are tuned sharp and flat respectively to what they would have been if all octaves were tuned pure with a frequency ratio of 2:1. From the figure it can be seen that this stretching effect amounts to approximately 35 cents sharp at C8 and 35 cents flat at C1 with respect to middle C.

The piano keyboard usually has 88 notes from A0 (27.5 Hz) to C8 (4186 Hz), giving it a playing range of just over seven octaves (see Figure 4.3). The use of thinner strings to help reduce inharmonicity means that less sound source energy is transferred to the body of the instrument, and over the majority of the piano's range multiple strings are used for each note. A concert grand piano can have over 240 strings for its 88 notes: single, wire-wrapped strings for the lowest notes, pairs of wire-wrapped strings for the next group of notes, and triplets of strings for the rest of the notes, the lower of which might be wire-wrapped. The use of multiple strings provides some control over the decay time of the note. If the multiple (2 or 3) strings of a particular note are exactly in-tune and beat free (see Section 2.2), the decay

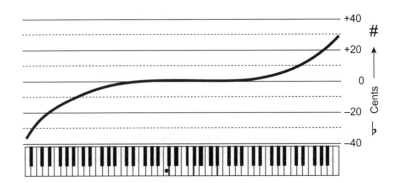

Approximate form of the average deviations from equal temperament due to inharmonicity in a small piano. Middle C marked with a spot. (Data from Martin and Ward, 1961.)

time is short as the exactly in-phase energy is transferred to the soundboard quickly. Appropriate tuning of multiple strings is a few cents apart, and this results in a richer sound which decays more slowly than exactly in-tune strings would. If the strings are out-of-tune by around 12 cents or more, then the result is the "pub piano" sound.

4.2.3 Sound source from a bowed string

The sound source that results from bowing a string is periodic and a continuous note can be produced while the bow travels in one direction. A bow supports many strands of hair, traditionally horsehair. Hair tends to grip in one direction but not in the other. This can be demonstrated with your own hair. Support the end of one hair firmly with one hand, and then grip the hair in the middle with the thumb and index finger of the other hand and slide that hand up and down the hair. You should feel the way the hair grips in one direction but slides easily in the other.

The bow is held at the end known as the "frog" or "heel," and the other end is known as the "point" or "tip." The hairs of the bow are laid out such that approximately half are laid one way round from heel to tip, and half are laid the other way round from tip to heel. In this way, about the same number of hairs are available to grip a string no matter in which direction the bow is moved (*up bow* or *down bow*). Rosin is applied to the hairs of a bow to increase its gripping ability. As the bow is moved across a string in either direction, the string is gripped and moved away from its rest position until the string releases itself, moving past its rest position until the bow hairs grip it again to repeat the cycle.

One complete cycle of the motion of the string immediately under a bow moving in one direction is illustrated in the graph on the right-hand side of Figure 4.8. (When the bow moves in the other direction, the pattern is reversed.) The string moves at a constant velocity when it is gripped by the bow hairs and then returns rapidly through its rest position until it is gripped by the bow hairs again. If the minute detail of the motion of the bowed string is observed closely, for example by means of stroboscopic illumination, it is seen to consist of two straight-line segments joining at a point which moves at a constant velocity around the dotted track as shown in the snapshot sequence in Figure 4.8.

The time taken to complete one cycle, or the fundamental period (T_0), is the time taken for the point joining the two line segments to travel twice the length of the string $(2L)$:

$$T_0 = \frac{2L}{v}$$

Substituting Equation 1.7 for the transverse wave velocity gives:

$$T_0 = 2L \sqrt{\frac{\mu}{T}}$$

The f_0 of vibration of the bowed string is therefore:

$$f_0 = \left[\frac{1}{T_0}\right] = \left[\frac{1}{2L}\right]\sqrt{\frac{T}{\mu}} \qquad (4.3)$$

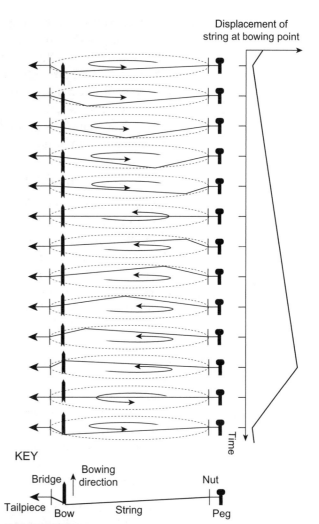

KEY

FIGURE 4.8 *One complete cycle of vibration of a bowed string and graph of string velocity at the bowing point as a function of time. (Adapted from Rossing, 2001.)*

Comparison with Equation 4.2a when $(n = 1)$ shows that this is the frequency of the first component mode of the string. Thus the f_0 for a bowed string is the frequency of the first natural mode of the string, and bowing is therefore an efficient way to excite the vibrational modes of a string.

The sound source from a bowed string is that of the waveform of string motion which excites the bridge of the instrument. Each of the snapshots in Figure 4.8 corresponds to equal time instants on the graph of string displacement at the bowing point in the figure, from which the resulting force acting on the bridge of the instrument can be inferred to be of a similar shape to that at the bowing point. In its ideal form, this is a sawtooth waveform (see Figure 4.9). The spectrum of an ideal sawtooth waveform contains all harmonics and their amplitudes decrease with ascending frequency as $(1/n)$, where n is the harmonic number. The spectrum of an ideal sawtooth waveform is plotted in Figure 4.9 and the amplitudes are shown relative to the amplitude of the f_0 component.

4.2.4 Sound modifiers in stringed instruments

The sound source provided by a plucked or bowed string is coupled to the sound modifiers of the instrument via a bridge. The vibrational properties of all elements of the body of the instrument play a part in determining the sound modification that takes place. In the case of the violin family, the components which contribute most significantly are the top plate (the plate under the strings that the bridge stands on and which has the f holes in it), the back plate (the back of the instrument), and the air contained within the main body of the instrument. The remainder of the instrument contributes to a much lesser extent to the sound-modification process, and there is still lively debate in some quarters about the importance or otherwise of the glues, varnish, choice of wood and wood treatment used by highly regarded violin makers of the past.

Two acoustic resonances dominate the sound modification due to the body of instruments in the violin family at low frequencies: the resonance of the air contained within the body of the instrument or the "air resonance," and the main resonance of the top plate or "top resonance." Hall (2001) summarizes the important resonance features of a typical violin as follows:

- practically no response below the first resonance at approximately 273 Hz (air resonance);
- another prominent resonance at about 473 Hz (top resonance);
- rather uneven response up to about 900 Hz, with a significant dip around 600–700 Hz;

FIGURE 4.9

Idealized sound source sawtooth waveform and its spectrum for a bowed string.

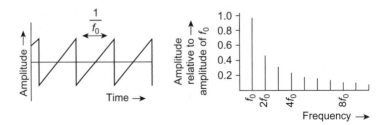

■ better mode overlapping and more even response (with some exceptions) above 900 Hz;

■ gradual decrease in response toward high frequencies.

Apart from the air resonance, which is defined by the internal dimensions of the instrument and the shape and size of the f holes, the detailed nature of the response of these instruments is related to the main vibrational modes of the top and back plates. As these plates are being shaped by detailed carving, the maker will hold each plate at particular points and tap it to hear how the so-called "tap tones" are developing to guide the shaping process. This ability is a vital part of the art of the experienced instrument maker in setting up what will become the resonant properties of the complete instrument when it is assembled.

The acoustic output from the instrument is the result of the sound input being modified by the acoustic properties of the instrument itself. Figure 4.10 (from Hall, 2001) shows the input spectrum for a bowed G3 ($f_0 = 196$ Hz) with a typical response curve for a violin, and the resulting output spectrum. Note that the frequency scales are logarithmic, and therefore the harmonics in the input and output spectra bunch together at high frequencies. The output spectrum is derived by multiplying the amplitude of each component of the input spectrum by the response of the body of the instrument at that frequency. In the figure, this multiplication becomes an addition since the amplitudes are expressed logarithmically as dB values, and adding logarithms of numbers is mathematically equivalent to multiplying the numbers themselves.

There are basic differences between the members of the orchestral string family (violin, viola, cello and double bass). They differ from each other acoustically in that the size of the body of each instrument becomes smaller relative to the f_0 values of the open strings (e.g., Hutchins, 1978). The air and tap resonances approximately coincide as follows: for the violin with f_0 of the D4 (2nd string) and A4 (3rd string) strings respectively, for the viola with f_0 values approximately

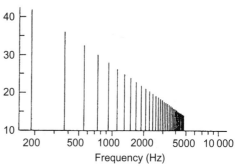

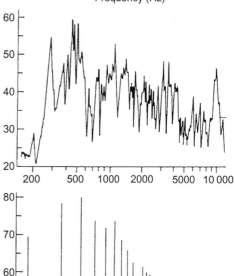

FIGURE 4.10 *Sound source spectrum for (from top to bottom): a bowed G3 ($f_0 = 196$ Hz), sound modifier response curve for a typical violin, and the resulting output spectrum. (From Figure 11.9 in* Musical Acoustics, *by Donald E. Hall, © 1991 Brooks/ Cole Publishing Company, Pacific Grove, CA 93950, by permission of the publisher.)*

midway between the G3 and D4 (2nd and 3rd strings) and D4 and A4 (3rd and 4th strings) strings respectively, for the cello with f_0 of the G2 string and approximately midway between the D3 and A3 (3rd and 4th strings) respectively, and for the double bass with f_0 of the D2 (3rd string) and G2 (4th string) strings respectively. Thus there is more acoustic support for the lower notes of the violin than for those of the viola or the double bass, and the varying distribution of these two resonances between the instruments of the string family is part of the acoustic reason why each member of the family has its own characteristic sound.

Figure 4.11 shows waveforms and spectra for notes played on two plucked instruments: C3 on a lute and F3 on a guitar. The decay of the note can be seen on the waveforms, and in each case the note lasts just over a second. The pluck position can be estimated from the spectra by looking for those harmonics which are reduced in amplitude and are integer

FIGURE 4.11

Waveforms and spectra for C3 played on a lute and F3 played on a six-string guitar.

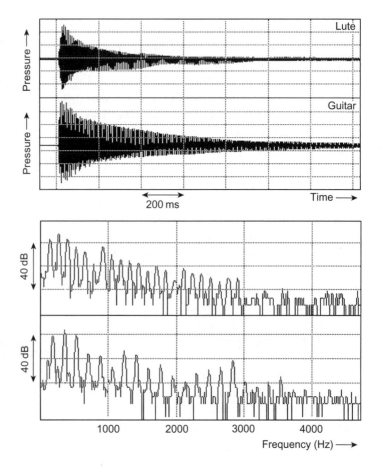

multiples of each other (see Equation 4.2a). The lute spectrum suggests a pluck point at approximately one-sixth of the string length due to the clear amplitude dips in the 6th and 12th harmonics, but there are also clear dips at the 15th and 20th harmonics.

An important point to note is that this is the spectrum of the output from the instrument, and therefore it includes the effects of the sound modifiers (e.g., air and plate resonances), so harmonic amplitudes are affected by the sound modifiers as well as the sound source. Also, the 15th and 20th harmonics are nearly 40 dB lower than the low harmonics in amplitude and therefore background noise will have a greater effect on their amplitudes. The guitar spectrum also suggests particularly clearly a pluck point at approximately one-sixth of the string length, given the dips in the amplitudes of the 6th, 12th and 18th harmonics.

Sound from stringed instruments does not radiate out in all directions to an equal extent and this can make a considerable difference if, for example, one is critically listening to or making recordings of members of the family. The acoustic output from any stringed instrument will contain frequency components across a wide range, whether it is plucked, struck or bowed. In general, low frequencies are radiated in essentially all directions, with the pattern of radiation becoming more directionally focused as frequency increases from the mid to high range. In the case of the violin, low frequencies in this context are those up to approximately 500 Hz, and high frequencies, which tend to radiate outwards from the top plate, are those above approximately 1000 Hz. The position of the listener's ear or a recording microphone is therefore an important factor in terms of the overall perceived sound of the instrument.

4.3 WIND INSTRUMENTS

The discussion of the acoustics of wind instruments involves similar principles to those used in the discussion of stringed instruments. However, the nature of the sound source in wind instruments is rather different but the description of the sound modifiers in wind instruments has much in common with that relating to possible modes on a string, but with a key difference that a string exhibits transverse wave motion, considered in terms of displacement modes, whereas in a pipe it is longitudinal wave motion, where considerations of the velocity and pressure modes are the key. This section concentrates on the acoustics of organ pipes to illustrate the acoustics of sound production in wind instruments. Some of the acoustic mechanisms basic to other wind instruments are given later in the section.

Wind instruments can be split into those with and those without reeds, and organ pipes can be split likewise, based on the sound source mechanism

> Wind instruments and the pipes of a pipe organ can be split into those without and those with reeds.

involved, into "flues" and "reeds" respectively. Organ pipes are used in this section to introduce the acoustic principles of wind instruments with and without reeds as the sound source. Figure 4.12 shows the main parts of flue and reed pipes. Each is constructed of a particular material, usually wood or a tin–lead alloy, and has a resonator of a particular shape and size depending on the sound that the pipe is designed to produce (e.g., Audsley, 1965; Sumner, 1975; Norman and Norman, 1980). The sources of sound in the flue and the reed pipe will be considered first, followed by the sound modification that occurs due to the resonator.

4.3.1 Sound source in organ flue pipes

The source of sound in flue pipes is described in detail in Hall (2001) and his description is as follows. The important features of a flue sound source are a narrow slit (the flue) through which air flows, and a wedge-shaped obstacle placed in the airstream from the slit. Figure 4.13 shows the detail of this mechanism for a wooden organ flue pipe (the similarity with a metal organ flue pipe can be observed in Figure 4.12). A narrow slit exists between the lower lip and the languid, and this is known as the "flue," and the wedge-shaped obstacle is the upper lip which is positioned in the airstream from the flue. This obstacle is usually placed off-center to the airflow.

FIGURE 4.12

The main parts of flue (open metal and stopped wood) and reed organ pipes.

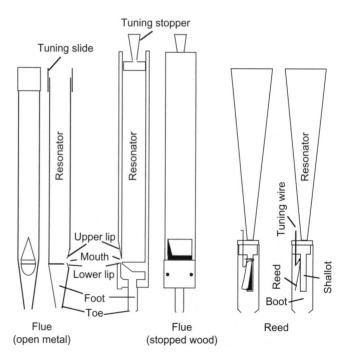

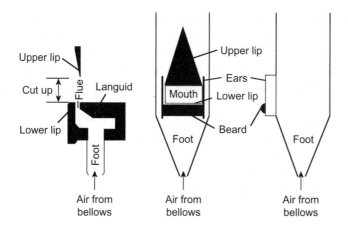

FIGURE 4.13

The main elements of the sound source in organ flue pipes based on a wooden flue pipe (left) and additional features found on some metal pipes (center and left).

Air enters the pipe from the organ bellows via the foot and a thin sheet of air emerges from the flue. If the upper lip were not present, the air emerging from the flue would be heard as noise. This indicates that the airstream is turbulent. A similar effect can be observed if you form the mouth shape for the "ff" in *far*, in which the bottom lip is placed in contact with the upper teeth to produce the "ff" sound. The airflow is turbulent, producing the acoustic noise which can be clearly heard. If the airstream flow rate is reduced, there is an air velocity below which turbulent flow ceases and acoustic noise is no longer heard. At this point the airflow has become smooth or "laminar." Turbulent airflow is the mechanism responsible for the non-pitched sounds in speech such as the "sh" in *shoe* and the "ss" in *sea*, for which waveforms and spectra are shown in Figure 3.9.

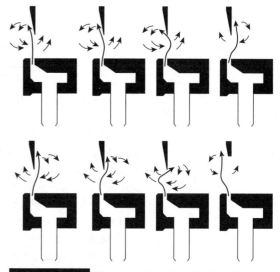

FIGURE 4.14 *Sequence of events to illustrate the sound source mechanism in a flue organ pipe.*

When a wedge-like obstruction is placed in the airstream emerging from the flue a definite sound is heard known as an "edgetone." Hall suggests a method for demonstrating this by placing a thin card in front of the mouth and blowing on its edge. Researchers are not fully agreed on the mechanism which underlies the sound source in flues. The preferred explanation is illustrated in Figure 4.14, and it is described in relation to the sequence of snapshots in the figure as follows. Air flows to one side of the obstruction, causing a local increase in pressure on that side of it. This local pressure increase causes air in its local vicinity to be moved out of the way, and some finds its way in a circular motion into the pipe via the mouth. This has the effect of "bending" the main stream of air increasingly, until it flips into the pipe. The process repeats itself, only this time the local pressure increase causes air to move in a circular motion out of the pipe via the mouth, gradually bending the main airstream until it flips outside the pipe again. The cycle then repeats providing a periodic sound source to the pipe itself. This process is sometimes referred to as a vibrating "air reed" due to the regular flipping to and fro of the airstream.

The f_0 of the pulses generated by this air reed mechanism in the *absence* of a pipe resonator is directly proportional to the airflow velocity from the flue, and inversely proportional to the length of the cut-up:

$$f_0 \propto \frac{v_j}{L_{\text{cut-up}}} \tag{4.4}$$

where $\propto$ means "is proportional to"
f_0 = fundamental frequency of air reed oscillation in absence of pipe resonator
v_j = airflow velocity
and $L_{\text{cut-up}}$ = length of cut-up

In other words, f_0 can be raised by either increasing the airflow velocity or reducing the cut-up. As the airflow velocity is increased or the cut-up size is decreased, there comes a point where the f_0 jumps up in value. This effect can be observed in the presence of a resonator with respect to increasing the airflow velocity by blowing with an increasing flow rate into a recorder (or if available, a flue organ pipe). It is often referred to as an "overblown" mode.

The acoustic nature of the sound source in flues is set by the pipe voicer, whose job it is to determine the overall sound from individual pipes and to establish an even tone across complete ranks of pipes. The following comments on the voicer's art in relation to the sound source in flue pipes are summarized from Norman and Norman (1980), who give the main modifications made by the voicer in order of application as:

- adjusting the cut-up;
- "nicking" the languid and lower lip;
- adjusting languid height with respect to that of the lower lip.

Adjusting the cut-up needs to be done accurately to achieve an even tone across a rank of pipes. This is achieved on metal pipes by using a sharp, short, thick-bladed knife. A high cut-up produces a louder and more "hollow" sound, and a lower cut-up gives a softer and "edgier" sound. The higher the cut-up, the greater the airflow required from the foot. However, the higher the airflow, the less prompt the speech of the pipe.

Nicking relates to a series of small nicks that are made in the approximating edges of the languid and the upper lip. This has the effect of reducing the high-frequency components in the sound source spectrum and giving the pipe a smoother, but slower, onset to its speech. More nicking is customarily applied to pipes which are heavily blown. A pipe which is not nicked has a characteristic consonantal attack to its sound, sometimes referred to as a "chiff." A current trend in organ voicing is the use of less or no nicking in order to take advantage of the onset chiff musically to give increased clarity to notes, particularly in contrapuntal music (e.g., Hurford, 1994).

The height of the languid is fixed at manufacture for wooden pipes, but it can be altered for metal pipes. The languid controls, in part, the direction of the air flowing from the flue. If it is too high, the pipe will be slow to speak and may not speak at all if the air misses the upper lip completely. If it is too low the pipe will speak too quickly, or speak in an uncontrolled manner. A pipe is adjusted to speak more rapidly if it is set to speak with a consonantal chiff by means of little or no nicking. Narrow scaled pipes (small diameter compared with the length) usually have a "stringy" tone color and often have ears added (see Figure 4.12) which stabilize air reed oscillation. Some bass pipes also have a wooden roller or "beard" placed between the ears to aid prompt pipe speech.

4.3.2 Sound modifiers in organ flue pipes

The sound modifier in an organ flue pipe is the main body of the pipe itself, or its "resonator" (see Figure 4.12). Organ pipe resonators are made in a variety of shapes developed over a number of years to achieve subtleties of tone color, but the most straightforward to consider are resonators whose dimensions do not vary along their length, or resonators of "uniform cross-section." Pipes made of metal are usually round in cross-section and those made of wood are generally square (some builders make triangular wooden pipes, partly to save on raw material). These shapes arise mainly from ease of construction with the material involved.

There are two basic types of organ flue pipe: those that are open and those that are stopped at the end farthest from the flue itself (see Figure 4.12). The flue end of the pipe is acoustically equivalent to an open end. Thus the open flue pipe is acoustically open at both ends, and the stopped flue pipe is acoustically open at one end and closed at the other. The air reed sound source mechanism in flue pipes as illustrated in Figure 4.14 launches a pulse of acoustic energy into the pipe. When a compression (positive amplitude) pulse of sound pressure energy is launched into a pipe, for example at the instant in the air reed cycle illustrated in the lower-right snapshot in Figure 4.14, it travels down the pipe at the velocity of sound as a compression pulse.

When the compression pulse reaches the far end of the pipe, it is reflected in one of the two ways described in the "standing waves" section of Chapter 1 (Section 1.5.7), depending on whether the end is open or closed. At a closed end there is a pressure antinode and a compression pulse is reflected back down the pipe. At an open end there is a pressure node and a compression pulse is reflected back as a rarefaction pulse to maintain atmospheric pressure at the open end of the pipe. Similarly, a rarefaction pulse arriving at a closed end is reflected back as a rarefaction pulse, but as a compression pulse when reflected from an open end. All four conditions are illustrated in Figure 4.15.

When the action of the resonator on the air reed sound source in a flue organ pipe is considered (see Figure 4.14), it is found that the f_0 of air reed vibration is entirely controlled by: (a) the length of the resonator, and (b) whether the pipe is open or stopped. This dependence of the f_0 of the air reed vibration can be appreciated by considering the arrival and departure of pulses at each end of the open and the stopped pipes.

Figure 4.16 shows a sequence of snapshots of pressure pulses generated by the air reed traveling down an open pipe of length L_o (left) and a stopped pipe of length L_s (right), and how they drive the vibration of the air reed. (Air reed vibration is illustrated in a manner similar to that used in Figure 4.14.) The figure shows pulses moving from left to right in the upper third of each pipe, those moving from right to left in the center third, and the summed pressure in the lower third. A time axis with arbitrary but equal units is marked in the figure to show equal time intervals. The pulses travel an equal distance in each frame of the figure since an acoustic pulse moves at a constant velocity. The flue end of the pipe acts as an open end in terms of the manner in which pulses are reflected (see Figure 4.15). At every instant when a pulse arrives and is reflected from the flue end, the air reed is flipped from inside to outside when a compression pulse arrives and is reflected as

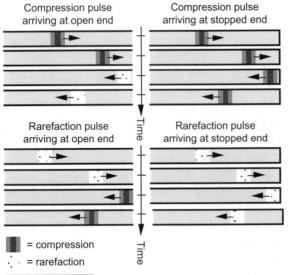

FIGURE 4.15 *The reflected pulses resulting from a compression (upper) and rarefaction (lower) pulse arriving at an open (left) and a stopped (right) end of a pipe of uniform cross-section. (Note: Time axes are marked in equal arbitrary units.)*

a rarefaction pulse, and vice versa when a rarefaction pulse arrives. This can be observed in Figure 4.16.

For the open pipe, the sequence in the figure begins with a compression pulse being launched into the pipe, and another compression pulse just leaving the open end (the presence of this second pulse will be explained shortly). The next snapshot (2) shows the instant when these two pulses reach the center of the pipe, their summed pressure being a maximum at this point. The pulses effectively travel through each other and emerge with their original identities due to "superposition" (see Chapter 1). In the third snapshot the compression pulse is being reflected from the open end of the pipe as a rarefaction pulse, and the air reed flips outside the pipe, generating a rarefaction pulse. (This may seem strange at first, but it is a necessary consequence of the event happening in the fifth snapshot.) The fourth snapshot shows two rarefactions at the center giving a summed pressure which is a minimum at this instant of twice the rarefaction pulse amplitude. In the fifth snapshot, when the rarefaction pulse is reflected from the flue end as a compression pulse, the air reed is flipped from outside to inside the pipe. One cycle is complete at this point since events in the fifth and first snapshots are similar. (A second cycle is illustrated on the right-hand side of Figure 4.1 to enable comparison with events in the stopped pipe.)

The fundamental period for the open pipe is the time taken to complete a complete cycle (i.e., the time between a compression pulse leaving the flue end of the pipe and the next compression pulse leaving the flue end of the pipe). In terms of Figure 4.16 it is four time frames (snapshot one

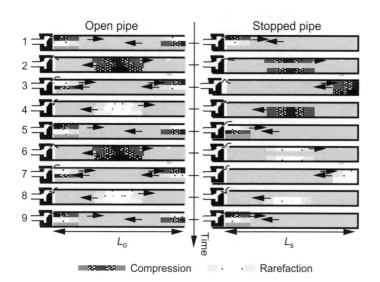

FIGURE 4.16

Pulses traveling in open (left) and stopped (right) pipes when they drive an air reed sound source. (Note: Time axis is marked in equal arbitrary time units; pulses traveling left to right are shown in the upper part of each pipe, those going right to left are shown in the center, and the sum is shown in the lower part.)

to snapshot five), being the time taken for the pulse to travel down to the other end and back (see Figure 4.15), or twice the open pipe length:

$$T_{0(open)} = \left[\frac{2L_o}{c}\right]$$

where $T_{0(open)}$ = fundamental period of open pipe
L_o = length of the open pipe
and c = velocity of sound

The f_0 value for the open pipe is therefore:

$$f_{0(open)} = \left[\frac{1}{T_{0(open)}}\right] = \left[\frac{c}{2L_o}\right] \tag{4.5}$$

In the stopped pipe, the sequence in Figure 4.15 again begins with a compression pulse being launched into the pipe, but there is no second pulse. Snapshot two shows the instant when the pulses reach the center of the pipe, and the third snapshot the instant when the compression pulse is reflected from the stopped end as a compression pulse (see Figure 4.15) and the summed pressure is a maximum for the cycle of twice the amplitude of the compression pulse. The fourth snapshot shows the compression pulse at the center and in the fifth, the compression pulse is reflected from the flue end as a rarefaction pulse, flipping the air reed from inside to outside the pipe. The sixth snapshot shows the rarefaction pulse halfway down the pipe and the seventh shows its reflection as a rarefaction pulse from the stopped end when the summed pressure there is the minimum for the cycle of twice the amplitude of the rarefaction pulse. The eighth snapshot shows the rarefaction pulse halfway back to the flue end and, by the ninth, one cycle is complete, since events in the ninth and first snapshots are the same.

It is immediately clear that one cycle for the stopped pipe takes twice as long as one cycle for the open pipe if the pipe lengths are equal (ignoring a small end correction which has to be applied in practice). Its fundamental period is therefore double that for the open pipe, and its f_0 is therefore half that for the open pipe, or an octave lower. This can be quantified by considering that the time taken to complete a complete cycle is the time required for the pulse to travel to the other end of the pipe and back twice, or four times the stopped pipe length (see Figure 4.15):

$$T_{0(stopped)} = \left[\frac{4L_s}{c}\right]$$

where $T_{0(stopped)}$ = fundamental period of stopped pipe
L_s = length of the stopped pipe
and c = velocity of sound

Therefore:

$$f_{0(stopped)} = \left[\frac{1}{T_{0(stopped)}} \right] = \left[\frac{c}{4L_s} \right]$$ (4.6)

EXAMPLE 4.2

If an open pipe and a stopped pipe are the same length, what is the relationship between their f_0 values?

Let ($L_s = L_o = L$) and substitute into Equations 4.5 and 4.6:

$$f_{0(open)} = \left[\frac{c}{2L} \right]$$

$$f_{0(stopped)} = \left[\frac{c}{4L} \right]$$

Therefore:

$$f_{0(stopped)} = \frac{1}{2}\left[\frac{c}{2L} \right] = \left[\frac{f_{0(open)}}{2} \right]$$

Therefore $f_{0(stopped)}$ is an octave lower than $f_{0(open)}$ (frequency ratio 1:2).

The natural modes of a pipe are constrained as described in the "standing waves" section of Chapter 1. Equation 1.35 gives the frequencies of the modes of an open pipe and Equation 1.36 gives the frequencies of the modes of a stopped pipe. In both equations, the velocity is the velocity of sound (c).

The frequency of the first mode of the open pipe is given by Equation 1.30 when ($n = 1$):

$$f_{open(1)} = \left[\frac{c}{2L_o} \right]$$ (4.7)

which is the same value obtained in Equation 4.5 by considering pulses in the open pipe. Using Equation 1.35, the frequencies of the other modes can be expressed in terms of its f_0 value as follows:

$$f_{open(2)} = \left[\frac{2c}{2L_o}\right] = 2f_{open(1)}$$

$$f_{open(3)} = \left[\frac{3c}{2L_o}\right] = 3f_{open(1)}$$

$$f_{open(4)} = \left[\frac{4c}{2L_o}\right] = 4f_{open(1)}$$

In general:

$$f_{open(n)} = nf_{open(1)} \tag{4.8}$$

The modes of the open pipe are thus all harmonically related and all harmonics are present. The musical intervals between the modes can be read from Figure 3.3.

The frequency of the fundamental mode of the stopped pipe is given by Equation 1.36 when $(n = 1)$:

$$f_{stopped(1)} = \left[\frac{c}{4L_s}\right] \tag{4.9}$$

This is the same value obtained in Equation 4.6 by considering pulses in the stopped pipe. The frequencies of the other stopped pipe modes can be expressed in terms of its $f_{stopped}$ using Equation 1.36 as follows:

$$f_{stopped(2)} = \left[\frac{3c}{4L_s}\right] = 3f_{stopped(1)}$$

$$f_{stopped(3)} = \left[\frac{5c}{4L_s}\right] = 5f_{stopped(1)}$$

$$f_{stopped(4)} = \left[\frac{7c}{4L_s}\right] = 7f_{stopped(1)}$$

In general:

$$f_{stopped(n)} = (2n - 1) f_{stopped(1)}$$ (4.10)

where $n = 1, 2, 3, 4, ...$

Thus the modes of the stopped pipe are harmonically related, but only the odd-numbered harmonics are present. The musical intervals between the modes can be read from Figure 3.3.

In open and stopped pipes the pipe's resonator acts as the sound modifier and the sound source is the air reed. The nature of the spectrum of the air reed source depends on the detailed shape of the pulses launched into the pipe, which in turn depends on the pipe's voicing summarized above. If a pipe is overblown, its f_0 jumps to the next higher mode that the resonator can support: up one octave to the second harmonic for an open pipe, and up an octave and a fifth to the third harmonic for the stopped pipe.

The length of the resonator controls the f_0 of the air reed (see Figure 4.15) and the natural modes of the pipe are the frequencies that the pipe can support in its output. The amplitude relationship between the pipe modes is governed by the material from which the pipe is constructed and the diameter of the pipe with respect to its length. In particular, wide pipes tend to be weak in upper harmonics. Organ pipes are tuned by adjusting the length of their resonators. In open pipes this is usually done nowadays by means of a tuning slide fitted round the outside of the pipe at the open end, and for stopped pipes by moving the stopper (see Figure 4.12).

A stopped organ pipe has an f_0 value which is an octave below that of an open organ pipe (Example 4.2), and, where space is limited in an organ, stopped pipes are often used in the bass register and played by the pedals. However, the trade-off is between the physical space saved and the acoustic result in that only the odd-numbered harmonics are supported. Figure 4.17 illustrates this with waveforms and spectra for middle C played on a gedackt 8″ and a principal 8′ (Section 5.4 describes organ stop footages: 8′, 4′, etc.). The gedackt stop has stopped wooden pipes, and the spectrum clearly shows the presence of odd harmonics only, in particular the first, third and fifth. The principal stop consists of open metal pipes, and odd and even harmonics exist in its output spectrum. Although the pitch of these stops is equivalent, and they are therefore both labeled 8′, the stopped gedackt pipe is half the length of the open principal pipe.

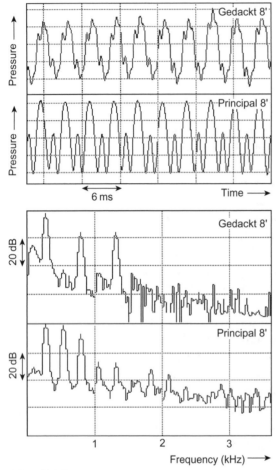

FIGURE 4.17 *Waveforms and spectra for middle C (C4) played on a gedackt 8′ (stopped flue) and a principal 8′ (open flue).*

4.3.3 Woodwind flue instruments

Other musical instruments which have an air reed sound source include the recorder and the flute. Useful additional material on woodwind flue instruments can be found in Benade (1976) and Fletcher and Rossing (1999). The air reed action is controlled by oscillatory changes in flow of air in and out of the flue (see Figure 4.16), often referred to as a "flow-controlled valve," and therefore there must be a velocity antinode and a pressure node. Hence the flue end of the pipe is acting as an open end, and woodwind flue instruments act acoustically as pipes open at both ends (see Figure 4.18).

Players are able to play a number of different notes on the same instrument by changing the effective acoustic length of the resonator. This can be achieved, for example, by means of the sliding piston associated with a swanee whistle or more commonly when particular notes are required, by covering and uncovering holes in the pipe walls known as "finger holes." A hole in a pipe will act in an acoustically similar manner to an open pipe end (pressure node, velocity antinode). The extent to which it does this is determined by the diameter of the hole with respect to the pipe diameter. When this is large with respect to the pipe diameter, as in the flute, the uncovered hole acts acoustically as if the pipe had an open end at that position. Smaller finger holes result acoustically in the effective open end being further down the pipe (away from the flue end). This is an important factor in the practical design of bass instruments with long resonators since it can enable the finger holes to be placed within the physical reach of a player's hands. It does, however, have an important consequence on the frequency relationship between the modes, and this is explored in detail below in connection with woodwind reed instruments. The other way to give a player control over finger holes which are out of reach, for example on a flute, is by providing each hole with a pad controlled by a key mechanism of rods and levers operated by the player's fingers to close or open the hole (depending on whether the hole is normally open or closed by default).

In general, a row of finger holes is gradually uncovered to effectively shorten the acoustic length of the resonator as an ascending scale is played. Occasionally some cross-fingering is used in instruments with small holes or small pairs of holes such as the recorder as illustrated in Figure 4.19. Here, the pressure node is further away from the flue than the first uncovered hole itself such that the state of other holes beyond it will affect its position. The figure shows typical fingerings used to play a two octave C major scale on a descant or tenor recorder. Hole fingerings are available to enable notes to be played which cover a full chromatic scale across one octave. To play a second octave on woodwind flue instruments, such as the recorder or flute, the flue is overblown. Since these instruments are acoustically open at both ends, the overblown flue jumps to the second mode which is one octave higher than the first (see Equation 4.8 and Figure 3.3). The finger holes can be reused to play the notes of the second octave.

Once an octave and a fifth above the bottom note has been reached, the flue can be overblown to the third mode (an octave and a fifth above the first mode) and the fingering can be started again to ascend higher. The fourth mode is available at the start of the third octave, and so on. Overblowing is supported in instruments

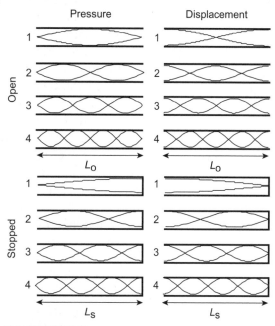

FIGURE 4.18 *The first four pressure and velocity modes of an open and a stopped pipe of uniform cross-section. (Note: The plots show maximum and minimum amplitudes of pressure and velocity.)*

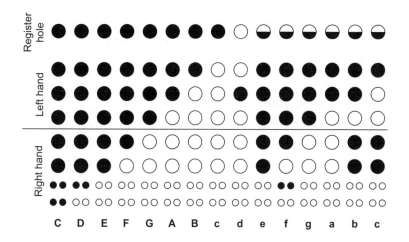

FIGURE 4.19

Fingering chart for recorders in C (descants and tenors).

such as the recorder by opening a small "register" or "vent" hole which is positioned such that it is at the pressure antinode for unwanted modes and these modes will be suppressed. The register hole marked in Figure 4.19 is a small hole on the back of the instrument which is controlled by the thumb of the left hand which either covers it completely, half covers it by pressing the thumb nail end-on against it, or uncovers it completely. To suppress the first mode in this way without affecting the second, this hole should be drilled in a position where the undesired mode has a pressure maximum. When all the tone holes are covered, this would be exactly halfway down the resonator—a point where the first mode has a pressure maximum and is therefore reduced, but the second mode has a pressure node and is therefore unaffected (see Figure 4.18). Register holes can be placed at other positions to enable overblowing to different modes. In practice, register holes may be set in compromise positions because they have to support all the notes available in that register, for which the effective pipe length is altered by uncovering tone holes.

A flute has a playing range between B3 and D7, and the piccolo sounds one octave higher between B4 and D8 (see Figure 4.3). Flute and piccolo players can control the stability of the overblown modes by adjusting their lip position with respect to the embouchure hole as illustrated in Figure 4.20. The air reed mechanism can be compared with that of flue organ pipes illustrated in Figures 4.13 and 4.14 as well as the associated discussion relating to organ pipe voicing. The flautist is able to adjust the distance between the flue outlet (the player's lips) and the edge of the mouthpiece, marked as the "cut-up" in the figure, a term borrowed from organ nomenclature (see Figure 4.13), by rolling the flute as indicated by the double-ended arrow. In addition, the airflow velocity can be varied as well as the fine detailed nature of the airstream dimensions by adjusting the shape, width and height of the opening between the lips. The flautist therefore has direct control over the stability of the overblown modes (Equation 4.4).

4.3.4 Sound source in organ reed pipes

The basic components of an organ reed pipe are shown in Figure 4.12. The sound source results from the vibrations of the reed, which is slightly larger than the shallot opening, against the edges of the shallot. Very occasionally, organ reeds make use of "free reeds," which are cut smaller than the shallot opening and move in and out of the shallot without coming into contact with its edges. In its rest position, as illustrated in Figure 4.12, there is a gap between the

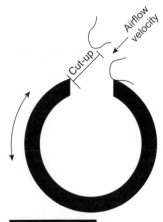

FIGURE 4.20 *Illustration of lip to embouchure adjustments available to a flautist.*

reed and shallot, enabled by the slight curve in the reed itself. The vibrating length of the reed is governed by the position of the "tuning wire," or "tuning spring," which can be nudged up or down to make the vibrating length longer or shorter, accordingly lowering or raising the f_0 of the reed vibration.

The reed vibrates when the stop is selected and a key on the appropriate keyboard is pressed. This causes air to enter the boot and flow past the open reed via the shallot to the resonator. The gap between the reed and shallot is narrow, and for air to flow there must be a higher pressure in the boot than in the shallot, which tends to close the reed fractionally, resulting in the gap between the reed and shallot being narrowed. When the gap is narrowed, the airflow rate is increased and the pressure difference which supports this higher airflow is raised. The increase in pressure difference exerts a slightly greater closing force on the reed, and this series of events continues, accelerating the reed towards the shallot until it hits the edge of the shallot, closing the gap completely and rapidly.

The reed is springy and once the gap is closed and the flow has dropped to zero, the reed's restoring force causes the reed to spring back towards its equilibrium position, opening the gap. The reed overshoots its equilibrium position, stops, and returns towards the shallot, in a manner similar to its vibration if it had been displaced from its equilibrium position and released by hand. Airflow is restored via the shallot and the cycle repeats.

In the absence of a resonator, the reed would vibrate at its natural frequency. This is the frequency at which it would vibrate if it were plucked. If a plucked reed continues to vibrate for a long time, then it has a strong tendency to vibrate at a frequency within a narrow range but, if it vibrates for a short time, there is a wide range of frequencies over which it is able to vibrate. This effect is illustrated in Figure 4.21. This difference is exhibited depending on the material from which the reed is made and how it is supported. A reed which vibrates over a narrow frequency range is usually made from brass and supported rigidly, and is known as a "hard" reed. A reed which vibrates over a wide range might be made from cane or plastic, held in a pliable support, and known as a "soft" reed. As shown in the figure, the natural period (T_N) is related to the natural frequency (F_N) as:

$$F_N = \frac{1}{T_N} \tag{4.11}$$

A reed vibrating against a shallot shuts off the flow of air rapidly and totally, and the consequent acoustic pressure variations are the sound source provided to the resonator. The rapid shutting off of the airflow

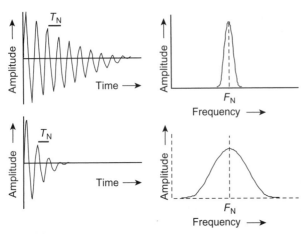

FIGURE 4.21 *Time (left) and frequency (right) responses of hard (upper) and soft (lower) reeds when plucked. Natural frequency (F_N) and natural period (T_N) are shown.*

produces a rapid, instantaneous drop in acoustic pressure within the shallot (as air flowing fast into the shallot is suddenly cut off). A rapid amplitude change in a waveform indicates a relatively high proportion of high harmonics are present. The exact nature of the sound source spectrum depends on the particular reed, shallot and bellows pressure being considered. Free reeds which do not make contact with a shallot, as found for example in a harmonica or harmonium, do not produce as high a proportion of high harmonics since the airflow is never completely shut off.

4.3.5 Sound modifiers in organ reed pipes

All reed pipes have resonators. The effect of a resonator has already been described and illustrated in Figure 4.16 in connection with air reeds. The same principles apply to reed pipes, but there is a major difference in that the shallot end of the resonator acts as a stopped end (as opposed to an open end as in the case of a flue). This is because during reed vibration, the pipe is either closed completely at the shallot end (when the reed is in contact with the shallot) or open with a very small aperture compared with the pipe diameter.

Organ reed pipes have hard reeds, which have a narrow natural frequency range (see Figure 4.21). Unlike the air reed, the presence of a resonator does not control the frequency of vibration of the hard reed. The sound-modifying effect of the resonator is based on the modes it supports (see Figure 4.18), bearing in mind the closed end at the shallot. Because the reed itself fixes the f_0 of the pipe, the resonator does not need to reinforce the fundamental and fractional length resonators are sometimes used to support only the higher harmonics. Figure 4.22 shows waveforms and spectra for middle C (C4) played on a hautbois 8', or oboe 8', and a trompette 8', or trumpet 8'. Both spectra exhibit an overall peak around the sixth/seventh harmonic. For the trompette this peak is quite broad with the odd harmonics dominating the even ones up to the tenth harmonic, probably a feature of its resonator shape. The hautbois spectrum exhibits more dips in the spectrum than the trompette—these are all features which characterize the sounds of different instruments as being different.

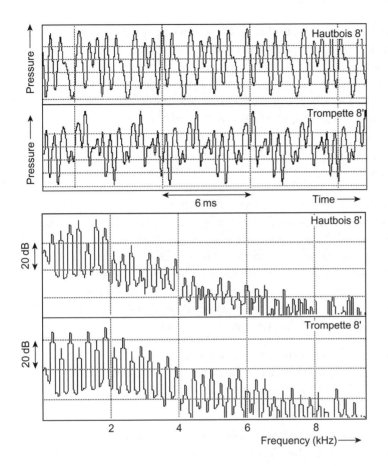

FIGURE 4.22

*Waveform and spectra
for middle C (C4) played
on a hautbois 8' and a
trompette 8'.*

4.3.6 Woodwind reed instruments

Woodwind reed instruments make use of either a single or a double vibrating reed sound source which controls the flow of air from the player's lungs to the instrument. The action of a vibrating reed at the end of a pipe is controlled as a function of the relative air pressure on either side of it in terms of when it opens and closes. It is therefore usually described as a *pressure-controlled valve*, and the reed end of the pipe acts as a *stopped end* (pressure antinode and velocity node—see Figure 4.18). Note that although the reed opens and closes such that airflow is not always zero, the reed opening is very much smaller than the pipe diameter elsewhere, making a stopped end reasonable. This is in direct contrast to the air reed in woodwind flue instruments such as the flute and recorder (see above), which, as a flow-controlled valve, provides a velocity antinode and a pressure node, and where the flue end of the pipe acts as an open end (see Figure 4.18).

Soft reeds are employed in woodwind reed instruments which can vibrate over a wide frequency range (see Figure 4.21). The reeds in clarinets and saxophones are single reeds which can close against the edge of the mouthpiece as in organ reed pipes where they vibrate against their shallots. The oboe and bassoon on the other hand use double reeds, but the basic opening and closing action of the sound source mechanism is the same.

Woodwind reed instruments have resonators whose modal behavior is crucial to the operation of these instruments and provide the sound modifier function. Woodwind instruments incorporate finger holes to enable chromatic scales to be played from the first mode to the second mode when the fingering can be used again as the reed excites the second mode. These mode changes continue up the chromatic scale to cover the full playing range of the instrument (see Figure 4.3). Clearly it is essential that the modes of the resonator retain their frequency ratios relative to each other as the tone holes are opened, or else the instrument's tuning will be adversely affected as higher modes are reached. Benade (1976) summarizes this effect and indicates the resulting constraint as follows:

Preserving a constant frequency ratio between the vibrational modes as the holes are opened is essential in all woodwinds and provides a limitation on the types of air column (often referred to as the bore) that are musically useful.

The musically useful bores in this context are based on tubing that is either cylindrical, as in the clarinet, or conical as in the oboe, cor Anglais, and members of the saxophone and bassoon families. The cylindrical resonator of a clarinet acts as a pipe that is stopped at the reed end (see above) but is open at the other. Odd numbered modes only are supported by such a resonator (see Figure 4.18), and its f_0 is an octave lower (see Example 4.2) than that of an instrument with a similar length pipe which is open at both ends, such as a flute (see Figure 4.3). The first overblown mode of a clarinet is therefore the third mode, an interval of an octave and a fifth (see Figure 3.3), and therefore, unlike a flute or recorder, it has to have sufficient holes to enable at least 19 chromatic notes to be fingered within the first mode prior to transition to the second.

Conical resonators that are stopped at the reed end and open at the other support all modes in a harmonically related manner. Taylor (1976) gives a description of this effect as follows:

Suppose by some means we can start a compression from the narrow end; the pipe will behave just as our pipe open at both ends until the rarefaction has returned to the start. Now, because the pipe has shrunk to a very small bore, the speed of the wave slows down and no real reflection occurs. . . . The result is that we need only consider one journey out and one back regardless of whether the pipe is open or closed at the narrow end. . . . The conical pipe will behave something like a pipe open at both ends as far as its modes are concerned.

The conical resonator therefore supports all modes, and the overblown mode of instruments with conical resonators, such as the oboe, cor Anglais, bassoon and saxophone family, is therefore to the second mode, or up an octave. Sufficient holes are therefore required for at least 12 chromatic notes to be fingered to enable the player to arrive at the second mode from the first.

The presence of a sequence of open tone holes in a pipe resonator of any shape is described by Benade (1976) as a *tone-hole lattice*. The effective acoustical end-point of the pipe varies slightly as a function of frequency when there is a tone-hole lattice, and therefore the effective pipe length is somewhat different for each mode. A pipe with a tone-hole lattice is acoustically shorter for low-frequency standing wave modes compared with higher-frequency modes, and therefore the higher-frequency modes are increasingly lowered slightly in frequency (lengthening the wavelength lowers the frequency). Above a particular frequency, described by Benade (1976) as the *open-holes lattice cut-off frequency* (given as around 350–500 Hz for quality bassoons, 1500 Hz for quality clarinets and between 1100 and 1500 Hz for quality oboes), sound waves are not reflected due to the presence of the lattice. Benade notes that this has a direct effect on the perceived timbre of woodwind instruments, correlating well with descriptions such as *bright* or *dark* given to instruments by players. It should also be noted that holes that are closed modify the acoustic properties of the pipe also, and this can be effectively modeled as a slight increase in pipe diameter at the position of the tone hole. The resulting acoustic change is considered below.

In order to compensate for these slight variations in the frequencies of the modes produced by the presence of open and closed tone holes, alterations can be made to the shape of the pipe. These might include flaring the open end, adding a tapered section, or small local voicing adjustments by enlarging or constricting the pipe, which on a wooden instrument can be achieved by reaming out or adding wax respectively (e.g., Nederveen, 1969). The acoustic effect on individual pipe mode frequencies of either enlarging or constricting the size of the pipe depends directly on the mode's distribution of standing wave pressure nodes and antinodes (or velocity antinodes and nodes respectively). The main effect of a constriction in relation to pressure antinodes (velocity nodes) is as follows (Kent and Read, 1992):

- A constriction near a pressure node (velocity antinode) lowers that mode's frequency.
- A constriction near a pressure antinode (velocity node) raises that mode's frequency.

A constriction at a pressure node (velocity antinode) has the effect of reducing the flow at the constriction since the local pressure difference across the constriction has not changed. Benade (1976) notes that this is equivalent to raising the local air density, and the discussion in Chapter 1 indicates that this will result in a lowering of the velocity of sound (see Equation 1.1) and therefore a lowering in the mode frequency (see Equations 4.7 and 4.9). A constriction at a pressure antinode (velocity node), on the other hand, provides a local rise in acoustic pressure which produces a greater opposition to local airflow of the sound waves that combine to produce the standing wave modes. This is equivalent to raising the local springiness in the medium (air), which is shown in Chapter 1 to be equivalent for air in Young's modulus (E_{gas}), which raises the velocity of sound (see Equation 1.5) and therefore raises the mode frequency (see Equations 4.7 and 4.9). By the same token, the effect of locally enlarging a pipe will be exactly opposite to that of constricting it.

Knowledge of the position of the pressure and velocity nodes and antinodes for the standing wave modes in a pipe therefore allows the effect on the mode frequencies of a local constriction or enlargement of a pipe to be predicted. Figure 4.23 shows the potential mode frequency variation for the first three modes of a cylindrical stopped pipe that could be caused by a constriction or enlargement at any point along its length. (The equivalent diagram for a cylindrical pipe open at both ends could be readily produced with reference to Figures 4.18 and 4.23; this is left as an exercise for the interested reader.)

The upper part of Figure 4.23 (taken from Figure 4.18) indicates the pressure and velocity node and antinode positions for the first three standing wave modes. The lower part of the figure exhibits plus and minus signs to indicate where that particular mode's frequency would be raised or lowered respectively by a local constriction or enlargement at that position in the pipe. The size of the signs indicates the sensitivity of the frequency variation based on how close the constriction is to the mode's pressure/velocity nodes and antinodes shown in the upper part of the figure. For example, a

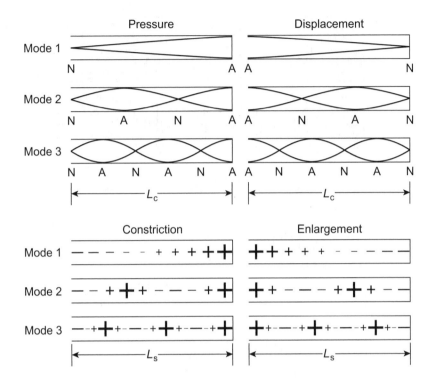

FIGURE 4.23

The effect of locally constricting or enlarging a stopped pipe on the frequencies of its first three modes: " + " indicates raised modal frequency, " − " indicates lowered modal frequency, and the magnitude of the change is indicated by the size of the " + " or " − " signs. The first three pressure and velocity modes of a stopped pipe are shown for reference: "N" and "A" indicate node and antinode positions respectively.

constriction close to the closed end of a cylindrical pipe will raise the frequencies of all modes since there is a pressure antinode at a closed end, whereas an enlargement at that position would lower the frequencies of all modes. However, if a constriction or enlargement were made one-third the way along a stopped cylindrical pipe from the closed end, the frequencies of the first and third modes would be raised somewhat, but that of the second would be lowered maximally. By creating local constrictions or enlargements, the skilled maker is able to set up a woodwind instrument to compensate for the presence of tone holes such that the modes remain close to being in integer frequency ratios over the playing range of the instrument.

Figure 4.24 shows waveforms and spectra for the note middle C played on a clarinet and a tenor saxophone. The saxophone spectrum contains all harmonics since its resonator is conical. The clarinet spectrum exhibits the odd harmonics clearly as its resonator is a cylindrical pipe closed at one end (see Figure 4.18), but there is also energy clearly visible in some of the even harmonics. Although the resonator itself does not support the even modes, the spectrum of the sound source does contain all harmonics (the saxophone

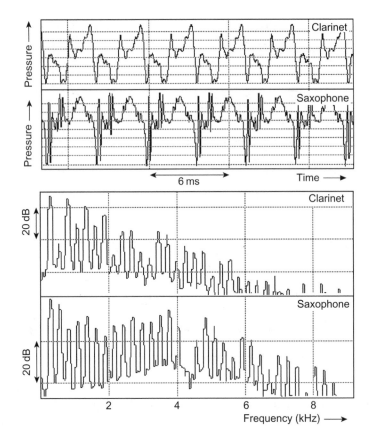

Waveforms and spectra for middle C (C4) played on a clarinet and a tenor saxophone.

and the clarinet are both single reed instruments). Therefore some energy will be radiated by the clarinet at even harmonics.

Sundberg (1989) summarizes this effect for the clarinet as follows:

> *This means that the even-numbered modes are not welcome in the resonator. . . . A common misunderstanding is that these partials are all but missing in the spectrum. The truth is that the second partial may be about 40 dB below the fundamental, so it hardly contributes to the timbre. Higher up in the spectrum the differences between odd- and even-numbered neighbors are smaller. Further . . . the differences can be found only for the instruments' lower tones.*

This description is in accord with the spectrum in Figure 4.24, where the amplitude of the second harmonic is approximately 40 dB below that of the fundamental, and the odd/even differences become less with increased frequency.

4.3.7 Brass instruments

The brass instrument family has an interesting history from early instruments derived from natural tube structures such as the horns of animals, seashells and plant stems, through a variety of wooden and metal instruments to today's metal brass orchestral family (e.g., Campbell and Greated, 1998; Fletcher and Rossing, 1999). The sound source in all brass instruments is the vibrating lips of the player in the mouthpiece. They form a double soft reed, but the player has the possibility of adjusting the physical properties of the double reed by lip tension and shape. The lips act as a pressure-controlled valve in the manner described in relation to the woodwind reed sound source, and therefore the mouthpiece end of the instrument acts acoustically as a stopped end (pressure antinode and velocity node—see Figure 4.18).

The double reed action of the lips can be illustrated if the lips are held slightly apart, and air is blown between them. For slow airflow rates nothing is heard, but, as the airflow is increased, acoustic noise is heard as the airflow becomes turbulent. If the flow is increased further, the lips will vibrate together as a double reed. This vibration is sustained by the physical vibrational properties of the lips themselves, and an effect known as the "Bernoulli effect."

As air flows past a constriction, in this case the lips, its velocity increases. The Bernoulli effect is based on the fact that at all points the sum of the energy of motion, or "kinetic" energy, plus the pressure energy, or "potential" energy, must be constant at all points along the tube. Figure 4.25 illustrates this effect in a tube with a flexible constriction. Airflow direction is represented by the lines with arrows, and the velocity of airflow is represented by the distance between these lines. Since airflow increases as it flows through the constriction, the kinetic energy increases. In order to satisfy the Bernoulli principle that the total energy remains constant, the potential energy or the pressure at the point of constriction must therefore reduce. This means that the force on the tube walls is lower at the point of constriction.

If the wall material at the point of constriction is elastic and the force exerted by the Bernoulli effect is sufficient to move the walls' mass (such as the brass player's lips) from its rest (equilibrium)

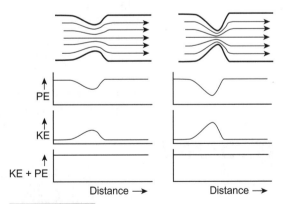

FIGURE 4.25 *An illustration of the Bernoulli effect (potential energy + kinetic energy = a constant) in a tube with a constriction. (Note: Lines with arrows represent airflow direction, and the distance between them is proportional to the airflow velocity. PE = potential energy; KE = kinetic energy.)*

position, then the walls are sucked together a little (compare the right- and left-hand illustrations in the figure). Now the kinetic energy (airflow velocity) becomes greater because the constriction is narrower; thus the potential energy (pressure) must reduce some more to compensate (compare the graphs in the figure), and the walls of the tube are sucked together with greater force. Therefore the walls are accelerated together as the constriction narrows until they smack together, cutting off the airflow. The air pressure in the tube tends to push the constriction apart, as does the natural tendency of the walls to return to their equilibrium position. Like two displaced pendulums, the walls move past their equilibrium position, stop and return towards each other, and the Bernoulli effect accelerates them together again. The oscillation of the walls will be sustained by the airflow, and the vibration will be regular if the two walls at the point of constriction have similar masses and tensions, such as the lips.

The lip reed vibration is supported by the resonator of the brass instrument formed by a length of tubing attached to a mouthpiece. Some mechanism is provided to enable the player to vary the length of the tube, which was done originally, for example, in the horn family by adding different lengths of tubing or "crooks" by hand. Nowadays this is accomplished by means of a sliding section as in the trombone or by adding extra lengths of tubing by means of valves. The tube profile in the region of the trombone slide or tunable valve mechanism has to be cylindrical in order for slides to function.

All brass instruments consist of four sections (see Figure 4.26): mouthpiece, a tapered mouthpipe, a main pipe fitted with slide or valves which is cylindrical (e.g., trumpet, French horn, trombone) or conical (e.g., cornet, flugelhorn, baritone horn, tuba), and a flared bell (Benade, 1976; Hall, 2001). If a brass instrument consisted only of a conical main pipe, all modes would be supported (see discussion on woodwind reed instruments above), but, if it were cylindrical, it acts as a stopped pipe due to the pressure-controlled action of the lip reed and therefore only odd-numbered modes would be supported (see Figure 4.18). However, instruments in the brass family support almost all

FIGURE 4.26

Basic sections of a brass instrument.

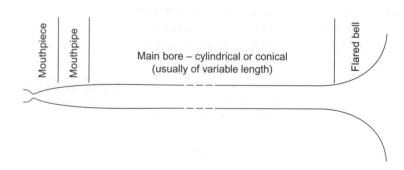

Mouthpiece · Mouthpipe · Main bore – cylindrical or conical (usually of variable length) · Flared bell

modes which are essentially harmonically related due to the acoustic action of the addition of the mouthpiece and bell.

The bell modifies as a function of frequency the manner in which the open end of the pipe acts as a reflector of sound waves arriving there from within the pipe. A detailed discussion is provided by Benade (1976) from which a summary is given here. Lower-frequency components are reflected back into the instrument from the narrower part of the bell while higher-frequency components are reflected from the wider regions of the bell. Frequencies higher than a cut-off frequency determined by the diameter of the outer edge of the bell (approximately 1500 Hz for a trumpet) are not reflected appreciably by the bell. Adding a bell to the main bore of the instrument has the effect of making the effective pipe length longer with increasing frequency. The frequency relationship between the modes of the stopped cylindrical pipe (odd-numbered modes only: $1f$, $3f$, $5f$, $7f$, etc.) will therefore be altered such that they are brought closer together in frequency. This effect is greater for the first few modes of the series.

The addition of a mouthpiece at the other end of the main bore also affects the frequency of some of the modes. The mouthpiece consists of a cup-shaped cavity which communicates via a small aperture with a short conical pipe. The mouthpiece has a resonant frequency associated with it, which is generally in the region of 850 Hz for a trumpet, which is otherwise known as the "popping frequency" since it can be heard by slapping its lip contact end on the flattened palm of one hand (Benade, 1976). The addition of a mouthpiece effectively extends the overall pipe length by an increasing amount. Benade notes that this effect "is a steady increase nearly to the top of the instrument's playing range," and that a mouthpiece with a "lower popping frequency will show a greater total change in effective length as one goes up in frequency" (Benade, 1976, p. 416). This pipe length extension caused by adding a mouthpiece therefore has a greater downwards frequency shifting effect on the higher compared with the lower modes.

In a complete brass instrument, it is possible through the use of an appropriately shaped bell, mouthpiece and mouthpipe to construct an instrument whose modes are frequency shifted from the odd only modes of a stopped cylindrical pipe to being very close to a complete harmonic series. In practice, the result is a harmonic series where all modes are within a few per cent of being integer multiples of a common lower-frequency value except for the first mode itself, which is well below that lower-frequency value common to the higher modes and therefore it is not harmonically related to them. The effects of the addition of the bell and mouthpiece/mouthpipe on the individual lowest six modes are broadly as summarized in Figure 4.27. Here the odd-numbered modal frequencies of the stopped cylindrical pipe are denoted as

FIGURE 4.27

Brass instrument mode
frequency modification to
stopped cylindrical pipe by
the addition of mouthpiece/
mouthpipe and bell.

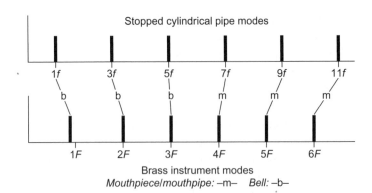

integer multiples of frequency "*f*," and the resulting brass instrument modal frequencies are shown as multiples of another frequency "*F*."

The second mode is therefore the lowest musically usable mode available in a brass instrument (note that the lowest mode does not correspond with 1*F*). Overblowing from the second mode to the third mode results in a pitch jump of a perfect fifth, or seven semitones. The addition of three valves to brass instruments (except the trombone), each of which adds a different length of tubing when it is depressed, enables six semitones to be played, sufficient to progress from the first to the second mode. Assuming this is from the written notes C4 to G4, the six required semitones are: C#4, D4, D#4, E4, F4, and F#4.

Figure 4.28 shows how this is achieved. The center (or second) valve lowers the pitch by one semitone, the first valve (nearest the mouthpiece) by two semitones, and the third valve by three semitones. Combinations of these valves therefore, in principle, enable the required six semitones to be played. It may at first sight seem odd that there are two valve fingerings for a lowering of three semitones (third valve alone or first and second valves together) as shown in the figure. This relates to a significant problem in relation to the use of valves for this purpose which is described below.

Assuming equal tempered tuning for the purposes of this section, it was shown in Chapter 3 that the frequency ratio for one semitone ($1/12$ of one octave) is:

$$r = \sqrt[12]{2} = 1.0595$$

The decrease in frequency required to lower a note by one semitone is therefore 5.95%, and this is also the factor by which a pipe should be lengthened by the second valve on a brass instrument. Depressing the first valve only should lower the f_0 and hence lengthen the pipe by 12.25% since

Semitones	−1	−2	−3	−3	−4	−5	−6
1st valve (−2 semitones)	○	●	●	○	○	●	●
2nd valve (−1 semitone)	●	○	●	○	●	○	●
3rd valve (−3 semitones)	○	○	○	●	●	●	●

FIGURE 4.28 *The basic valve combinations used on brass instruments to enable 6 semitones to be fingered. (Note: Black circle = valve depressed; white circle = valve not depressed; on a trumpet, first valve is nearest mouthpiece, second in the middle and third nearest the bell.)*

the frequency ratio for two semitones is the square of that for one semitone $(1.0595^2 = 1.1225)$. Depressing the first and second valve together will lengthen the pipe by 18.2% (12.25% + 5.95%), which is not sufficient for three semitones since this requires the pipe to be lengthened by 18.9% $(1.0595^3 = 1.1893)$. The player must lip notes using this valve combination down in pitch. The third valve is also set nominally to lower the f_0 by three semitones but, because of the requirement to add a larger length the further down that is progressed, it is set to operate with the first valve to produce an accurate lowering of five semitones. Five semitones is equivalent to 33.51% $(1.0595^5 = 1.3351)$, and subtracting the lowering produced by the first valve gives the extra pipe length required from the third valve as 21.26% (33.51−12.25%), which is rather more than both the 18.2% available from the combination of the first and second valves and the 18.9% required for an accurate three-semitone lowering.

In practice, on a trumpet, for example, the third valve is often fitted with a tuning slide so that the player can alter the added pipe length while playing. No such issues arise for the trombonist, who can alter the slide position accurately to ensure the appropriate additional pipe lengths are added for accurate tuning of the intervals.

Figure 4.29 shows waveforms and spectra for the note C3 played on a trombone and a tuba. The harmonics in the spectrum of the trombone extend far higher in frequency than those of the tuba. This effect can be seen by comparing the shape of their waveforms where the trombone has many more oscillations during each cycle than the tuba. In these examples, the first three harmonics dominate the spectrum of the tuba in terms of amplitude and eight harmonics can be readily seen, whereas the fifth harmonic dominates the spectrum of the trombone, and harmonics up to about the 29th can be identified.

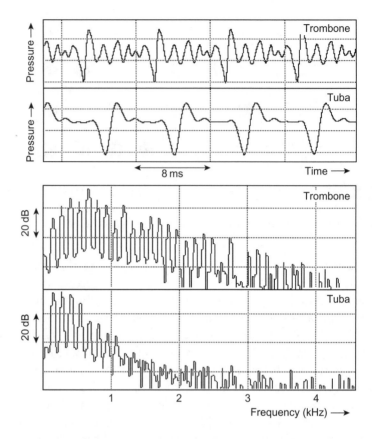

FIGURE 4.29

Waveforms and spectra for C3 played on a trombone and a tuba.

4.4 PERCUSSION INSTRUMENTS

The percussion family is an important body of instruments which can also be described acoustically in terms of the "black box" model. Humans have always struck objects, whether to draw attention to or to imbue others and themselves with rhythm. Rhythm is basic to all forms of music in all cultures and members of the percussion family are often used to support it. Further reading in this area can be found in Benade (1976); Rossing (2001); Hall (2001); and Fletcher and Rossing (1999).

4.4.1 Sound source in percussion instruments

The sound source in percussion instruments usually involves some kind of striking. This is most often by means of a stick or mallet which may be made of wood or metal and may have a plastic or cloth-covered, padded striking end. However, this is not the case in a cymbal crash when two cymbals are struck together. Such a sound source is known as an "impulse."

The spectrum of a single impulse is continuous since it is non-periodic (i.e., it never repeats), and all frequency components are present. Therefore any instrument which is struck is excited by an acoustic sound source of short duration in which all frequencies are present. All modes that the instrument can support will be excited, and each will respond in the same way that the plucked reed vibrates as illustrated in Figure 4.21. The narrower the frequency band of the mode, the longer it will "ring" for. (One useful analogy is the impulse provided if a parent pushes a child on a swing just once. The child will swing back and forth at the natural frequency of the swing and child, and the amplitude of the swinging will gradually diminish. A graph of swing position against time would be similar to the time response for the hard reed plotted in Figure 4.21.)

4.4.2 Sound modifiers in percussion instruments

Percussion instruments are characterized acoustically by the modes of vibration they are able to support, and the position of the strike point with respect to the node and antinode points of each mode (e.g., see the discussion on plucked and struck strings earlier in this chapter). Percussion instruments can be considered in three classes: those that make use of bars (e.g., xylophone, glockenspiel, Celeste, triangle); membranes (e.g., drums); or plates (e.g., cymbals). In each case, the natural mode frequencies are not harmonically related, with the exception of longitudinal modes excited in a bar which is stimulated by stroking with a cloth or glove coated with rosin whose mode frequencies are given by Equation 1.35 if the bar is free to move (unfixed) at both ends, and equation 1.36 if it is supported at one end and free at the other.

Percussion instruments using bars or plates are known as "idiophones" and those using membranes are known as "membrano-phones."

Transverse modes are excited in bars that are struck, as, for example, when playing a xylophone or triangle, and these are not harmonically related. The following equations (adapted from Fletcher and Rossing, 1999) relate the frequencies of higher modes to that of the first mode.

For transverse modes in a bar resting on supports (e.g., glockenspiel, xylophone):

$$f_n = 0.11030 \, ((2n) + 1)^2 \, f_1 \qquad (4.12)$$

where n = mode numbers from 2 (i.e., 2, 3, 4, ...)
and f_1 = frequency of first mode

For transverse modes in a bar clamped at one end (e.g., celeste):

$$f_2 = 0.70144 \, (2.988)^2 \, f_1$$

$$f_n = 0.70144\,(2n + 1)^2\,f_1 \qquad (4.13)$$

where n = mode numbers from 3 (i.e., 3, 4, 5, ...)
and f_1 = frequency of first mode

The frequencies of the transverse modes in a bar are inversely proportional to the square of the length of the bar:

$$f_{\text{transverse}} \propto \left[\frac{1}{L^2}\right] \qquad (4.14)$$

whereas those of the longitudinal modes are inversely proportional to the length (from Equations 1.35 and 1.36):

$$f_{\text{longitudinal}} \propto \left[\frac{1}{L}\right] \qquad (4.15)$$

Therefore halving the length of a bar will raise its transverse mode frequencies by a factor of four, or two octaves, whereas the longitudinal modes will be raised by a factor of two, or one octave. The transverse mode frequencies vary as the square of the mode number, apart from the second mode of the clamped bar (see Equation 4.13) whose factor (2.988) is very close to (3). Table 4.1 shows the frequencies of the first five modes relative to the frequency of the first mode as a ratio and in equal tempered semitones (Appendix 2 gives a frequency ratio to semitone conversion equation) for a bar resting on supports (Equation 4.12) and one clamped at one end (Equation 4.13).

Table 4.1 Frequency ratios (Equations 4.20 and 4.21) and semitone spacings (see Appendix 2) of the first five theoretical modes relative to the first mode for a bar clamped at one end and a bar resting on supports

Transverse mode of bar	Bar resting on supports		Bar clamped at one end	
	Ratio	Semitones	Ratio	Semitones
1 (rel. 1st mode)	1.000	0.00	1.000	0.00
2 (rel. 1st mode)	2.758	17.56	6.267	31.77
3 (rel. 1st mode)	5.405	29.21	17.536	49.58
4 (rel. 1st mode)	8.934	37.91	34.371	61.23
5 (rel. 1st mode)	13.346	44.86	56.817	69.93

The first three modes of a bar resting on supports and a bar clamped at one end are shown in the upper and lower parts respectively of Figure 4.30 along with the appropriate frequency ratio to the first mode (see Table 4.1). Note that the clamped modes are those found for a tuning fork which can be considered as a pair of bars clamped at one end.

None of the higher modes is a whole number of equal tempered semitones above the fundamental and none forms an interval available within a musical scale. The intervals between the modes are very wide compared with harmonic spacing as they are essentially related by the square of odd integers (i.e., $3^2, 5^2, 7^2, 9^2, ...$). The relative excitation strength of each mode is in part governed by the point at which the bar is hit.

Benade (1976) notes that often the measured frequencies of the vibrating modes of instruments which use bars differ somewhat from the theoretical modes (in Table 4.1) due to the effect of "mounting hole(s) drilled in the actual bar and the grinding away of the underside of the center of the bar which is done for tuning purposes."

In order that notes can be played which have a clearly perceived pitch on percussion instruments such as the xylophone, marimba, and vibraphone (with playing ranges from C5 (523 Hz) to C8 (4186 Hz), A2 (110 Hz) to C7 (2093 Hz), and F3 (175 Hz) to F6 (1397 Hz) respectively), the bars are shaped with an arch on their undersides to tune the modes to be close to harmonics of the first mode. In the marimba and vibraphone the second mode is tuned to two octaves above the first mode, and in the xylophone it is tuned to a twelfth above the first mode. These instruments have resonators, which consist of a tube closed at one end, mounted under each bar. The first mode of these resonators is tuned to the f_0 of the bar to enhance its loudness, and therefore the length of the resonator is a quarter of the wavelength of f_0 (see Equation 1.36).

In percussion instruments which make use of membranes and plates, the modal patterns which can be adopted by the membranes or plates themselves govern the frequencies of the modes that are supported. The membrane in a drum and the plate of a cymbal are circular, and the first 10 mode patterns which they can adopt in terms of where displacement nodes and antinodes can occur are shown in the upper part of Figure 4.31. Displacement nodes occur in circles and/or diametrically across and these are shown in the figure. They are identified by the numbers given in brackets as follows: (number of diametric modes, number of circular modes). The drum membrane always has at least one circular mode where there is a displacement node, which is the clamped edge.

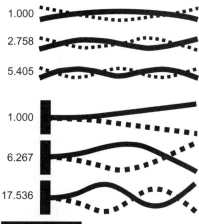

FIGURE 4.30 *The first three modes of a free bar (upper) and a clamped bar (lower).*

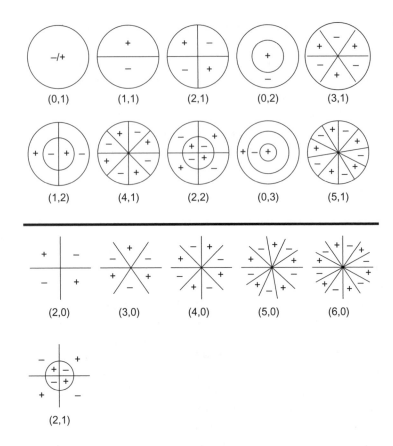

FIGURE 4.31

The first ten modes of a stretched drum membrane (upper) and the first six modes of a plate cymbal (lower). The mode numbers are given in brackets as (number of diametric nodes, number of circular nodes). The plus and minus signs show the relative phasing of the vibration of different parts of the structure within each mode. (Adapted from Fletcher and Rossing, 1999.)

The frequencies of the modes can be calculated mathematically, but the result is rather more complicated than for the bars. Table 4.2 gives the frequencies of each mode relative to the first mode (Fletcher and Rossing, 1999) and the equivalent number of semitones (calculated using the equation given in Appendix 3). As with the bars, none of the modes is an exact number of equal tempered semitones apart or in an integer ratio and therefore they are not harmonically related.

Drums consist of membranes or "drum heads," which are either made of a synthetic material or animal skin, stretched across a supporting frame which is usually round. A small hand-drum, such as the tabor which is commonly used in early music, consists of a cylindrical ring with the drum head stretched across one end, the other end being open. This construction is also used for the tambourine, which is like a tabor but with small cymbal-like disks, or "jingles," mounted in pairs on pins in slots around the cylindrical ring. These rattle together when the instrument is mechanically excited, either by striking it with the knuckles or fingertips to set the jingles

Table 4.2 Modes and frequency ratios (from Fletcher and Rossing, 1999) as well as semitone (ST) distances (see Appendix 3 for details of converting frequency ratios to semitones) between each of the first 10 theoretical modes relative to the first mode for an ideal circular membrane but without a bowl (left), a circular membrane mounted on a bowl as a timpani (center) and an ideal circular plate (right). The asterisks indicate those modes that tend not to be excited strongly when a timpani is struck at the normal playing position (see text)

Circular membrane			Circular plate	
Mode	Ideal ratio (ST)	Timpani ratio (ST)	Mode	Cymbal ratio (ST)
(0,1)	1.000 (0.0)	1.70 (9.19)*	(2,0)	1.00 (0.0)
(1,1)	1.59 (8.1)	2.00 (12.0)	(3,0)	1.94 (11.5)
(2,1)	2.14 (13.1)	3.00 (19.0)	(4,0)	3.42 (21.3)
(0,2)	2.30 (14.4)	3.36 (21.0)*	(5,0)	5.08 (28.1)
(3,1)	2.65 (16.9)	4.00 (24.0)	(6,0)	6.90 (22.4)
(1,2)	2.92 (18.6)	4.18 (24.7)*	(2,1)	8.63 (37.3)
(4,1)	3.16 (19.9)	4.98 (27.8)	–	–
(2,2)	3.50 (21.7)	5.34 (29.0)*	–	–
(0,3)	3.60 (22.2)	5.59 (29.8)*	–	–
(5,1)	3.65 (22.4)	5.96 (30.9)	–	–

ringing briefly, or by rubbing a thumb around the head near the edge to set the jingles ringing continuously. Larger drums exist which have a single head stretched over a long cylinder, such as congas and bongos, and these are usually struck with the player's hands.

The drums found in a drum kit are the bass drum, snare drum and two or more tom-toms of different sizes. These, along with the orchestral bass drum, have two drum heads—one mounted on each side of their cylindrical ring. Bass and snare drums are set up so as to be essentially non-pitched by setting a different tension on each of the drum heads, and therefore spreading the non-harmonic modes produced by the two heads widely in frequency. When they are struck the result is not pitch specific, and therefore they can be used to provide rhythm for music in any key. The snare drum (or "side drum" in marching bands) has a set of metal wires stretched across its lower head (or "snare head"), which vibrate against the snare head when the drum's upper head or "batter head" is struck. The snares can be dropped from the snare head by means of a lever to allow the drum to be used without the vibrating snare. Tom-toms on the other hand are often tuned so as to provide a more definite sense of pitch by matching more closely the tuning of the upper and lower heads on each drum. They are often used in fills to give rising or, more commonly, falling groups of multiple strikes during a fill.

While the tom–tom can produce a more definite sense of pitch, it is not the clearest that can be achieved from a drum. The main orchestral drum is the kettledrum, and there will usually be between two and five kettle-drums used in today's orchestras with a single player devoted to playing them. Such a group of kettledrums is referred to as "timpani" (the *Oxford Companion to Music* notes that "tympani" is a common misspelling of "timpani"); therefore a single kettledrum would be a "timpano." As the music modulates from one key to another, composers can continue to make use of the timpani in the new key since they can be retuned rapidly.

A kettledrum has a single membrane which is stretched over a metal bowl (or kettle) that is suspended on a supporting frame. Absolute tuning of a kettledrum head is set by means of adjusting screws (usually eight) around its rim, to enable it to be tuned to the appropriate absolute pitch reference. During a performance, its tuning can be changed in semitone steps by means of a lever tensioning system operated by a pedal, which typically enables a pitch variation over five semitones. The head tension varies as the lever is moved from note to note, and, if the kettledrum is struck immediately prior to the lever being moved, a rising (or falling) pitch is heard as a glide or glissando. This is sometimes used for musical effect.

The modes produced when a kettledrum is struck are the same shape as those given in Figure 4.31, but their mode frequencies are different from those of an unmounted head (shown as "ideal" in Table 4.2), because of the effect of the air in the bowl over which the head is stretched and the position at which it is struck. Hall (2001) describes this as follows: "When sections of the head move into and out of the bowl, other sections move in the opposite direction (out of and into the bowl respectively)." These are marked in Figure 4.31 with plus and minus signs. The (0,1) mode is an exception to this, and it is marked with "−/+" since it involves the whole head moving either into or out of the bowl, as it attempts to compress and rarefy the trapped air respectively.

In practice, this mode is damped by means of a vent hole in the bowl which allows air to move out of and into the bowl in response to this mode, thereby absorbing its energy. This vent hole has no such effect on the other modes since they all involve compensating movements by sections of the head as indicated by the equal numbers of plus and minus signs in Figure 4.31. Hall notes that the usual strike position for a kettledrum is half to three quar-ters of the distance from the center to the rim, and that this is reasonably close to the circular node positions for all modes that have two or three circu-lar nodes: {(0,2), (1,2), (2,2) and (0,3)} in Figure 4.31. These modes will not be greatly excited since they will be unable to realize strong circular nodes due to the strike producing a significant velocity in their circular nodal regions (see Section 4.2.1 on modes not excited when a string is plucked).

In summary, only modes with one circular node (the one they must all have at the rim) *except* the first mode (0,1) contribute significantly to the sound produced by a kettledrum, bearing in mind that this depends on the strike position being between a half and three quarters of the distance from the drum center to the rim. The presence of the bowl lowers the frequencies of these contributing modes since the head is effectively made more massive due to the presence of the air trapped in the bowl which loads it. The extent of the pitch flattening of the modes is dependent on the shape of the bowl itself, and the aim for the kettledrum maker is to achieve modal frequencies for the modes with one circular mode that are close to being members of the same harmonic series.

Table 4.2 shows the frequency ratios for the lowest 10 modes of an ideal supported membrane (these are the modes shown in Figure 4.31), and those for a kettledrum (adapted from Rossing, 2001). The modes that do not contribute significantly to the overall output from a kettledrum {(0,1), (0,2), (1,2), (2,2) and (0,3)} are indicated with an asterisk. It can be seen that the frequency ratios of the remaining modes are very close to being the second, third, fourth, fifth and sixth harmonics—making the kettledrum an orchestral instrument whose output has a "missing fundamental"—but this does not affect our ability to perceive its pitch as being associated with the missing fundamental as discussed in Section 3.2.1. A kettledrum then will output a strongly pitched note, provided it is struck in a position that tends not to favor those modes marked with an asterisk in Table 4.2.

The player does then have some control over the output spectrum from a kettledrum depending on where it is struck. Note that a strike in the center is not very satisfactory in terms of the resulting sound because almost all of the first 10 modes have nodes in the center, and therefore they will tend not to be excited. Another form of control is from the use of different mallets. Small hard mallets produce a large excitation that is focused over a small area and therefore can excite a number of modes extending to high frequencies, while the excitation from large soft mallets is somewhat muffled, duller and less strong due to it being spread over a larger area, thereby tending to excite more strongly the lower frequency modes. This muffling effect was produced on early kettledrums by placing a cloth over the drum head. The normal orchestral playing effects for timpani are the roll and repeated notes.

The lower part of Figure 4.31 shows the first six modes of a cymbal (Rossing, 2001). It should be noted that unlike the stretched membrane a metal plate is not supported around its outer edge, and therefore the low-frequency modes have no circular nodes. Rossing notes that the modes above the sixth tend to be combinations of more than one individual mode, and are therefore rather difficult to pin down in terms of their modal patterns, but they do tend to have at least one circular node. The mode frequencies do not approximate closely to a harmonic series and therefore no strong pitch is produced by cymbals. Cymbals are struck in a variety of ways, either as a pair of orchestral crash cymbals or as a "hi-hat" in a drum kit or with hard or soft beaters. Crash and hi-hat cymbals excite all modes since all modes have antinodes around the edges of the cymbals. The use of beaters gives the player some control of the output spectrum in the same way that is described for a kettledrum above.

4.5 THE SPEAKING AND SINGING VOICE

The singing voice is probably the most versatile of all musical instruments. Anyone who can speak is capable of singing, but we are not all destined to be opera or pop stars. While considerable mystique surrounds the work of some singing teachers and how they achieve their results, the acoustics of the singing voice is now established as a research topic in its own right. Issues such as the following are being considered:

- the differences between adult male and adult female voices;
- the effects of singing training;
- the development of pitching skills by children;
- "tone deafness;"
- the acoustic nature of different singing styles;
- the effect of different acoustics on choral singing;
- electronic synthesis of the singing voice;
- choral blend in choirs;
- solo voice.

Knowledge of the acoustics of the singing and speaking voice can be helpful to music technologists when they are developing synthetic sounds since humans are remarkably good at vocalizing the sound they desire. In such cases, knowledge of the acoustics of the singing and speaking voice can help in the development of synthesis strategies. This section discusses the human singing voice in terms of the input/system/output model and points to some of the key differences between the speaking and singing voice. The discussion presented in this section is necessarily brief. A number of texts are available which consider the acoustics of the speaking voice (e.g., Fant, 1960; Fry, 1979; Borden and Harris, 1980; Baken, 1987; Baken and Danilof, 1991; Kent and Read, 1992; Howard, 1998; Howard and Angus, 1998), and the acoustics of the singing voice (e.g., Benade, 1976; Sundberg, 1987; Bunch, 1993; Dejonckere *et al.*, 1995; Howard, 1999).

4.5.1 Sound source in singing

The sound source in singing is the acoustic result of the vocal folds vibrating in the larynx which is sustained by air flowing from the lungs. The sound modifiers in singing are the spaces between the larynx and the lips and nostrils, known as the "vocal tract," which can be changed in shape and size by moving the "articulators," for example the jaw, tongue and lips (see Figure 4.32). As we sing or speak, the shape of the vocal tract is continually changing to produce different sounds. The soft palate acts as a valve to shut off and open the nasal cavity (nose) from the airstream.

Vocal fold vibration in a healthy larynx is a cyclic sequence in which the vocal folds close and open regularly when a note is being sung. Thus the vocal folds of a soprano singing A4 ($f_0 = 440.0\,\text{Hz}$) will complete this vocal fold closing and opening sequence 440 times a second. Singers have two methods by which they can change the f_0 of vocal fold vibration: they alter the stiffness of the folds themselves by changing the tension of the fold muscle tissue, or by altering the vibrating mass by supporting an equal portion of each fold in an immobile position. Adjustments of the physical properties of the folds themselves allow many trained singers to sing over a pitch range of well over two octaves.

The vocal folds vibrate as a result of the Bernoulli effect in much the same way as the lips of a brass player. A consequence of this is that the folds close more rapidly than they open. An acoustic pressure pulse is generated at each instant when the vocal folds snap together, rather like a hand clap. As these closures occur regularly during singing, the acoustic input to the vocal tract consists of a regular series of pressure pulses (see Figure 4.33), the note being sung depending on the number per second. The pressure pulses are shown as negative-going in the figure since the rapid closure of the vocal folds suddenly causes the airflow from the lungs to stop, resulting in a pressure *drop* immediately above the vocal folds. The time between each pulse is the

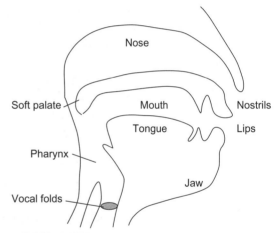

FIGURE 4.32   *A cross-section of the vocal tract.*

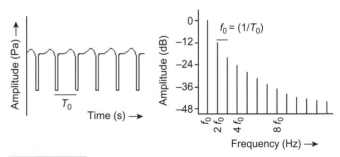

FIGURE 4.33   *Idealized waveform (left) and spectrum (right) of acoustic excitation due to normal vocal fold vibration. (Note that T_0 indicates the period of the acoustic excitation waveform, and f_0 indicates the fundamental frequency.)*

fundamental period. Benade (1976) notes though that the analogy between the lip vibration of brass players and vocal fold vibration speakers and singers should not be taken too far because the vocal folds can vibrate with little influence being exerted by the presence of the vocal tract, whereas the brass player's lip vibration is very strongly influenced by the presence of the instrument's pipe.

Figure 4.34 shows a schematic vocal fold vibration sequence as if viewed from the front associated with an idealized airflow waveform between the vibrating vocal folds. This is referred to as "glottal" airflow since the space between the vocal folds is known as the "glottis." Three key phases of the vibration cycle are usefully identified: closed phase (vocal folds together), opening phase (vocal folds parting), and closing phase (vocal folds coming together). The opening and closing phases are often referred to as the "open phase" as shown in the figure, because this is the time during which air flows. It should also be noted that airflow is not necessarily zero during the closed phase since there are vocal fold vibration configurations for which the vocal folds do not come together over their whole length (e.g., Sundberg, 1987; Howard, 1998, 1999).

The nature of vocal fold vibration changes with voice training, whether for oratory, acting or singing. The time for which the vocal folds are in contact in each cycle, known as "larynx closed quotient" or "CQ," has been investigated as a possible means by which trained adult male (Howard *et al.*, 1990) and female (Howard, 1995) singers can be helped in producing a more efficient acoustic output. Experimental measurements on trained and untrained singers suggest that CQ is higher at all pitches for trained adult males, and that it tends to increase with pitch for trained adult females in a patterned manner.

Howard *et al.* suggest that the higher CQ provides the potential for a more efficient voice output by three means: (i) the time in each cycle during

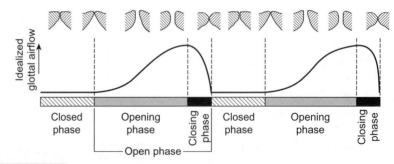

FIGURE 4.34 *Schematic sequence for two vocal fold vibration cycles to illustrate vocal fold vibration sequence as if viewed from the front, and idealized glottal airflow waveform. Vocal fold opening, closing, open and closed phases are indicated.*

which there is an acoustic path via the open vocal folds to the lungs, where sound is essentially completely absorbed, is reduced, (ii) longer notes can be sustained since less air is lost via the open vocal folds in each cycle, and (iii) the voice quality is less breathy since less air flows via the open vocal folds in each cycle.

The frequency spectrum of the regular pressure pulses generated by the vibrating vocal folds during speech and singing consists of all harmonics with an amplitude change on average of -12 dB per octave rise in frequency (see the illustration on the right in Figure 4.33). Thus for every doubling in frequency, equivalent to an increase of one octave, the amplitude reduces by 12 dB. The amplitudes of the first, second, fourth and eighth harmonics (which are separated by octaves) in the figure illustrate this effect.

The shape of the acoustic excitation spectrum remains essentially constant while singing, although the amplitude change of -12 dB per octave is varied for artistic effect and singing style and to aid voice projection by professional singers (e.g., Sundberg, 1987). The spacing between the harmonics will change as different notes are sung; Figure 4.38 shows three input spectra for sung notes an octave apart. Trained singers, particularly those with Western operatic voices, exhibit an effect known as "vibrato" in which their f_0 is varied at a rate of approximately 5.5–7.5 Hz with a range of between ± 0.5 and ± 2 semitones (Dejonckere *et al.*, 1995).

4.5.2 Sound modifiers in singing

The regular series of pulses from the vibrating vocal folds are modified by the acoustic properties of the vocal tract (see Figure 4.32). In acoustic terms, the vocal tract can be considered as a stopped tube (closed at the larynx, which operates as a *flow-controlled reed*, and open at the lips) which is approximately 17.5 cm in length for an adult male. When the vowel at the end of *announcer* is produced, the vocal tract is set to what is referred to as a "neutral" position, in which the articulators are relaxed, and the soft palate (see Figure 4.32) is raised to cut off the nose; the vowel is termed "non-nasalized." The neutral vocal tract approximates quite closely to a

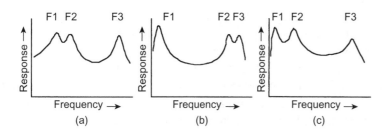

FIGURE 4.35

Idealized vocal tract response plots for the vowels in the words fast *(a),* feed *(b), and* food *(c).*

tube of constant diameter throughout its length and therefore the equation governing modal frequencies in a cylindrical stopped pipe can be used to find the vocal tract standing wave mode frequencies for this vowel.

EXAMPLE 4.3

Calculate the first three mode frequencies of the neutral adult male vocal tract. (Take the velocity of sound in air as 344 ms^{-1}.)

The vocal tract length is 17.5 cm, or 0.175 m.

From Equation 4.9, the fundamental or first mode:

$$F_{stopped(1)} = \left[\frac{c}{4L_s}\right] = \left[\frac{344}{4 \times 0.175}\right] 491.4 \text{ Hz}$$

From Equation 4.10, the higher mode frequencies are:

$$f_{stopped(n)} = (2n - 1) \, f_{stopped(1)}$$
$$\text{where } n = 1, \, 2, \, 3, \, 4, \ldots$$

Thus the second mode frequency ($n = 2$) is: $3 \times 491.4 = 1474$ Hz
and the third mode frequency ($n = 3$) is: $5 \times 491.4 = 2457$ Hz.

Example 4.3 gives the frequencies for the neutral vowel, and these are often rounded to 500 Hz, 1500 Hz and 2500 Hz for convenience. When considering the acoustics of speech and singing, the standing wave modes are generally referred to as "formants." Idealized frequency response curves for a vocal tract set to produce the vowels in the words *fast*, *feed* and *food* are shown in Figure 4.35 and the center frequency of each formant is labeled starting with "F1" or "first formant" for the peak that is lowest in frequency, continuing with "F2" (second formant) and "F3" (third formant) as shown in the figure. The formants are acoustic resonances of the vocal tract itself resulting from the various dimensions of the vocal tract spaces. These are modified during speech and singing by movements of the articulators.

When considering the different sounds produced during speech, usually just the first, second and third formants are considered since these are the only formants whose frequencies tend to vary. Six or seven formants can often be identified in the laboratory and the higher formants are thought to contribute to the individual identity of a speaking or singing voice. However, in singing, important contributions to the overall projection of sound are believed to be made by formants higher than the third.

In order to produce different sounds, the shape of the vocal tract is altered by means of the articulators to change its acoustic properties. The *perturbation theory* principles explored in the context of woodwind reed instruments (see Figure 4.23) can be employed here also (Kent and Read, 1992). Figure 4.36 shows the velocity nodes and antinode positions for the first three formants of the vocal tract during a neutral non-nasalized vowel, which can be confirmed with reference to the upper-right-hand part of Figure 4.23. Following the same line of reasoning as that presented in the context of Figure 4.23, the effect of constrictions (and therefore enlargements) on the first three formants of the vocal tract can be predicted as shown in Figure 4.37. For example, all formants have a volume velocity antinode at the lips, and a lip constriction therefore lowers the frequencies of all formants. (It should be noted that there are two other means of lowering all formant frequencies by means of vocal tract lengthening either by protruding the lip or by lowering the larynx.)

A commonly referenced set of average formant frequency values for men, women and children for a number of vowels, taken from Peterson and Barney (1952), is shown in Table 4.3. Formant frequency values for these vowels can be predicted with reference to their articulation. For example, the vowel in *beat* has a constriction towards the front of the tongue in the region of both N2 and N3 (see Figure 4.36), and reference to Figure 4.37 suggests that F1 is lowered in frequency and F2 and F3 are raised from the values one would expect for the neutral vowel. The vowel in *part*, on the other hand, has a significant constriction in the region of both A2 and A3 (see Figure 4.36) resulting in a raising of F1, and a lowering of both F2 and F3 from their neutral vowel values. The vowel in *boot* has a constriction at the lips,

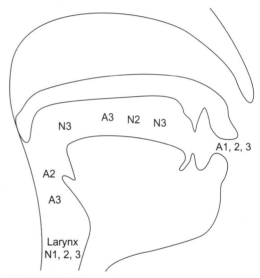

FIGURE 4.36 *Velocity nodes and antinode positions for the first three modes (or formants: F1, F2, F3) of the vocal tract during a neutral non-nasalized vowel.*

FIGURE 4.37

Formant frequency modification with position of vocal tract of constriction.

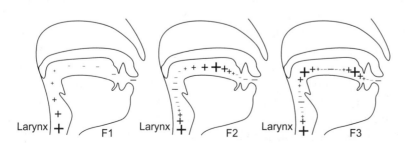

Table 4.3	Average formant frequencies in Hz for men, women and children for a selection of vowels. (From Peterson and Barney, 1952.)								
Vowel in	Men			Women			Children		
	F1	F2	F3	F1	F2	F3	F1	F2	F3
beat	270	2300	3000	300	2800	3300	370	3200	3700
bit	400	2000	2550	430	2500	3100	530	2750	3600
bet	530	1850	2500	600	2350	3000	700	2600	3550
bat	660	1700	2400	860	2050	2850	1000	2300	3300
part	730	1100	2450	850	1200	2800	1030	1350	3200
pot	570	850	2400	590	900	2700	680	1050	3200
boot	440	1000	2250	470	1150	2700	560	1400	3300
book	300	850	2250	370	950	2650	430	1150	3250
but	640	1200	2400	760	1400	2800	850	1600	3350
pert	490	1350	1700	500	1650	1950	560	1650	2150

which are also rounded so as to extend the length of the vocal tract, and thus all formant frequencies are lowered from their neutral vowel values. These changes can be confirmed from Table 4.3.

The input/system/output model for singing consists of the acoustic excitation due to vocal fold vibration (input) and the vocal tract response (system) to give the output. These are usually considered in terms of their spectra, and both the input and system change with time during singing. Figure 4.38 shows the model for the vowel in *fast* sung on three different notes. This is to allow one of the main effects of singing at different pitches to be illustrated.

The input in each case is the acoustic spectrum resulting from vocal fold vibration (see Figure 4.33). The output is the result of the response of the vocal tract for the vowel in *fast* acting on the input vocal fold vibration. The effect of this is to multiply the amplitude of each harmonic of the input by the response of the vocal tract at that frequency. This effectively imparts the formant peaks of the vocal tract response curve onto the harmonics of the input spectrum. In this example, there are three formant peaks shown, and it can be seen that in the cases of the lower two notes the formant structure can be readily seen in the output, but that in the case of the highest note the formant peaks cannot be identified in the output spectrum because the harmonics of the input are too far apart to represent clearly the formant structure.

The representation of the formant structure in the output spectrum is important if the listener is to identify different vowels. Figure 4.38 suggests that somewhere between the G above middle C and the G an octave above,

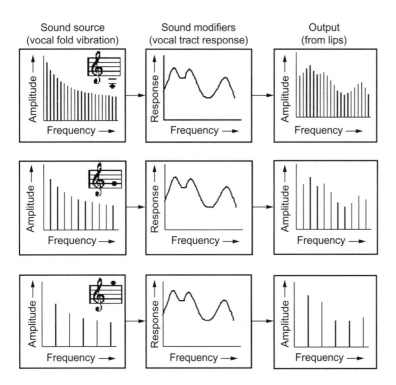

FIGURE 4.38

Singing voice input/system/ output model idealized for the vowel in fast sung on three notes an octave apart.

vowel identification will become increasingly difficult. This is readily tested by asking a soprano to sing different vowels on mid and top G, as shown in the figure, and listening to the result. In fact, when singing these higher notes, professional sopranos adopt vocal tract shapes which place the lower formants over individual harmonics of the excitation so that they are transmitted via the vocal tract with the greatest amplitude. In this way, sopranos can produce sounds of high intensity which will project well. This effect is used from approximately the C above middle C where the vocal tract is, in effect, being "tuned in" to each individual note sung, but at the expense of vowel clarity.

This tuning-in effect is not something that tenors need to do since the ratio between the formant frequencies and the f_0 of the tenor's range is higher than that for sopranos. However, all singers who do not use amplification need to project above accompaniment, particularly when this is a full orchestra and the performance is in a large auditorium. The way in which professional opera singers achieve this can be seen with reference to Figure 4.39, which shows idealized spectra for the following:

- a professional opera singer speaking the text of an operatic aria;
- the orchestra playing the accompaniment to the aria; and
- the aria sung by the singer with the orchestral accompaniment.

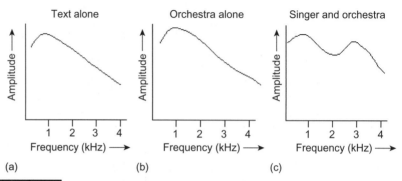

FIGURE 4.39 *Idealized spectra: (a) a singer speaking the text of an opera aria, (b) the orchestra playing the accompaniment to the aria, (c) the aria being sung with orchestral accompaniments (adapted from Sundberg, 1987).*

It should be noted that the amplitude levels cannot be directly compared between (a) and (b) in the figure (i.e. the singer does not speak as loudly as the orchestral accompaniment!) since they have been normalized for comparison.

The idealized spectrum for the text read alone has the same general shape as that for the orchestra playing alone. When the professional singer sings the aria with orchestral accompaniment, it can be seen that this combined response curve has a shape similar to both the speech and orchestral accompaniment at low frequencies, but with an additional broad peak between approximately 2.5 kHz and 4 kHz and centered at about 3 kHz. This peak relates to the acoustic output from the singer when singing but not when speaking, since it is absent for the read text and also in the orchestral accompaniment alone.

This peak has similar characteristics to the formants in the vocal tract response, and for this reason it is known as the "singer's formant." The presence of energy in this peak enables the singer to be heard above an accompanying orchestra because it is a section of the frequency spectrum in which the singer's output prevails. This is what gives the professional singing voice its characteristic "ring," and it is believed to be the result of lowering the larynx and widening the pharynx (see Figure 4.32) which is adopted by trained Western operatic singers. (The lower plot in Figure 5.5 in Chapter 5 is an analysis of a CD recording of a professional tenor whose singer's formant is very much in evidence.)

Singing teachers set out to achieve these effects from pupils by suggesting that pupils: "sing on the point of the yawn," or "sing as if they have swallowed an apple which has stuck in their throat." Sundberg (1987) discusses the articulatory origin of the singer's formant as follows: "it shows a strong dependence on the larynx tube," concluding that: "it is necessary, however, that the pharynx tube be lengthened and that the cross-sectional area in the pharynx at the level of the larynx tube opening be more than six times the area of that opening."

Professional singing is a complex task which extends the action of the instrument used for speech. It is salutary to note that the prime function of the vocal folds is to act as a valve to protect the lungs, and not to provide the sound source basic to human communication by means of speech and song.

4.5.3 Tuning in a capella (unaccompanied) singing

Experienced singers are able to vary their intonation (sung pitch) to a very fine degree with respect either to the pitch of any accompanying instrument(s) or, if singing a capella (unaccompanied), to the pitches of the other singers. If singing with accompaniment that is being provided by a modern piano or pipe organ, then the tuning system will be equal tempered where all the semitones have a twelfth root of two frequency ratio (see Section 3.4.3).

When singing a capella though, there is no requirement to stick to equal tempered tuning and the possibility exists for singers to make use of the more consonant just intonation where the intervals are maximally consonant (see Section 3.3.3), for example with perfect fifths (frequency ratio 3:2) and just major and minor thirds (frequency ratios 5:4 and 6:5 respectively). Bearing in mind that just intonation cannot be used to tune a 12-note chromatic musical keyboard if the octaves are to remain in the ratio 2:1 (see Section 3.4), there could be musical situations where singing in just intonation causes the overall pitch to drift as a piece modulates through different keys and back to the starting key.

When singers take care with their listening, they will tend away from non-equal tempered tuning towards just tuning (Helmholtz, 1954; Bohrer, 2002). A note sung by one singer has many harmonics whose relative amplitudes depend mainly on the setting of the sound modifiers and therefore the formant frequencies (see Section 4.5.2). When singing a chord, each harmonic of the note produced by one singer will have a nearest neighbor in the set of harmonics produced by another singer and the overall consonance of the result will depend heavily on the tuning accuracy. This is maximized when the fundamental frequencies are in an integer relationship and are harmonics of each other (see Section 3.3.3).

When a capella singing group adopts the non-equal temperament of just tuning, there is the question of whether the group's pitch remains in-tune or not if the music modulates away from the starting key and back again. In order to test this, some musical exercises were written by Howard (2007a) for a four-part (soprano, alto, tenor, bass or SATB) quartet which involved a four-part block chord movement that visited a large number (compared with the total number) of key chords. An example (Exercise 3) is shown in Figure 4.40 where the first and last chords are spaced by one octave. This exercise was one of a set that has been used to test the hypothesis that an SATB a capella vocal quartet will drift in pitch with key modulation if it tends to non-equal temperament.

The f_0 of each of the four singers was measured when they sang the set of exercises by means of an electrolaryngograph (Abberton *et al.*, 1989), which monitors vocal fold vibration directly. It makes use of two electrodes that are placed externally on the neck at the level of the larynx. A small high-frequency electrical current flows between the electrodes that is higher when the vocal folds are in contact than when they are apart. Note that acoustic measurements could not be used for f_0 measurement in this experiment for four singers since the recordings of each singer would be contaminated with the sound from the other singers.

The predicted pitch drift for just tuning of the chords in the exercise shown in Figure 4.40 is shown in Figure 4.41. It is based on the tuning of the just scale (see Section 3.4.2), and it can be seen that the prediction is for the pitch to drift a semitone flat (rather surprising for an exercise consisting of only 13 chords!).

Figure 4.42 shows the measured results for a quartet where the average f_0 for each sung note of each chord has been used (Howard, 2007a).

FIGURE 4.40 *Tuning exercise to test for pitch drift in a capella singing (from Howard, 2007a).*

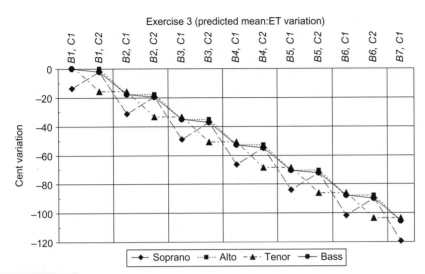

FIGURE 4.41 *Predicted pitch drift based on just tuning for the exercise shown in Figure 4.40. The 13 chords of the exercise are predicted for each part separately and therefore indicated as "note number" on the X axis (adapted from Howard, 2007a).*

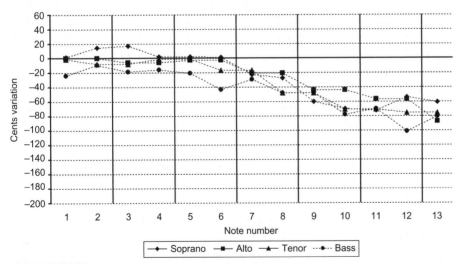

FIGURE 4.42 *Measured pitch drift for a quartet singing the exercise shown in Figure 4.40. The 13 chords of the exercise (see Figure 4.40) are analyzed for each part separately and therefore indicated as "note number" on the X axis (adapted from Howard, 2007a).*

The measured results have a similar trend to that predicted, but with some variation in detail. The singers tend to drift flat more gently than the prediction in the early stages then more rapidly afterwards, with individual singers generally moving together except for the bass whose pitch swings are quite wide from chord to chord. The soprano and to some extent the alto in this quartet seem to be reluctant to allow the pitch to drift any further flat towards the end, almost maintaining it in the position it has reached somewhat against the efforts of the tenor and to some extent the bass to shift it further in line with the prediction. The overall measured pitch change is quite close to the semitone flat shift predicted. The experiment was repeated with this quartet and another quartet sang the exercise twice (Howard, 2007b); the results in each case were similar.

This experiment confirms that singers do tend towards just tuning and that they maintain this overall tuning even at the expense of the overall pitch of the piece. Thus a dichotomy is presented to choral singers and their conductors in terms of whether to keep the overall pitch in-tune and make some compromise to the tuning of individual chords or whether to allow the overall pitch to drift. Since equal temperament is at the heart of music heard by most people brought up within the Western musical tradition, its compromised intervals in terms of their lack of consonance (all intervals except the octave are out-of-tune) form the basis of how musical intervals are learned.

It is therefore interesting to note the tendency of a capella choral singers towards the more consonant intervals of just intonation which are based on minimum beats, whilst at the same time noting the difficulty of tuning in equal temperament for which there is no readily available physical guide such as the absence of beating. Tuning notes in equal temperament is therefore likely to rely on memories of music from an early age rather than relying on any physical attributes of the relations between the notes themselves.

Howard (2007b) lists a number of basic points in relation to singing a capella in just intonation including:

- Singers do seem able to change their tuning, even subconsciously.
- Singers are not completely "locked" to equal temperament.
- Some consistency in approach has been demonstrated.
- Natural shift is towards just intonation.
- Present one note to start a piece—not an equal tempered chord.
- Have good listening skills.

In addition, he notes some points which tend to work against achieving just tuning including:

- choir practice with a piano or other keyboard tuned in equal temperament;
- listening to recorded and live music in equal temperament;
- working with choral conductors who are unaware of the implications;
- presenting an equal tempered chord to start a piece rather than one note;

- being trained to remain in-tune;
- poor listening skills.

This aspect of a capella singing is one that many choral conductors are unaware of: knowledge of acoustics and psychoacoustics provides the basis for a proper understanding of its underlying causes. Conductors and directors should be aware of the effect and then they can make their own musical judgment in terms of what to do about pitch drift when it arises.

REFERENCES

Abberton, E.R.M., Howard, D.M., Fourcin, A.J., 1989. Laryngographic assessment of normal voice: a tutorial. Clin. Linguist. Phon. 3, 281–296.

Askenfelt, A. (ed.), 1990. Five Lectures on the Acoustics of the Piano, with compact disc, Publication No. 64. Royal Swedish Academy of Music, Stockholm.

Audsley, G.A., 1965. The Art of Organ-Building 2 Vols. Dover, New York (reprint of 1905 edition, New York: Dodd, Mead and Company).

Baken, R.J., 1987. Clinical Measurement of Speech and Voice. Taylor and Francis, London.

Baken, R.J., Danilof, R.G., 1991. Readings in Clinical Spectrography of Speech. Singular Publishing Group, San Diego.

Benade, A.H., 1976. Fundamentals of Musical Acoustics. Oxford University Press, New York.

Bohrer, J.C.S., 2002. Intonational Strategies in Ensemble Singing [Doctoral thesis]. University of London, London.

Borden, G.J., Harris, K.S., 1980. Speech Science Primer. Williams and Wilkins, Baltimore.

Bunch, M., 1993. Dynamics of the Singing Voice. Springer-Verlag, New York.

Campbell, M., Greated, C., 1998. The Musician's Guide to Acoustics. Oxford University Press, Oxford.

Dejonckere, P.H., Hirano, M., Sundberg, J. (eds), 1995. Vibrato. Singular Publishing Group, San Diego.

Fant, C.G.M., 1960. Acoustic Theory of Speech Production. Mouton, The Hague.

Fletcher, N.H., Rossing, T.D., 1999. The Physics of Musical Instruments, Second edn. Springer-Verlag, New York.

Fry, D.B., 1979. The Physical of Speech. Cambridge University Press, Cambridge.

Hall, D.E., 2001. Musical Acoustics: An Introduction, Third edn. Wadsworth Publishing Company, Belmont, CA.

Helmholtz, H., 1954. On the Sensations of Tone, 2nd edn. (1885 translation by A. J. Ellis of the 1877 4th edn.), Dover, New York.

Howard, D.M., 1995. Variation of electrolaryngographically derived closed quotient for trained and untrained adult female singers. J. Voice 9, 163–172.

Howard, D.M., 1998. Practical voice measurement. In: Harris, T., Harris, S., Rubin, J.S., Howard, D.M. (eds), The Voice Clinic Handbook. Whurr Publishing Company, London, 323–382.

Howard, D.M., 1999. The human singing voice. In: Day, P. (Ed.), Killers in the Brain. Oxford University Press, Oxford, pp. 113–134.

Howard, D.M., 2007a. Intonation drift in *a capella* Soprano, Alto, Tenor, Bass quartet singing with key modulation. J. Voice 21 (3), 300–315.

Howard, D.M., 2007b. Equal or non-equal temperament in a capella SATB singing. Logop. Phoniatr. Vocol. 32 (2), 87–94.

Howard, D.M., Angus, J.A.S., 1998. Introduction to human speech production, human hearing and speech analysis. In: Westall, F.A., Johnson, R.D., Lewis, A.V. (Eds), Speech Technology for Telecommunications. Chapman and Hall, London, pp. 30–72.

Howard, D.M., Lindsey, G.A., Allen, B., 1990. Towards the quantification of vocal efficiency. J. Voice 4, 205–221 (See also errata: (1991). J. Voice, **5**, 93–95.)

Hurford, P., 1994. Making Music on the Organ, sixth edn. Oxford University Press, Oxford.

Hutchins, C.M. (ed.), 1975a. Musical Acoustics, Part I: Violin Family Components. Dowden, Hutchinson and Ross Inc, Pennsylvania.

Hutchins, C.M. (ed.), 1975b. Musical Acoustics, Part II: Violin Family Functions. Dowden, Hutchinson and Ross Inc, Pennsylvania.

Hutchins, C.M. (ed.), 1978. The Physics of Music. Reprints from Scientific American, W.H. Freeman and Company, San Francisco.

Kent, R.D., Read, C., 1992. The Acoustic Analysis of Speech. Singular Publishing Group, San Diego.

Martin, D., Ward, D., 1961. Subjective evaluation of musical scale temperament in pianos. J. Acoust. Soc. Am. 33, 582–585.

Nederveen, C.J., 1969. Acoustical Aspects of Woodwind Instruments. Frits Knuf, Amsterdam.

Norman, H., Norman, H.J., 1980. The Organ Today. David and Charles, London.

Peterson, G.E., Barney, H.E., 1952. Control methods used in the study of vowels. J. Acoust. Soc. Am. 24, 175–184.

Rossing, T.D., 2001. The Science of Sound, third edition. Addison-Wesley, New York.

Sumner, W.L., 1975. The Organ, Fifth edn. Macdonald and Company, London.

Sundberg, J., 1987. The Science of the Singing Voice. Illinois University Press, DeKalb.

Sundberg, J., 1989. The Science of Musical Sounds. Academic Press, San Diego.

Taylor, C.A., 1976. Sounds of Music. Butler and Tanner Ltd, London.

Hearing Timbre and Deceiving the Ear

5.1 WHAT IS TIMBRE?

Pitch and loudness are two of three important descriptors of musical sounds commonly used by musicians, the other being "timbre." Pitch relates to issues such as notes on a score, key, melody, harmony, tuning systems, and intonation in performance. Loudness relates to matters such as musical dynamics (e.g., pp, p, mp, mf, f, ff) and the balance between members of a musical ensemble (e.g., between individual parts, choir and orchestra, or soloist and accompaniment). Timbre to sound quality descriptions include: mellow, rich, covered, open, dull, bright, dark, strident, grating, harsh, shrill, sonorous, somber, colorless and lackluster. Timbral descriptors are therefore used to indicate the perceived quality or tonal nature of a sound which can have a particular pitch and loudness also.

There is no subjective rating scale against which timbre judgments can be made, unlike pitch and loudness which can, on average, be reliably rated by listeners on scales from "high" to "low." The commonly quoted American National Standards Institute formal definition of timbre reflects this: "Timbre is that attribute of auditory sensation in terms of which a listener can judge two sounds similarly presented and having the same loudness and pitch as being dissimilar" (ANSI, 1960). In other words, two sounds that are perceived as being different but which have the same perceived loudness and pitch differ by virtue of their timbre (consider the examples on track 62 of the accompanying CD). The timbre of a note is the aspect by which a listener recognizes the instrument which is playing a note when, for example, instruments play notes with the same pitch, loudness and duration.

The definition given by Scholes (1970) encompasses some timbral descriptors: "Timbre means tone quality – coarse or smooth, ringing or more subtly penetrating, 'scarlet' like that of a trumpet, 'rich brown' like that of a cello, or 'silver' like that of the flute. These color analogies come naturally to every mind. . . . The one and only factor in sound production which conditions timbre is the presence or absence, or relative strength or weakness, of overtones." (In Chapter 3, Table 3.1 gives the relationship between overtones and harmonics.) While his color analogies might not come naturally to every mind, Scholes' later comments about the acoustic nature of sounds which have different timbres are a useful contribution to the acoustic discussion of the timbre of musical sounds.

When considering the notes played on pitched musical instruments, timbre relates to those aspects of the note which can be varied without affecting the pitch, duration or loudness of the note as a whole, such as the spectral components present and the way in which their frequencies and amplitudes vary during the sound. In Chapter 4 the acoustics of musical instruments are considered in terms of the output from the instrument as a consequence of the effect of the sound modifiers on the sound input (e.g., Figure 4.2). What is not considered, due to the complexity of modeling, is the acoustic development from silence at the start of the note and back to silence at the end. It is then convenient to consider a note in terms of three phases: the "onset" or "attack" (the build-up from silence at the start of the note), the "steady state" (the main portion of the note), and the "offset" or "release" (the return to silence at the end of the note after the energy source is stopped). The onset and offset portions of a note tend to last for a short time, of the order of a few tens of milliseconds (or a few hundredths of a second). Changes that occur during the onset and offset phases, and in particular during the onset, turn out to have a very important role in defining the timbre of a note.

In this chapter, timbre is considered in terms of the acoustics of sounds which have different timbres, and the psychoacoustics of how sounds are perceived. Finally, the pipe organ is reviewed in terms of its capacity to synthesize different timbres.

5.2 ACOUSTICS OF TIMBRE

The description of the acoustics of notes played on musical instruments presented in Chapter 4 was in many cases supported by plots of waveforms and spectra of the outputs from some instruments (Figures 4.17, 4.22, 4.24 and 4.29). Except in the plots for the plucked notes on the lute and guitar (Figure 4.11) where the waveforms are for the whole note and the spectra are for a single spectral analysis, the waveform plots show a few cycles from the steady-state phase of the note concerned and the spectral plots are based on averaging together individual spectral measurements taken during the steady-state phase. The number of spectra averaged together depends on how long the steady-state portion of the note lasts. For the single notes illustrated in Chapter 4, spectral averaging was carried out over approximately a quarter to three quarters of a second, depending on the length of the note available.

An alternative way of thinking about this is in terms of the number of cycles of the waveform over which the averaging takes place, which would be 110 cycles for a quarter of a second to 330 cycles for three quarters of a second for A4 ($f_0 = 440\,\mathrm{Hz}$), or 66 cycles to 198 cycles for C4 ($f_0 = 261.6\,\mathrm{Hz}$). Such average spectra are commonly used for analyzing the frequency components of musical notes, and they are known as "long-term average spectra" or "LTAS." One main advantage of using LTAS is that the spectral features of interest during the steady-state portion of the note are enhanced in the resulting plot by the averaging process with respect to competing acoustic sounds, such as background noise, which change over the period of the LTAS and thus average towards zero.

LTAS cannot, however, be used to investigate acoustic features that change rapidly such as the onset and offset of musical notes, because these will also tend to average towards zero. In terms of the timbre of the note, it is not only the variations that occur during the onset and offset that are of interest, but also how they change with time. Therefore an analysis method is required in which the timing of acoustic changes during a note is preserved in the result. One analysis technique commonly used for the acoustic analysis of speech is a plot of amplitude, frequency and time known as a "spectrogram." Frequency is plotted on the vertical scale, time on the horizontal axis and amplitude is plotted as the darkness on a gray scale, or in some cases the color, of the spectrogram.

The upper plot in Figure 5.1 shows a spectrogram and acoustic pressure waveform of C4 played on a principal 8′ (open flue), the same note for which an LTAS is presented in Figure 4.17. The LTAS plot in Figure 4.17 showed that the first and second harmonics dominate the spectrum, the amplitude of the third harmonic being approximately 8 dB lower than the first harmonic, and with energy also clearly visible in the fourth, fifth, seventh and eighth harmonics whose amplitudes are at least 25 dB lower than that of the first harmonic.

A spectrogram shows which frequency components are present (measured against the vertical axis), at what amplitude (blackness of marking) and when (measured against the horizontal axis). Thus harmonics are represented on spectrograms as horizontal lines, where the vertical position of the line marks the frequency and the horizontal position shows the time for which that harmonic lasts. The amplitudes of the harmonics are plotted as the blackness of marking of the lines. The frequency and time axes on the spectrogram are marked and the amplitude is shown as the blackness of the marking.

The spectrogram shown in Figure 5.1 shows three black horizontal lines which are the first three harmonics of the principal note (since the frequency axis is linear, they are equally spaced). The first and second harmonics are slightly blacker (and thicker) than the third, reflecting the amplitude difference as shown in Figure 4.17. The fourth, fifth and seventh harmonics are visible and their amplitude relative to the first harmonic is reflected in the blackness with which they are plotted.

5.2.1 Note envelope

The onset, steady-state, and offset phases of the note are indicated above the waveform in the figure, and these are determined mainly with reference to the spectrogram because they relate to the changes in spectral content at the start and end of the note, leaving the steady portion in between. However, "steady state" does not mean that no aspect of the note varies.

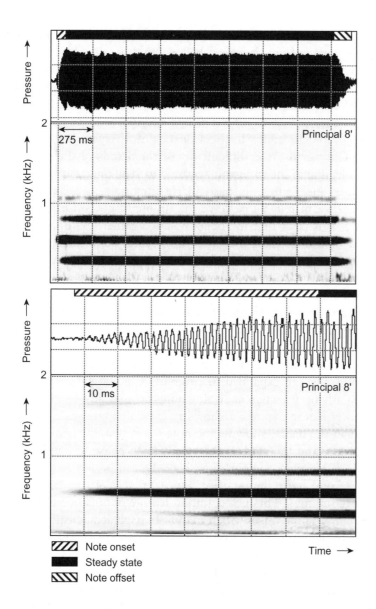

Pressure →

Frequency (kHz) →

2

275 ms

1

Principal 8'

Pressure →

2

10 ms

1

Principal 8'

Time →

| Note onset |
| Steady state |
| Note offset |

FIGURE 5.1

Waveform and spectrogram of whole note (upper) and onset phase (lower) for C4 played on the principal 8' organ stop (open flue) for which an LTAS is shown in Figure 4.17. (Note that onset, steady-state and offset phases are marked.)

The timbre of a principal organ stop sounds "steady" during a prolonged note such as that plotted, which lasts for approximately 2 seconds, but it is clear from the acoustic pressure waveform plot in Figure 5.1 that the amplitude, or "envelope," varies even during the so-called "steady-state" portion of this note. This is an important aspect of musical notes to be aware of when, for example, synthesizing notes of musical instruments, particularly if using looping techniques on a sampling synthesizer.

For the principal pipe, the end of the note begins when the key is released and the air flowing from the organ bellows to drive the air reed is stopped. In the note offset for this example, which lasts approximately 200 ms, the high harmonics die away well before the first and second. However, interpretation of note offsets is rather difficult if the recording has been made in an enclosed environment as opposed to free space (see Chapter 4), since any reverberation due to the acoustics of the space is also being analyzed (see Chapter 6). It is difficult to see the details of the note onset in this example, due to the timescale required to view the complete note.

The note onset phase is particularly important to perceived timbre. Since listeners can reliably perceive the timbre of notes *during* the steady-state phase, it is clear that the offset phase is rather less important to the perception of timbre than the onset and steady-state phases. The onset phase is also more acoustically robust from the effects of the local environment in which the notes are played, since coloration of the direct sound by the first reflection (see Chapter 6) may occur *after* the onset phase is complete (and therefore transmitted uncolored to the listener). By definition the first reflection certainly occurs after part of the note onset has been heard uncolored. The onset phase is therefore a vital element and the offset phase an important factor in terms of timbre perception. Spectrograms whose timescales are expanded to cover the time of the note onset phase are particularly useful when analyzing notes acoustically.

The lower plot in Figure 5.1 shows an expanded timescale version of the upper plot in the figure, showing the note onset phase which lasts approximately 70 ms, and the start of the steady-state phase. It can be seen that the detail of the onset instant of each of the harmonics is clearly visible, with the second harmonic starting approximately 30 ms before the first and third harmonics. This is a common feature of organ pipes voiced with a chiff or consonantal onset which manifests itself acoustically in the onset phase as a initial jump to the first, or sometimes higher, overblown mode.

The first overblown mode for an open flue pipe is to the second harmonic (see Chapter 4). Careful listening to pipes voiced with a chiff will reveal that open pipes start briefly an octave high since their first overblown mode is the second harmonic, and stopped pipes start an octave and a fifth high since their first overblown mode is the third harmonic. The fourth harmonic in the figure starts with the third and its amplitude briefly drops 60 ms into the sound when the fifth starts, and the seventh starts almost with the second and its amplitude drops 30 ms later. The effect of the harmonic build-up on the acoustic pressure waveform can be observed in Figure 5.1 in terms of the changes in its shape, particularly the gradual increase in amplitude during the onset phase. The onset phase for this principal organ pipe is a complex series of acoustic events, or acoustic "cues," which are available as potential contributors to the listener's perception of the timbre of the note.

5.2.2 Note onset

In order to provide some data to enable appreciation of the importance of the note onset phase for timbre perception, Figures 5.2 to 5.4 are presented in which the note onset and start of the steady-state phases for four organ stops, four woodwind instruments and four brass instruments respectively are presented for the note C4 (except for the trombone and tuba for which the note is C3). By way of a caveat it should be noted that these figures are presented to provide only examples of the general nature of the acoustics of the note onset phases for these instruments. Had these notes been played at a different loudness, by a different player, on a different instrument, in a different environment, or even just a second time by the same player on the same instrument in the same place while attempting to keep all details constant, the waveforms and spectra would probably be noticeably different.

The organ stops for which waveforms and spectra are illustrated in Figure 5.2 are three reed stops: hautbois and trompette (LTAS in Figure 4.22) and a regal, and gedackt (LTAS in Figure 4.17), which is an example of a stopped flue pipe. The stopped flue supports only the odd modes (see Chapter 4), and, during the onset phase of this particular example, the fifth harmonic sounds first, which is the second overblown mode sounding two octaves and a major third above the fundamental (see Figure 3.3), followed by the fundamental and then the third harmonic, giving a characteristic chiff to the stop.

The onset phase for the reed stops is considerably more complicated since many more harmonics are present in each case. The fundamental for the hautbois and regal is evident first, and the second harmonic for the trompette. In all cases, the fundamental exhibits a frequency rise at onset during the first few cycles of reed vibration. The staggered times of entry of the higher harmonics form part of the acoustic timbral characteristic of that particular stop, the trompette having all harmonics present up to the 4 kHz upper frequency bound of the plot, the hautbois having all harmonics up to about 2.5 kHz, and the regal exhibiting little or no evidence (at this amplitude setting of the plot) of the fourth or eighth harmonics.

Figure 5.3 shows plots for four woodwind instruments: clarinet, oboe, tenor saxophone and flute. For these particular examples of the clarinet and tenor saxophone, the fundamental is apparent first and the oboe note begins with the second harmonic, followed by the third and fourth harmonics after approximately 5 ms, and then the fundamental some 8 ms later. The higher

FIGURE 5.2

Waveform (upper) and spectrogram (lower) of the note onset phase for C4 played on the following pipe organ stops: hautbois 8' (reed), trompette 8' (reed), gedackt 8' (stopped flue) and regal 8' (reed). LTAS for the hautbois and trompette notes are shown in Figure 4.22, and for the gedackt in Figure 4.17.

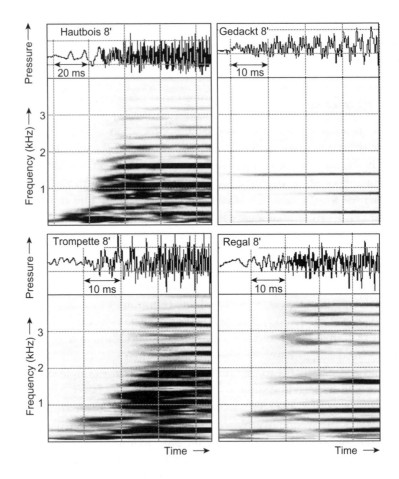

harmonics of the clarinet are apparent nearly 30 ms after the fundamental; the dominance of the odd harmonics is discussed in Chapter 4.

This particular example of C4 from the flute begins with a notably "breathy" onset just prior to and as the fundamental component starts. This can be seen in the frequency region of the spectrogram that is above 2 kHz lasting some 70 ms. The higher harmonics enter one by one approximately 80 ms after the fundamental starts. The rather long note onset phase is characteristic of a flute note played with some deliberation. The periodicity in the waveforms develops gradually, and in all cases there is an appreciable time over which the amplitude of the waveform reaches its steady state.

Figure 5.4 shows plots for four brass instruments. The notes played on the trumpet and French horn are C4, and those for the trombone and tuba are C3. The trumpet is the only example with energy in high harmonic components in this particular example, with the fourth, fifth and sixth harmonics

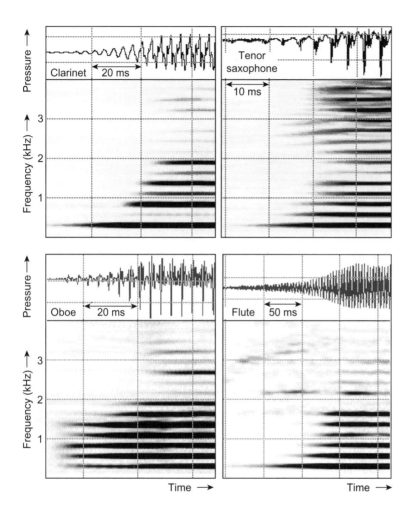

FIGURE 5.3

Waveform (upper) and spectrogram (lower) of the note onset phase for C4 played on a clarinet, flute, oboe and tenor saxophone. LTAS for the clarinet and tenor saxophone are shown in Figure 4.24.

having the highest amplitudes. The other instruments in this figure do not have energy apparent above approximately the fourth harmonic (French horn and tuba) or the sixth harmonic for the trombone. The note onset phase for all four instruments starts with the fundamental (noting that this is rather weak for the trombone), followed by the other harmonics. The waveforms in all cases become periodic almost immediately.

Waveforms and spectrograms are presented in the upper plot of Figure 5.5 for C4 played with a bow on a violin. Approximately 250 ms into the violin note, vibrato is apparent as a frequency variation particularly in the high harmonics. This is a feature of using a linear frequency scale, since a change of x Hz in f_0 will manifest itself as a change of $2x$ Hz in the second harmonic, $3x$ Hz in the third harmonic, and so on. In general the frequency change in the

FIGURE 5.4

Waveform (upper) and spectrogram (lower) of the note onset phase for C4 played on a trumpet and French horn, and C3 played on a trombone and tuba. LTAS for the trombone and tuba are shown in Figure 4.29.

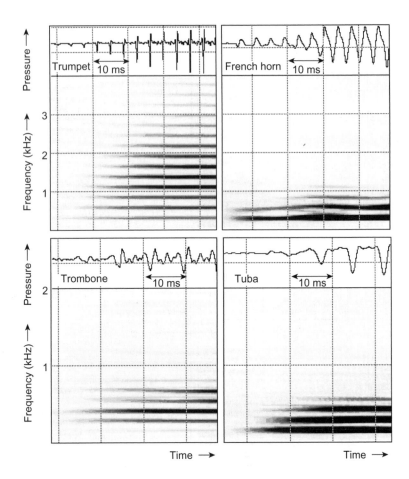

nth harmonic will be n_x Hz; therefore the frequency variation in the upper harmonics during vibrato will be greater than that for the lower harmonics when frequency is plotted on a linear scale as in the figure. Vibrato often has a delayed start, as in this example, as the player makes subtle intonation adjustments to the note. This particular bowed violin note has an onset phase of approximately 160 ms and an offset phase of some 250 ms.

Finally in this section, a note is analyzed from a CD recording of a professional tenor singing the last three syllables of the word *Vittoria* (i.e., "toria") on B♭4 from the second act of *Tosca* by Puccini (lower plot in Figure 5.5). This is a moment in the score when the orchestra stops playing and leaves the tenor singing alone. The orchestra stops approximately 500 ms into this example: its spectrographic record can be seen particularly in the lower left-hand corner of the spectrogram, where it is almost impossible to analyze any detailed acoustic patterning. This provides just a hint at

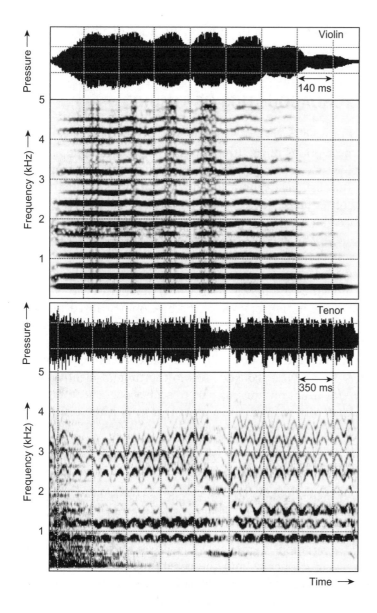

FIGURE 5.5

UPPER: *Waveform and spectrogram of C4 (262 Hz) played on a violin. LOWER: Waveform and spectrogram of the last three syllables of the word* Vittoria *from the second act of* Tosca *by Puccini sung by a professional tenor (f₀ = B♭4) from a CD recording.*

the real acoustic analysis task facing the hearing system when listening to music.

The spectrogram of the tenor shows the harmonics and the extent of the vibrato very clearly, and his singer's formant (compare with Figure 4.38) can be clearly seen in the frequency region between 2.4 kHz and 3.5 kHz. The first and third of the three syllables ("toria") are long, and the second ("ri") is considerably shorter in this particular tenor's interpretation. The second syllable manifests itself as the dip in amplitude of all harmonics just over halfway through the spectrogram.

5.3 PSYCHOACOUSTICS OF TIMBRE

A number of psychoacoustic experiments have been carried out to explore listeners' perceptions of the timbre of musical instruments and the acoustic factors on which it depends. Such experiments have demonstrated, for example, that listeners cannot reliably identify musical instruments if the onset and offset phases of notes are removed. For example, if recordings of a note played on a violin open string and the same note played on a trumpet are modified to remove their onset and offset phases in each case, it becomes very difficult to tell them apart.

The detailed acoustic nature of a number of example onset phases is provided in Figures 5.1 to 5.5, from which differences can be noted. Thus, for example, the initial scraping of the bow on a stringed instrument, the consonant-like onset of a note played on a brass instrument, the breath noise of the flautist, the initial flapping of a reed, the percussive thud of a piano hammer and the final fall of the jacks of a harpsichord back onto the strings are all vital acoustic cues to the timbral identity of an instrument. Careful attention must be paid to such acoustic features, for example when synthesizing acoustic musical instruments if the resulting timbre is to sound convincingly natural to listeners.

5.3.1 Critical bands and timbre

A psychoacoustic description of timbre perception must be based on the nature of the critical bandwidth variation with frequency since this describes the nature of the spectral analysis carried out by the hearing system. The variation in critical bandwidth is such that it becomes wider with increasing frequency, and the general conclusion was drawn in the section on pitch perception in Chapter 3 (Section 3.2) that no harmonic above about the fifth to seventh is resolved no matter what the value of f_0. Harmonics below the fifth to seventh are therefore resolved separately by the hearing system (e.g., see Figure 3.11), which suggests that these harmonics might play a distinct and individual role in timbre perception. Harmonics above the fifth or seventh, on the other hand, which are not separately isolated by the hearing system are not likely to have such a strong individual effect on timbre perception, but could affect it as groups that lie within a particular critical band.

Based on this notion, the perceived timbre is reviewed of instruments for which the results of acoustic analysis are presented in this book, bearing in mind that these analyses are for single examples of notes played on these instruments by a particular player on a particular instrument at a particular loudness and pitch in a particular acoustic environment.

Instruments amongst those for which spectra have been presented, which have significant amplitudes in harmonics above the fifth or seventh during their steady-state phases, include organ reed stops (see Figures 4.22 and 5.2), the tenor saxophone (see Figures 4.24 and 5.3), the trumpet (see Figure 5.4), the violin and professional singing voice (see Figure 5.5). The timbres of such instruments might be compared with those of other instruments using descriptive terms such as "bright," "brilliant," or "shrill."

Instruments which do not exhibit energy in harmonics above the fifth or seventh during their steady-state phases include the principal 8' (see Figures 4.17 and 5.1), the gedackt 8'(see Figures 4.17 and 5.2), the clarinet, oboe and flute (see Figures 4.24 and 5.3), and the trombone, French horn and tuba (see Figures 4.29 and 5.4). In comparison with their counterpart organ stops or other instruments of their category (woodwind or brass), their timbres might be described as being: "less bright" or "dark," "less brilliant" or "dull," or "less shrill" or "bland."

Within this latter group of instruments there is an additional potential timbral grouping between those instruments which exhibit all harmonics up to the fifth or seventh, such as the clarinet, oboe, and flute, compared with those which just have a few low harmonics such as the principal 8', gedackt 8', trombone, French horn and tuba. It may come as a surprise to find the flute in the same group as the oboe and clarinet, but the lack of the seventh harmonic in the flute spectrum compared with the clarinet and oboe (see Figure 5.3) is crucial.

Notes excluding the seventh harmonic sound considerably less "reedy" than those with it; the seventh harmonic is one of the lowest which is not resolved by the hearing system (provided the sixth and/or eighth are/is also present). This last point is relevant to the clarinet where the seventh harmonic is present but both the sixth and eighth are weak. The clarinet has a particular timbre of its own due to the dominance of the odd harmonics in its output, and it is often described as being "nasal." Organists who are familiar with the effect of the tierce $(1\frac{3}{5})$ and the rarely found septième $(1\frac{1}{7})$ stops (see Section 5.4) will appreciate the particular timbral significance of the fifth and seventh harmonics respectively and the "reediness" they tend to impart to the overall timbre when used in combination with other stops.

Percussion instruments that make use of bars, membranes or plates as their vibrating system (described in Section 4.4), which are struck, have a distinct timbral quality of their own. This is due to the non-harmonic relationship between the frequencies of their natural modes which provides a clear acoustic cue to their family identity. It gives the characteristic "clanginess" to this class of instruments which endows them with a timbral quality of their own.

5.3.2 Acoustic cues and timbre perception

Timbre judgments are highly subjective and therefore individualistic. Unlike pitch or loudness judgments, where listeners might be asked to rate sounds on scales of low to high or soft to loud respectively, there is no "right" answer for timbre judgments. Listeners will usually be asked to compare the timbre of different sounds and rate each numerically between two opposite extremes of descriptive adjectives, for example on a one to ten scale between "bright" (1)—"dark" (10) or "brilliant" (1)—"dull" (10), and a number of such descriptive adjective pairs could be rated for a particular sound. The average of ratings obtained from a number of listeners is often used to give a sound an overall timbral description. Hall (1991) suggests that it is theoretically possible that one day up to five specific rating scales could be "sufficient to accurately identify almost any timbre."

Researchers have attempted to identify relationships between particular features in the outputs from acoustic musical instruments and their perceived timbre. A significant experiment in this field was conducted by Grey (1977). Listeners were asked to rate the similarity between recordings of pairs of synthesized musical instruments on a numerical scale from 1 to 30. All sounds were equalized in pitch, loudness and duration. The results were analyzed by "multidimensional scaling" which is a computational technique that places the instruments in relation to each other in a multi-dimensional space based on the similarity ratings given by listeners.

In Grey's experiment, a three-dimensional space was chosen and each dimension in the resulting three-dimensional representation was then investigated in terms of the acoustic differences between the instruments lying along it "to explore the various factors which contributed to the subjective distance relationships." Grey identified the following acoustic factors with respect to each of the three axes: (1) "spectral energy distribution" observed as increasing high-frequency components in the spectrum; (2) "synchronicity in the collective attacks and decays of upper harmonics" from sounds with note onsets in which all harmonics enter in close time alignment to those in which the entry of the harmonics is tapered; and (3) from sounds with "precedent high-frequency, low-amplitude energy, most often inharmonic energy, during the attack phase" to those without high-frequency attack energy.

These results serve to demonstrate that (a) useful experimental work can and has been carried out on timbre, and (b) that acoustic conclusions can be reached which fit in with other observations, for example the emphasis of Grey's axes (2) and (3) on the note onset phase.

The sound of an acoustic musical instrument is always changing, even during the rather misleadingly so-called "steady-state" portion of a note. This is clearly shown, for example, in the waveforms and spectrograms for the violin and sung notes in Figure 5.5. Pipe organ notes are often presented as being "steady" due to the inherent airflow regulation within the instrument, but Figure 5.1 shows that even the acoustic output from a single organ pipe has an amplitude envelope that is not particularly steady. This effect manifests itself perceptually extremely clearly when attempts are made to synthesize the sounds of musical instruments electronically and no attempt is made to vary the sound in any way during its steady state.

Variation of some kind is needed during any sound in order to hold the listener's attention. The acoustic communication of new information to a listener, whether speech, music, environmental sounds or warning signals from a natural or person-made source, requires that the input signal varies in some way, with time. Such variation may be of the pitch, loudness or timbre of the sound. The popularity of post-processing effects, particularly chorus (see Chapter 7), either as a feature on synthesizers themselves or as a studio effects unit reflects this. However, while these can make sounds more interesting to listen to by time variation imposed by adding post-processing, such an addition rarely does anything to improve the overall *naturalness* of a synthesized sound.

A note from any acoustic musical instrument typically changes dynamically throughout in its pitch, loudness and timbre. Pitch and loudness have one-dimensional subjective scales from "low" to "high" that can be related fairly directly to physical changes which can be measured, but timbre has no such one-dimensional subjective scale. Methods have been proposed to track the dynamic nature of timbre based on the manner in which the harmonic content of a sound changes throughout. The "tristimulus diagram" described by Pollard and Jansson (1982) is one such method in which the time course of individual notes is plotted on a triangular graph such as the example plotted in Figure 5.6.

The graph is plotted based on the proportion of energy in: (1) the second, third and fourth harmonics or "mid" frequency components (Y axis); and (2) the high-frequency partials, which here are the fifth and above, or "high" frequency components (X axis); and (3) the fundamental or f_0 (where X and Y tend towards zero). The corners of the plot in Figure 5.6 are marked: "mid", "high" and "f_0" to indicate this. A point on a tristimulus

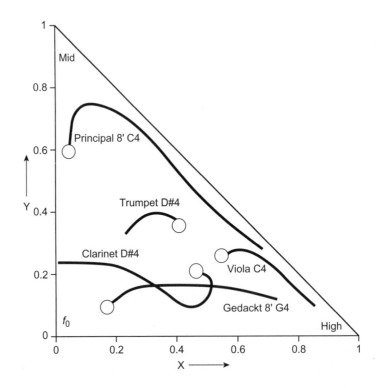

FIGURE 5.6

Approximate timbre representation by means of a tristimulus diagram for note onsets of notes played on a selection of instruments. In each case, the note onset tracks along the line towards the open circle which represents the approximate steady-state position. "Mid" represents "strong mid-frequency partials"; "High" represents "strong high-frequency partials"; "f_0" represents "strong fundamental". (Data from Pollard and Jansson, 1982.)

diagram therefore indicates the relationship between f_0, harmonics which are resolved, and harmonics which are not resolved.

The tristimulus diagram enables the dynamic relationship between high, mid and f_0 to be plotted as a line, and several are shown in the figure for the note onset phases of notes from a selection of instruments (data from Pollard and Jansson, 1982). The time course is not even and is not calibrated here for clarity. The approximate steady-state position of each note is represented by the open circle, and the start of the note is at the other end of the line. The note onsets in these examples lasted as follows: gedackt (10–60 ms); trumpet (10–100 ms); clarinet (30–160 ms); principal (10–150 ms); and viola (10–65 ms). The track taken by each note is very different and the steady-state positions lie in different locations. Pollard and Jansson present data for additional notes on some of these instruments which suggest that each instrument maintains its approximate position on the tristimulus representation as shown in the figure. This provides a method for visualizing timbral differences between instruments which is based on critical band analysis. It also provides a particular representation which gives an insight as to the nature of the patterns which could be used to represent timbral differences perceptually.

The tristimulus representation essentially gives the relative weighting between the f_0 component, those harmonics other than the f_0 component that are resolved, and those that are not resolved. The thinking underlying the tristimulus representation is itself not new in regard to timbre. In his seminal work, *On the Sensations of Tone as a Physiological Basis for the Theory of Music*—first published in 1877—Helmholtz (1877, translated into English in 1954) deduced four general rules in regard to timbre, which he presented "to shew the dependence of quality of tone from the mode in which a musical tone is compounded" (Helmholtz, 1877). It should though be remembered that for Helmholtz, no notion of critical bands or today's models of pitch and loudness perception were known at that time to support his thinking. Despite this, the four *general rules* for timbre that Helmholtz developed demonstrate insights that are as relevant today as they were then in the context of our understanding of the objective nature of timbre. The four *rules* (Helmholtz [1877], translation [1954] pp. 118–119) are as follows (track 63 on the accompanying CD illustrates these).

1. *Simple tones*, like those of tuning forks applied to resonance chambers and wide stopped organ pipes, have a very soft, pleasant sound, free from all roughness, but wanting in power, and dull at low pitches.

2. *Musical tones*, which are accompanied by a moderately loud series of the lower partial tones up to about the sixth partial, are more harmonious and musical. Compared with simple tones they are rich and splendid, while they are at the same time perfectly sweet and soft if the higher upper partials are absent. To these belong the musical tones produced by the pianoforte, open organ pipes, the softer piano tones of the human voice and the French horn.

3. If only the unevenly numbered partials are present (as in narrow stopped organ pipes, pianoforte strings struck in their middle points, and clarinets), the quality of the tone is *hollow*, and, when a large number of such partials are present, *nasal*. When the prime tone predominates the quality of the tone is *rich*; but when the prime tone is not sufficiently superior in strength to the upper partials, the quality of tone is *poor*.

4. When partials higher than the sixth or seventh are very distinct, the quality of the tone is *cutting or rough*. . . . The most important musical tones of this description are those of the bowed instruments and of most reed pipes, oboe, bassoon, harmonium and the human voice. The rough, braying tones of brass instruments are extremely penetrating, and hence are better adapted to give the impression of great power than similar tones of a softer quality.

These *general rules* provide a basis for considering changes in timbre with respect to sound spectra which link directly with today's understanding of the nature of the critical bandwidth. His second and fourth general rule make a clear distinction between sounds in which partials higher than the sixth or seventh are distinct or not, and his first rule gives a particular importance to the f_0 component. It is worth noting the similarity with the axes of the tristimulus diagram as well as axis (I) resulting from Grey's multidimensional experiment. It is further worth noting which words Helmholtz chose to describe the timbre of the different sounds (bearing in mind that these have been translated by Alexander Ellis in 1954 from the German original). As an aside on the word *timbre* itself, Alexander Ellis translated the German word *klangfarbe* as "quality of tone" which he argued is "familiar to our language". He notes that he could have used one of the following three words: *register, color,* or *timbre,* but he chose not to because of the following common meanings that existed for these words at the time.

1. Register—"has a distinct meaning in vocal music which must not be disturbed."

2. Timbre—means kettledrum, helmet, coat of arms with helmet; in French it means "postage stamp." Ellis concluded that: "timbre is a foreign word, often mispronounced and not worth preserving."

3. Color—"is used at most in music as a metaphorical expression."

Howard and Tyrrell (1997) describe a frequency domain manifestation of the four general rules based on the use of spectrograms which display the output of a human hearing modeling analysis system. This is described in the next few paragraphs, and it is based on Figure 5.7 which shows human hearing modeling spectrograms for the following synthesized sounds: (a) a sinewave, (b) a complex periodic tone consisting of the first five harmonics, (c) a complex periodic tone consisting of the odd harmonics only from the first to the nineteenth inclusive, and (d) a complex periodic tone consisting of the first 20 (odd and even) harmonics. In each case, the f_0 varies between 128 Hz and 160 Hz. The sounds (a–d) were synthesized to represent the sounds described by Helmholtz for each of the general rules (1–4). The key feature to note about the hearing modeling spectrogram is that the sound is analyzed via a bank of filters that model the auditory filters in terms of their shape and bandwidth. As a result, the frequency axis is based on the ERB (equivalent rectangular bandwidth) scale itself, since the output from each filter is plotted with an equal distance on the frequency axis of the spectrogram.

Howard and Tyrrell (1997) note the following. The human hearing modeling spectrogram for the sinewave exhibits a single horizontal band of energy (the f_0 component) which rises in frequency (from 128 to 160 Hz)

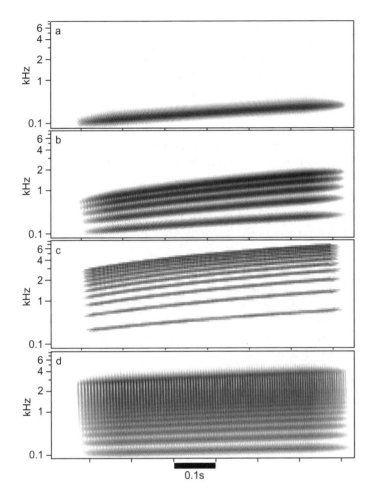

FIGURE 5.7

Gamma tone spectrograms of synthesized sounds with the f_0 varying between 128 Hz and 160 Hz: (a) a sinewave, (b) a complex periodic tone consisting of the first five harmonics, (c) a complex periodic tone consisting of the odd harmonics only from the first to the nineteenth inclusive, and (d) a complex periodic tone consisting of the first 20 (odd and even) harmonics.

during the course of the sound. It is worth noting the presence of the f_0 component in all four spectrograms since its variation is the same in each case. For the sound consisting of the first five harmonics, it can be seen that these are isolated separately as five horizontal bands of energy. These harmonic bands of energy become closer together with increasing frequency on the spectrogram due to the nature of the ERB frequency scale.

In the spectrogram of the fourth sound, where all harmonics are present up to the twentieth, the lowest six or seven harmonics appear as horizontal bands of energy and are therefore resolved, but the energy in the frequency region above the seventh harmonic is plotted as vertical lines, known as "striations," which occur once per cycle (recall the discussion about the temporal theory of pitch perception). This is because all filters with center frequencies above the seventh harmonic have bandwidths that are greater than the f_0 of the sound, and therefore at least two adjacent harmonics are captured by each of these filters.

In the spectrogram of the third sound, where only the odd harmonics up to the nineteenth are present, there are approximately seven resolved horizontal energy bands, but in this case these are the lowest seven *odd* harmonics, or the first to the thirteenth harmonic inclusive. The point at which the odd harmonics cease to be resolved occurs where the spacing between them ($2f_0$—the spacing between the odd harmonics) is less than the ERB. This will occur at about the position of the fifth to seventh harmonic of a sound with double the f_0: somewhere between the tenth and the fourteenth harmonic of this sound, which concurs with the spectrogram.

Table 5.1, which has been adapted from Howard and Tyrrell (1997), provides a useful summary of the key features of the human hearing spectrogram

Table 5.1	Summary of the frequency domain properties as exemplified by the human hearing modeling spectrograms, example timbre descriptions and example acoustic instruments which fall under each of the four rules introduced by Helmholtz (1877). (Table adapted from Howard and Tyrrell, 1997)		
Helmholtz rule	**Human hearing modeling spectrogram**	**Example timbre descriptors**	**Example acoustic instruments**
1	f_0 dominates	Pure Soft Simple Pleasant Dull at low pitch Free from roughness	Tuning fork Wide stopped organ flues Baroque flute
2	Harmonics dominate	Sweet and soft Rich Splendid Dark Dull Less shrill Bland	French horn, tuba Modern flute Recorder Open organ flues Soft sung sounds
3	Odd harmonics dominate	Hollow Nasal	Clarinet Narrow stopped organ flues
4	Striations dominate	Cutting Rough Bright Brilliant Shrill Brash	Oboe, bassoon Trumpet, trombone Loud sung sounds Bowed instruments Harmonium Organ reeds

example timbre descriptors (including those proposed by Helmholtz) and example acoustic instruments for each of the four Helmholtz general rules. Howard and Tyrrell note, though, "that the grouping of instruments into categories is somewhat generalized, since it may well be possible to produce sounds on these instruments where the acoustic output would fall into another timbral category based on the frequency domain [hearing modeling] description, for example by the use of an extended playing technique." Human hearing modeling spectrographic analysis has also been applied to speech (Brookes *et al.*, 2000), forensic audio (Howard *et al.*, 1995) and singing (Howard, 2005), where it is providing new insights into the acoustic content that is relevant to human perception of sound.

Audio engineers need to understand timbre to allow them to communicate successfully with musicians during recording sessions in a manner that everyone can understand. Many musicians do not think in terms of frequency but they are well aware of the ranges of musical instruments which they will discuss in terms of note pitches. Sound engineers who take the trouble to work in these terms will find the communication process both easier and more rapid with the likely result that the overall quality of the final product is higher than it might have been otherwise. Reference to Figures 3.21 and 4.3 will provide the necessary information in terms of the f_0 values for the equal tempered scale against a piano keyboard and the overall frequency ranges of common musical instruments.

Timbral descriptions also have their place in musician/sound engineer communication, but this is not a trivial task since many of the words employed have different meanings depending on who is using them. Figure 5.8 shows a diagram adapted from Katz (2007) in which timbral descriptors are presented that relate to boosting or reducing energy in the spectral regions indicated set against a keyboard, frequency values spanning the normally quoted human hearing range of 20 Hz to 20 kHz and terms ("sub bass," "bass," "midrange" and "treble") commonly used to describe spectral regions. The timbral descriptors are separated by a horizontal line which indicates whether the energy in the frequency region indicated has to be increased or decreased to modify the sound as indicated by the timbral term. For example, using an equalizer a sound can be made *thinner* by decreasing the energy in the frequency range between about 100 Hz and 500 Hz (note that "this" is under the horizontal line in the "energy down" part of the diagram). The sound will become *brighter, more sibilant, harsher,* and *sweeter* by increasing the energy around 2 kHz and 8 kHz.

It is curious to note that a sound can be made *warmer* by either increasing the energy between approximately 200 and 600 Hz or by reducing the energy in the 3–7 kHz region. Increasing *projection* approximately in the

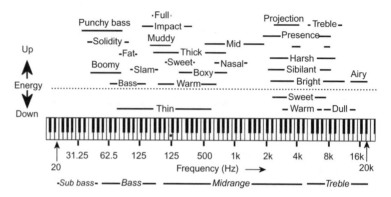

FIGURE 5.8 *Timbral descriptors used to describe the effect of boosting (above) and reducing (below) the spectral energy in various frequency regions set against a keyboard (middle C marked with a spot). Frequency values spanning the normally quoted human hearing range of 20 to 20 kHz, and terms commonly used to describe spectral regions are shown. (Adapted from Katz, 2007.)*

2.5–4 kHz region relates primarily to voice production and specifically to providing a hint of singer's formant (see Section 4.5.2), which has the effect of bringing a singer out towards the front of a mix. This must, however, be done judiciously using a boost of just a few dB otherwise the sound will tend towards being *edgy* and *harsh*; a trained singer modifies other aspects of his or her vocal output such as vibrato.

In an experiment to establish which adjectives were commonly used in a similar manner by musicians, listening tests were carried out with musicians (Howard *et al.*, 2007) who were asked to rate a set of sounds using various adjectives. They found that the following adjectives had the highest agreement between listeners when describing the sounds: *bright, percussive, gentle, harsh* and *warm*, whilst those with the least were: *nasal, ringing, metallic, wooden* and *evolving*. Listeners were also asked to indicate their confidence in assigning the adjectives: the one that elicited the least confidence was *evolving*, whilst the highest confidence ratings were given to *clear, percussive, ringing* and *bright*.

There is still much work to be done on timbre to understand it more fully. While there are results and ideas which indicate what acoustic aspects of different instruments contribute to the perception of their timbre differences, such differences are far too coarse to explain how the experienced listeners are able to tell apart the timbre differences between, for example, violins made by different makers. The importance of timbre in music performance has been realized for many hundreds of years as manifested in the so-called "king" of instruments—the pipe organ—well before there was any detailed knowledge of the function of the human hearing system.

5.4 THE PIPE ORGAN AS A TIMBRAL SYNTHESIZER

A pipe organ is one of the earliest forms of "additive synthesizer" for creating different sound timbres. In additive synthesis, the output sound is manipulated by means of adding harmonics together, and the stops of a pipe organ provide the means for this process. There are references to the existence of some form of pipe organ going back at least to 250 BC (Sumner, 1975).

A modern pipe organ usually has two or more keyboards, or "manuals," each with up to five octaves (61 notes from C2 to C7) and a pedal board with up to two and a half octaves (32 note from C1 to G3). Very large organs might have up to six manuals, and small "chamber" organs might have one manual and no pedals. Each stop on a five octave manual has 61 pipes, apart from stops known as "mixtures" which have a number of pipes per note. An organ stop which has the same f_0 values as on a piano (i.e., f_0 for the note A4 is 440 Hz—see Figure 3.21) is known as an "eight foot" or 8′ rank on the manuals and "sixteen foot" or 16′ rank on the pedals. Eight and sixteen feet are used because they are the approximate lengths of open pipes of the bottom note of a manual (C2) and the pedals (C1) respectively. The pedals thus sound an octave below the note written on the score. A 4′ rank and a 2′ rank would sound one and two octaves higher than an 8′ rank respectively, and a 32′ rank would sound one octave lower than a 16′ rank.

It should be noted that this footage terminology is used to denote the sounding pitch of the rank and gives no indication as to whether open or stopped pipes are employed. Thus a stopped rank on a manual sounding a pitch equivalent to a rank of 8′ open pipes would have a four foot long bottom stopped pipe but its stop knob would be labeled as 8′.

Organs have a number of stops on each manual of various footages, most of which are flues. Some are voiced to be used alone as solo stops, usually at 8′ pitch, but the majority are voiced to blend together to allow variations in loudness and timbre to be achieved by acoustic harmonic synthesis by drawing different combinations of stops. The timbral changes are controlled by reinforcing the natural harmonics of the 8′ harmonic series on the manuals (16′ harmonic series for the pedals). The following equation relates the footage of a stop to the member of the harmonic series which its fundamental reinforces:

$$\text{(stop footage)} = \text{(footage of first harmonic)}/N \qquad (5.1)$$
$$\text{where } N = \text{harmonic number (1, 2, 3, 4 ...)}$$

Thus, for example, the footage of the stop where the fundamental reinforces the sixth harmonic on the manuals (which is related to the 8' harmonic series) is given by:

$$\left[\frac{8}{6}\right] = 1\frac{2'}{6} = 1\frac{1b'}{3}$$

The footage of the stop where the fundamental reinforces the third harmonic on the pedals (which is related to the 16' harmonic series) is given by:

$$\left[\frac{16}{3}\right] = 5\frac{1'}{3}$$

EXAMPLE 5.1

Find the footage of pipe organ stops which reinforce the third and seventh natural harmonics of the 8' harmonic series.

The third harmonic is reinforced by a stop of $\left[\frac{8}{3}\right] = 2\frac{2}{3}'$

The seventh harmonic is reinforced by a stop of $\left[\frac{8}{7}\right] = 1\frac{1}{7}'$

Footages of organ stops which might be found on pipe organs are given in Table 5.2, along with a selection of names used for members of the flute family, other flue stops and reeds. The interval of each footage shown with respect to an 8' stop is given in degrees of a major scale (note that all the sevenths are flat as indicated by "(b)"), those footages that are members of the natural 8', 16', 32'and 64'harmonic series are given as harmonic numbers, and the f_0 values in Hz are provided for the bottom C (manuals or pedals). Those stops which reinforce harmonics that are *neither* in unison (1:1) with, *nor* a whole number of octaves away (i.e., 2:1; 4:1; 8:1; . . . or 2^n:1) from, the first harmonic are known as "mutation" stops. The stops shown include those found on manuals as well as the pedals. Only the largest cathedral organs contain the majority of footages as shown; many small parish church organs will have a small selection, usually including flues at 8', 4', 2' and perhaps $2\frac{2}{3}'$ as well as one or two reed stops.

The organist can build up the overall sound of the organ by drawing stops which build on the harmonic series of 8' on the manuals (consider tracks 64 and 65 on the accompanying CD) and 16' on the pedals. It is important to note, however, that a single 8' open diapason or principal stop,

Table 5.2 Example non-mixture organ stop names, their footages, f_0 values for bottom C, sounding interval to 8' in degrees of the major scale, and their harmonic number in the 8', 16', 32' and 64' harmonic series. Flues are indicated in bold and reeds in italic. The sevenths are all indicated with "(b)" to show that they are flat

Example stop names (flute, other flues, *reeds*)	Stop footage	Interval to 8'	Harmonic to				f_0 bottom C
			8'	16'	32'	64'	
gravissima, resultant bass, *diaphone dulzian, contra trombone dulzian*	64'	22nd below	—	—	—	1	8.2 Hz
double bourdon, acoustic bass	42 2/3'	18th below	—	—	—	—	12.2 Hz
double open diapason, contra violone, diaphone *contra trombone, contra bombarde, contra cor Anglais*	32'	15th below	—	—	1	2	16.4 Hz
tibia clausa	21 1/3'	11th below	—	—	—	3	24.6 Hz
bourdon, open bass, open wood, quintadena, quintaton, melodia, grosse geigen, double diapason, *regal, posaune, ophicleide, bombarde, fagotto, trombone, double trumpet*	16'	8th below	—	1	2	4	32.7 Hz
tierce	12 4/5'	6th below	—	—	—	5	40.9 Hz
double quint, sub quint, *trumpet profunda*	10 2/3'	4th below	—	—	3	6	49.1 Hz
septième	9 1/7'	1st below (b)	—	—	—	7	57.3 Hz
stopped diapason, gedackt, claribel, open diapason, principal, dulciana, gamba, salicional, *tuba, trumpet, tromba, horn, oboe, cornopean, krummhorn, clarinet, posaune*	8'	unison	1	2	4	8	65.4 Hz
voix celeste, vox angelica, viol celeste, unda maris (2 ranks: 8' & 8'#)	8' — #	unison	1	2	4	8	65.4 Hz
third, tierce, *tromba tierce*	6 2/5'	3rd above	—	—	5	10	81.75 Hz
quint	5 1/3'	5th above	—	3	6	12	98.1 Hz
septième	4 4/7'	7th above (b)	—	—	7	14	114.5 Hz
rohrflüte, spitzflüte, principal, gemshorn, octave, salicet, *clarion, octave tuba, schalmei*	4'	8th above	2	4	8	16	130.8 Hz
double tierce, tenth	3 1/5'	10th above	—	5	10	20	163.5 Hz
rohrquint, nazard, twelfth, octave quint	2 2/3'	12th above	3	6	12	24	196.2 Hz
seventh, septième	2 2/7'	14th above (b)	—	7	14	28	228.9 Hz
piccolo, principal, fifteenth, super octave	2	15th above	4	8	16	32	261.6 Hz
tierce, seventeenth	1 3/5'	17th above	5	10	20	40	327.0 Hz
larigot, nineteenth, nasat	1 1/3'	19th above	6	12	24	48	392.4 Hz
septième	1 1/7'	21st above (b)	7	14	28	56	457.8 Hz
sifflöte, spitzflöte	1'	22nd above	8	16	32	64	523.2 Hz
none, dulciana twenty-third	8/9'	23rd above	9	18	36	72	588.6 Hz
third	4/5'	24th above	10	20	40	80	654.0 Hz
eleventh	8/11'	25th above	11	22	44	88	719.4 Hz
fifth	2/3'	26th above	12	24	48	96	784.8 Hz
Gemshorn	1/2'	29th above	16	32	64	128	1046.4 Hz
French thirty-first	2/5'	31st above	20	40	80	160	1308.0 Hz
gemshorn	1/4'	36th above	32	64	128	256	2092.8 Hz

	Table 5.3	Example mixture organ stop names, the number of ranks and typical intervals as degrees of the major scale to the first harmonic of the natural series (8' for manuals and 16' for pedals). Note that these intervals change across the keyboard/pedal board as discussed in the text

Mixture stop name	No. ranks	Typical intervals to first harmonic
Sesquialtera	II	12, 17
Tertian	II	17, 19
Mixture	II	19, 22
Zimbal	III	15, 17, 19
Mixture	III	19, 22, 26
Cymbal	III	29, 33, 36
Scharff	III	26, 29, 33
Plein jeu/mixture	IV	19, 22, 26, 29
Mixture	V	15, 19, 22, 26, 29
Pedal mixtur	V	12, 17, 19, 21(flat), 22
Kornet/cornet	V	1, 8, 12, 15, 17

which produces the foundation 8' tone of the organ, consists of open flue pipes, which themselves are rich in harmonics (see Figure 5.1). Adding an open flue 4' principal will not only enhance the second harmonic of the 8' stop in particular, but also all other even harmonics. In general, when a stop is added that is set to reinforce a member ($n = 1, 2, 3, 4, \ldots$) of the natural harmonic series, it also enhances the ($2 * n, 3 * n, 4 * n, \ldots$) members.

There is a basic pipe organ timbral issue when tuning the instrument to equal temperament (see Chapter 3). Mutation stops have to be tuned in their appropriate integer frequency ratio (see Figure 3.3) to reinforce harmonics appropriately, but as a result of this they are not tuned in equal temperament and therefore they introduce beats when chords are played. For example, if a C is played with 8' and 2/3' stops drawn, the f_0 of the C on the $2\frac{2}{3}'$ rank will be exactly in tune with the third harmonic of the C on the 8' rank. If the G above is now played also to produce a two-note chord, the first harmonic of the G on the 8' rank will beat with the f_0 of the C on the $2\frac{2}{3}'$ rank (and the third harmonic of the C on the 8' rank).

Equal tempered tuning thus colors with beats the desired effect of adding mutation stops, and therefore the inclusion of mutation stops tended to go out of fashion with the introduction of equal temperament tuning (Padgham, 1986). Recent movements to revive the performance of authentic early music have extended to the pipe organ, and the resulting use of non-equal tempered tuning systems and inclusion of more mutation stops is giving new life particularly to contrapuntal music.

Often, ranks which speak an octave below the foundation pitch of 8′ (manuals) and below the foundation pitch of 16′ (pedals) are provided, particularly on large instruments. Thus a 16′ bourdon, double open diapason or double trumpet might be found on a manual, and a 32′ contra open bass or double open diapason might be found on the pedals. Extremely occasionally, stops which sound two octaves below might be found, such as a 32′ contra violone on a manual or a 64′ gravissima or contra trombone on the pedals as found on the John Wanamaker organ in Philadelphia and the Centennial Hall in Sydney, respectively.

Higher members of the series are more rarely found on pipe organs. The septième and none are the most uncommon and are used to enhance "reedy" timbres. The tierce produces a particularly reedy timbre, and it is a stop commonly found on French organs. In order to give reasonable control to the build-up of organ loudness and timbre, pipes enhancing higher harmonics are grouped together in "mixture" stops. These consist of at least two ranks of pipes per note, each rank reinforcing a high member of the natural harmonic series. Table 5.3 gives a selection of typical mixture stops.

Mixture stops have another important role to play in the synthesis process. If a chromatic scale is played on a mixture stop alone, the scale would not be continuous and jumps in pitch would be heard at approximately every octave or twelfth, keeping the overall frequency span of the mixture much narrower than the five octave range of a manual. In this way, the mixture adds brilliance to the overall sound by enhancing high harmonics at the bass end of the keyboard, but these would become inaudible (and the pipe length would be too short to be practical) if continued to the treble. A mixture IV might consist of 19, 22, 26, 29 from C1 to C2; 15, 19, 22, 26 from C2 to C3; 12, 15, 19, 22 from C3 to C4; 8, 12, 15, 19 from C4 to C5; and 8, 12, 15, 15 from C5 to C6. The values in the table show the typical content at the bass end for each mixture stop in terms of their intervals to the first harmonic expressed as note numbers of the major scale. A mixture such as the "pedal mixtur" shown in the table would produce a strongly reed-like sound due to the presence of the flattened twenty-first (the seventh harmonic).

The sesquialtera and terzian are mainly used to synthesize solo stops, since each contains a tierce (fifth harmonic). The cornet is also usually used as a solo stop, and is particularly associated with "cornet voluntaries." When it is not provided as a stop in its own right, it can be synthesized (if the stops are available) using: 8′, 4′, 2⅔′, 2′ and 1⅗′, or 8′, 4′, 2′ and sesquialtera.

5.5 DECEIVING THE EAR

This section concerns sounds which in some sense could be said to "deceive" the ear. Such sounds have a psychoacoustic realization which is not what might be expected from knowledge of their acoustic components. In other words, the subjective and objective realizations of sounds cannot be always directly matched up. While some of the examples given may be of no obvious musical use to the performer or composer, they may in the future find musical application in electronically produced music for particular musical effects where control over the acoustic components of the output sound is exact.

5.5.1 Perception of pure tones

When two pure tones are played simultaneously, they are not always perceived as two separate pure tones. The discussion relating to Figure 2.7 introducing critical bandwidth in Chapter 2 provides a first example of sounds which in some sense deceive the ear. These two pure tones are only perceived as separate pure tones when their frequency difference is greater than the critical bandwidth. Otherwise they are perceived as a single fused tone which is "rough" or as beats depending on the frequency difference between the two pure tones.

When two pure tones are heard together, other tones with frequencies lower than the frequencies of either of the two pure tones themselves may be heard also. These lower tones are not acoustically present in the stimulating signal and they occur as a result of the stimulus consisting of a "combination" of at least two pure tones; they are known as "combination tones." The frequency of one such combination tone which is usually quite easily perceived is the difference (higher minus the lower) between the frequencies of the two tones; this is known as the "difference tone":

$$f_d = f_h - f_l \qquad (5.2)$$

where f_d = frequency of the difference tone
f_h = frequency of the higher frequency pure tone
and f_l = frequency of the lower frequency pure tone

Notice that this is the beat frequency when the frequency difference is less than approximately 12.5 Hz (see Chapter 2). The frequencies of other possible combination tones that can result from two pure tones sounding simultaneously can be calculated as follows:

$$f_{(n)} = f_l - [n(f_h - f_l)] = f_l - [nf_d] \qquad (5.3)$$

where $f_{(n)}$ = frequency of the nth combination tone
n = (1, 2, 3, 4, ...)
f_l = frequency of the pure tone with the lower frequency
and f_h = frequency of the pure tone with the higher frequency

These tones are always below the frequency of the lower pure tone, and occur at integer multiples of the difference tone frequency below the lower tone. No listeners hear all and some hear none of these combination tones. The difference tone and the combination tones for $n = 1$ and $n = 2$, known as the "second-order difference tone" and the "third-order difference tone," are those that are perceived most readily (e.g., Rasch and Plomp, 1982).

EXAMPLE 5.2

Calculate the difference tone and first four combination tones which occur when pure tones of 1200 Hz and 1100 Hz sound simultaneously.

Equation 5.2 gives the difference tone frequency = $f_h - f_l = 1200 - 1100 = 100$ Hz.
    Equation 5.3 gives combination tone frequencies, and the first four are for $n = 1, 2,$ 3 and 4.

for $n = 1$: $f_{(1)} = 1100 - (1 * 100) = 1000$ Hz
for $n = 2$: $f_{(2)} = 1100 - (2 * 100) = 900$ Hz
for $n = 3$: $f_{(3)} = 1100 - (3 * 100) = 800$ Hz
for $n = 4$: $f_{(4)} = 1100 - (4 * 100) = 700$ Hz

When the two pure tone frequencies are both themselves adjacent harmonics of some f_0 (in Example 5.2 the tones are the 11th and 12th harmonics of 100 Hz), then the difference tone is equal to f_0 and the other combination tones form "missing" members of the harmonic series. When the two tones are not members of a harmonic series, the combination tones have no equivalent f_0, but they will be equally spaced in frequency.

Combination tones are perceived quite easily when two musical instruments produce fairly pure tone outputs, such as the descant recorder, baroque flute or piccolo, whose f_0 values are high and close in frequency.

When the two notes played are themselves both exact and adjacent members of the harmonic series formed on their difference tone, the combination tones will be consecutive members of the harmonic series adjacent and below the lower played note (i.e., the f_0 values of both notes and their combination tones would be exact integer multiples of the difference frequency between the notes themselves). The musical relationship of combination tones to notes played therefore depends on the tuning system in use. Two notes played using a tuning system which results in the interval between the notes never being pure, such as the equal tempered system, will produce combination tones which are close but not exact harmonics of the series formed on the difference tone.

EXAMPLE 5.3

If two descant recorders are playing the notes A5 and B5 simultaneously in equal tempered tuning, which notes on the equal tempered scale are closest to the most readily perceived combination tones?

The most readily perceived combination tones are the difference tone and the combination tones for $n = 1$ and $n = 2$ in Equation 5.3. Equal tempered f_0 values for notes are given in Figure 3.21. Thus A5 has an f_0 of 880.0 Hz and for B5, $f_0 = 987.8$ Hz.

The difference tone frequency = $987.8 - 880.0 = 107.8$ Hz; closest note is A2 ($f_0 = 110.0$ Hz).

The combination tones are:

for $n = 1$: $880.0 - 107.8 = 772.2$ Hz; closest note is G5 ($f_0 = 784.0$ Hz)
for $n = 2$: $880.0 - 215.6 = 664.4$ Hz; closest note is E5 ($f_0 = 659.3$ Hz)

These combination tones would beat with the f_0 component of any other instruments in an ensemble playing a note close to a combination tone. This will not be as marked as it might appear at first, due to an effect known as "masking," which is described in the next section.

5.5.2 Masking of one sound by another

When we listen to music, it is very rare that it consists of just a single pure tone. While it is possible and relatively simple to arrange to listen to a pure tone of a particular frequency in a laboratory, or by means of an electronic synthesizer (a useful, important and valuable experience), such a sound would not sustain any prolonged musical interest. Almost every sound we hear in music consists of at least two frequency components.

When two or more pure tones are heard together an effect known as "masking" can occur, where each individual tone can become more difficult or impossible to perceive, or it is partially or completely "masked," due to the presence of another tone. In such a case the tone which causes the masking is known as the "masker" and the tone which is masked is known as the "maskee." These tones could be individual pure tones, but, given the rarity of such sounds in music, they are more likely to be individual frequency components of a note played on one instrument which mask either other components in that note, or frequency components of another note.

The extent to which masking occurs depends on the frequencies of the masker and maskee and their amplitudes.

As is the case with most psychoacoustic investigations, masking is usually discussed in terms of the masking effect one pure tone can have on another, and the result is extended to complex sounds by considering the masking effect in relation to individual components. (This is similar, for example, to the approach adopted in the section on consonance and dissonance in Chapter 3, Section 3.3.2.) In psychoacoustic terms, the threshold of hearing of the maskee is shifted when in the presence of the masker. This provides the basis upon which masking can be quantified as the shift of a listener's threshold of hearing curve when a masker is present.

The dependence of masking on the frequencies of masker and maskee can be illustrated by reference to Figure 2.9 in which an idealized frequency response curve for an auditory filter is plotted. The filter will respond to components in the input acoustic signal which fall within its response curve, whose bandwidth is given by the critical bandwidth for the filter's center frequency. The filter will respond to components in the input whose frequencies are lower than its center frequency to a greater degree than components which are higher in frequency than the center frequency due to the asymmetry of the response curve.

Masking can be thought of as the filter's effectiveness in analyzing a component at its center frequency (maskee) being reduced to some degree by the presence of another component (masker) whose frequency falls within the filter's response curve. The degree to which the filter's effectiveness is reduced is usually measured as a shift in hearing threshold, or "masking level," as illustrated in Figure 5.9(a). The figure shows that the asymmetry of the response curve results in the masking effect being considerably greater for maskees which are above rather than those below the frequency of the masker. This effect is often referred to as:

- the upward spread of masking; or
- low masks high.

The dependence of masking on the amplitudes of masker and maskee is illustrated in Figure 5.9(b) in which idealized masking level curves

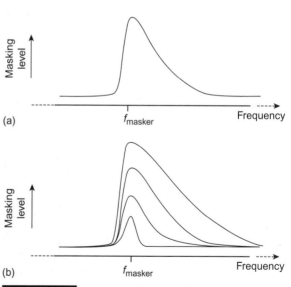

FIGURE 5.9 *(a) Idealized masking level to illustrate the "low masks high," or "upward spread of masking effect" for a masker of frequency f_{masker} Hz. (b) Idealized change in masking level with different levels of masker of frequency f_{masker} Hz.*

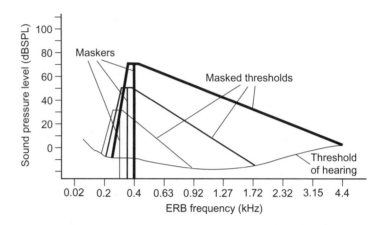

FIGURE 5.10

Idealized masked thresholds for masker pure tones at 300 Hz, 350 Hz and 400 Hz at 50 dBSPL, 70 dBSPL and 90 dBSPL respectively, plotted on a critical band spaced frequency scale. (From Sundberg, 1991.)

are plotted for different amplitude levels of a masker of frequency f_{masker}. At low amplitude levels, the masking effect tends to be similar for frequencies above and below f_{masker}. As the amplitude of the masker is raised, the low masks high effect increases and the resulting masking level curve becomes increasingly asymmetric. Thus the masking effect is highly dependent on the amplitude of the masker. This effect is illustrated in Figure 5.10, which is taken from Sundberg (1991). The frequency scale in this figure is plotted such that each critical bandwidth occupies the same distance. Sundberg summarizes this figure in terms of a three straight-line approximation to the threshold of hearing in the presence of the masker, or "masked threshold," as follows:

- the masked threshold above the critical band in which the masker falls off at about 5–13 dB per critical band;

- the masked threshold in the critical band in which the masker falls, approximately 20 dB below the level of the masker itself;

- the masked threshold below the critical band in which the masker falls off considerably more steeply than it does above the critical band in which the masker falls.

The masking effect of individual components in musical sounds, which are complex with many spectral components, can be determined in terms of the masking effect of individual components on other components in the sound. If a component is completely masked by another component in the sound, the masked component makes no contribution to the perceived nature of the sound itself and is therefore effectively ignored. If the masker is broadband noise, or "white noise," then components at all frequencies are masked in an essentially linear fashion (i.e., a 10 dB increase in the level of the noise increases the masking effect by 10 dB at all frequencies). This can be the case, for example, with background noise or a brushed snare drum (see Figure 3.6) which has spectral energy generally spread over a wide frequency range which can mask components of other sounds that fall within that frequency range.

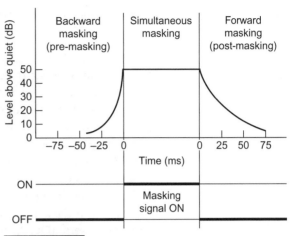

FIGURE 5.11 *Idealized illustration of simultaneous and non-simultaneous masking.*

The masking effects considered so far are known as "simultaneous masking" because the masking effect on the maskee by the masker occurs when both sound together (or simultaneously). Two further masking effects are important for the perception of music where the masker and maskee are not sounding together; these are referred to as "non-simultaneous masking." These are "forward masking" or "post-masking," and "backward masking" or pre-masking. In forward masking, a pure tone masker can mask another tone (maskee) which starts after the masker itself has ceased to sound. In other words the masking effect is "forward" in time from the masker to the maskee. Forward masking can occur for time intervals between the end of the masker and the start of the maskee of up to approximately 30 ms. In backward masking a maskee can be masked by a masker which follows it in time, starting up to approximately 10 ms after the maskee itself has ended. It should be noted, however, that considerable variation exists between listeners in terms of the time intervals over which forward and backward masking takes place.

Simultaneous and non-simultaneous masking are summarized in an idealized graphical format in Figure 5.11, which gives an indication of the masking effect in the time domain. The instant at which the masker starts and stops is indicated at the bottom of the figure, and it is assumed that the simultaneous masking effect is such that the threshold is raised by 50 dB. The potential spreading in time of masking as non-simultaneous pre- and post-masking effects is also shown. Moore (1996) makes the following observations about non-simultaneous masking:

- Backward masking is considerably lessened (to zero in some cases) with practice.
- Recovery rate from forward masking is greater at higher masking levels.
- The forward masking effect disappears 100–200 ms after the masker ceases.

- The forward masking effect increases for masker durations up to about 50 ms.

Masking is exploited practically in digital systems that store and transmit digital audio in order to reduce the amount of information that has to be handled, and therefore reduces the transmission resource, or bandwidth, and memory, hard disk or other storage medium required. Such systems are generally referred to as "perceptual coders" because they exploit knowledge of human perception. For example, perceptual coding is the operating basis of MP3 systems. It is also used to transmit music over the Internet, in MP3 players that store many hours of such music in a pocket-sized device, in multi-channel sound in digital audio broadcasting and in satellite television systems, MiniDisk recorders (Maes, 1996), and the now obsolete digital compact cassette (DCC). Audio coding systems are discussed in more detail in Chapter 7.

Demonstrations of masking effects are available on the CD recording of Houtsma *et al.* (1987).

5.5.3 Note grouping illusions

There are some situations when the perceived sound is unexpected, as a result of either what amounts to an acoustic illusion or the way in which the human hearing system analyzes sounds. While some of these sounds will not be found in traditional musical performances using acoustic instruments, since they can only be generated electronically, some of the effects have a bearing on how music is performed. The nature of the illusion and its relationship with the acoustic input which produced it can give rise to new theories of how sound is perceived, and, in some cases, the effect might have already or could in the future be used in the performance of music.

Diana Deutsch describes a number of note grouping acoustic illusions, some of which are summarized below with an indication of their manifestation in music perception and/or performance. Deutsch (1974) describes an "octave illusion" in which a sequence of two tones an octave apart with high (800 Hz) and low (400 Hz) f_0 values are alternated between the ears as illustrated in the upper part of Figure 5.12.

Most listeners report hearing a high tone in the right ear alternating with a low tone in the left ear as illustrated in the figure, no matter which way round the headphones are placed. She further notes that right-handed listeners tend to report hearing the high tone in the right ear alternating with a low tone in the left ear, while left-handed listeners tend to hear a

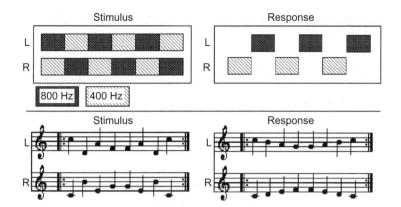

FIGURE 5.12

A schematic representation of the stimulus for, and most common response to, the "octave illusion" (upper) described by Deutsch (1974), and scale illusion (lower) described by Deutsch (1975).

high tone alternating with a low tone but it is equally likely that the high tone is heard in the left or right ear. This illusion persists when the stimuli are played over loudspeakers. This stimulus is available on the CD recording of Houtsma *et al.* (1987).

In a further experiment Deutsch (1975) played an ascending and descending C major scale simultaneously with alternate notes being switched between the two ears as shown in the lower part of Figure 5.12. The most commonly perceived response is also shown in the figure. Once again the high notes tend to be heard in the right ear and the low notes in the left ear, resulting in a snippet of a C major scale being heard in each ear. Such effects are known as "grouping" or "streaming" and by way of explanation, Deutsch invokes some of the grouping principles of the "Gestalt school" of psychology known as "good continuation," "proximity" and "similarity." She describes these as follows (Deutsch, 1982):

■ Grouping by good continuation: "Elements that follow each other in a given direction are perceived as blending together."
■ Grouping by proximity: "Nearer elements are grouped together in preference to elements that are spaced farther apart."
■ Grouping by similarity: "Like elements are grouped together."

In each case the "elements" referred to are the individual notes in these stimuli. Applying these principles to the stimuli shown in the figure, Deutsch suggests that the principle of proximity is important, grouping the higher tones (and lower tones) together, rather than good continuation, which would suggest that complete ascending and/or descending scales of C major would be perceived. Deutsch (1982) describes other experiments which support this view.

FIGURE 5.13

Bars 45 to 50 of the Preludio from Partita III in E major for solo violin by J. S. Bach showing the notes scored for the violin (upper stave) and the three parts normally perceived by streaming (lower three staves).

Music in which grouping of notes together by frequency proximity produces the sensation of a number of parts being played, even though only a single line of music is being performed, includes works for solo instruments such as the Partitas and Sonatas for solo violin by J. S. Bach. An example of this effect is shown in Figure 5.13 from the *Preludio* from Partita number III in E major for solo violin by J. S. Bach (track 66 on the accompanying CD). The score (upper stave) and three parts usually perceived (lower stave) are shown, where the perceived parts are grouped by frequency proximity.

The rather extraordinary string part writing in the final movement of Tchaikovsky's 6th symphony in the passage shown in Figure 5.14 is also often

FIGURE 5.14

Snippet of the final movement of Tchaikovsky's 6th symphony showing the notes scored for the strings and the four parts normally perceived.

noted in this context because it is generally perceived as the four-part passage shown (track 67 on the accompanying CD). This can again be explained by the principle of grouping by frequency proximity. The effect would have been heard in terms of stereo listening by audiences of the day, since the strings were then positioned in the following order (audience view from left to right): first violins, double basses, cellos, violas and second violins. This is as opposed to the more common arrangement today (audience view from left to right): first violins, second violins, violas, double basses and cellos.

Other illusions can be produced which are based on timbral proximity streaming. Pierce (1992) describes an experiment "described in 1978 by David L. Wessel" and illustrated in Figure 5.15. In this experiment the rising arpeggio shown as response (A) is perceived as expected for stimulus

FIGURE 5.15 *Stimulus and usually perceived responses for Wessel's timbral streaming experiment described by Pierce (1992). Different timbres in (B) are represented by the noteheads and plus signs.*

(A) when all the note timbres are the same. However, the response changes to two separate falling arpeggii, shown as response (B), if note timbres are alternated between two timbres represented by the different notehead shapes, and "the difference in timbres is increased" as shown for stimulus (B). This is described as timbral streaming (e.g., Bregman, 1990).

A variation on this effect is shown in Figure 5.16 in which the pattern of notes shown is produced with four different timbres represented by the different notehead shapes. (This forms the basis of one of our laboratory exercises for music technology students.) The score is repeated indefinitely and the speed can be varied. Ascending or descending scales are perceived depending on the speed at which this sequence is played. For slow speeds (less than one note per second) an ascending sequence of scales is perceived (stave B in the figure). The streaming is based on "note order." When the speed is increased, for example to greater than 10 notes per second, a descending sequence of scales of different timbres is perceived (staves C–F in the figure). The streaming is based on timbre. The ear can switch from one descending stream to another between those shown in staves (C–F) in the figure by concentrating on another timbre in the stimulus.

The finding that the majority of listeners to the stimuli shown in Figure 5.12 hear the high notes in the right ear and the low notes in the left ear may have some bearing on the natural layout of groups of performing musicians. For example, a string quartet will usually play with the cellist sitting on the left of the viola player who is sitting on the left of the second violinist, who in turn is sitting on the left of the first violinist as illustrated in Figure 5.17. This means that each player has the instruments playing parts lower than their own on their left-hand side, and those instruments playing higher parts on their right-hand side.

Vocal groups tend to organize themselves such that the sopranos are on the right of the altos, and the tenors are on the right of the basses if they

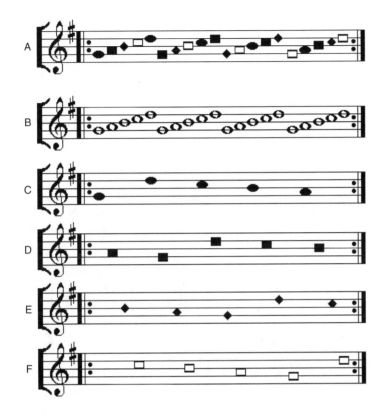

FIGURE 5.16

Stimuli (stave A) used in timbre and note-order streaming experiment in which notehead shapes represent different timbres. At low speeds, note-order streaming is perceived (stave B), and at higher speeds timbre streaming is perceived (staves C–F).

are in two or more rows. Small vocal groups such as a quartet consisting of a soprano, alto, tenor and bass will tend to be in a line with the bass on the left and the soprano on the right. In orchestras, the treble instruments tend to be placed with the highest pitched instruments within their section (first violin, piccolo, trumpet, etc.) on the left and bass instruments on the right. Such layouts have become traditional and moving players or singers around such that they are not in this physical position with respect to other instruments or singers is not welcomed. This tradition of musical performance layout may well be in part due to a right-ear preference for the higher notes.

However, while this may work well for the performers, it is back-to-front for the audience. When an audience faces a stage to watch a live performance (see Figure 5.17), the instruments or singers producing the treble notes are on the left and the bass instruments or singers are on the right. This is the wrong way round in terms of the right-ear treble preference, but the correct way round for observing the performers themselves. It is interesting to compare the normal concert hall layout as a listener with the experience of sitting in the audience area behind the orchestra, which is

FIGURE 5.17

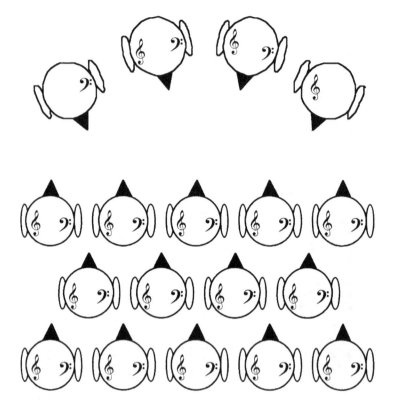

Traditional performer and audience layout in a concert situation showing treble/bass bias in the ears of performers and listeners.

possible in halls such as the Royal Festival Hall in London. Unfortunately this is not a test that can be carried out very satisfactorily since it is not usually possible to sit far enough behind the players to gain as good an overall balance as can be obtained from the auditorium in front of the orchestra. It is, however, possible to experience this effect by turning round when listening to a good stereo recording over loudspeakers or by reversing the left and right channels.

5.5.4 Pitch illusions

A pitch illusion, which has been compared with the continuous staircase pictures of Maurits Escher, has been demonstrated by Shepherd (1964) and is often referred to as a "Shepherd tone." This illusion produces the sensation of an endless scale which ascends in semitone steps. After 12 semitone steps when the pitch has risen by one octave, the spectrum is identical to the starting spectrum so the scale ascends but never climbs acoustically more than one octave. This stimulus is available on the CD recording of

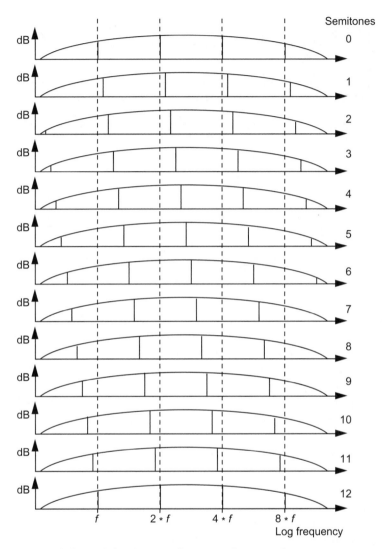

FIGURE 5.18

Illustration of the spectra of stimuli which would produce the Shepherd continuous ascending scale illusion.

Houtsma *et al.* (1987) (and as tracks 68 and 69 on the accompanying CD). Figure 5.18 illustrates the spectral nature of the Shepherd tone stimuli. Only the fundamental and harmonics which are multiple octaves above the fundamental are employed in the stimuli. The component frequencies of the Shepherd tone can be represented as:

$$f_{(\text{Shepherd})} = (2_n f_0) \qquad (5.4)$$

where $f_{(\text{Shepherd})}$ = frequencies of Shepherd tone components
and $n = (0, 1, 2, 3, ...)$

FIGURE 5.19

An extract from bar 31 of the Fantasia in G minor (BWV 542) for organ by J. S. Bach.

The amplitudes of the components are constrained within the curved envelope shown. Each time the tone moves up one semitone, the partials all move up by a twelfth of an octave, or one semitone. The upper harmonics become weaker and eventually disappear and new lower harmonics appear and become stronger.

A musical example relating to the Shepherd tone effect in which some pitch ambiguity is perceived by some listeners can be found in the pedal line starting at bar 31 of the *Fantasia in G minor* (BWV 542) for organ by J. S. Bach (consider tracks 70–72 on the accompanying CD). These bars are reproduced in Figure 5.19 as an organ score in which the lower of the three staves is played on the pedals while the upper two staves are played with the left and right hands.

The pedal line consists of a sequence of five descending scales with eight notes in each except the last. Each scale ends with an upward leap of a minor seventh and the exact moment where the upward leap occurs is often perceived with some ambiguity, even when listeners have the score in front of them. The strength of this effect depends on the particular stops used. This ambiguity is particularly common amongst listeners in the third bar of the extract where the upward leap is often very strongly perceived as occurring one or even two notes later. This could be due to the entry of a new part in the left hand playing F3 which starts as the pedal part jumps up to written Bb3. (Reference to "written" Bb3 is made since the 16′ rank provides the fundamental on the pedals which sounds an octave lower than written pitch as discussed in Section 5.4.)

The f_0 components of these two notes, i.e., F3 and B♭2, form the second and third harmonics of B♭1, which would have been the next sounding note of the pedal descending scale had it not jumped up an octave. At all the upward leaps in the descending pedal scales the chord in the manual part changes from minor to major.

In bar 32 the left hand change from E♭ to E natural adds a member of the harmonic series (see Figure 3.3) of what would have been the next note (written C3) in the pedal scale had it not risen up the octave. E♭ is not a member of that harmonic series. The case is similar for the D natural in the right hand in the third bar of the extract with the entry of the left hand F3, and the left hand C natural in the fourth bar. These entries of notes which are members of the harmonic series of what would have been the next note in the descending pedal scale had it not jumped up the octave serve to provide the perceived ambiguity in definition as to the exact instant at which the upward leap occurs.

The illusion produced by the combination of organ pipes to produce a sensation of pitch lower than any note actually sounding is also used in organ pedal resultant bass stops. These sound at 32′ (and very occasionally 64′), and their f_0 values for bottom C are 16.25 Hz and 8.175 Hz respectively. A resultant bass at 32′ pitch is formed by sounding together stops of 16′ and $10\frac{2}{3}$′ which form the second and third harmonics of the 32′ harmonic series (see Section 5.4). A 32′ stop perhaps labeled "acoustic bass" is a mutation stop of $10\frac{2}{3}$′, which when sounded with a 16′ rank produces a perceived pitch at 32′ (place theory of pitch perception from the second and third harmonics—see Chapter 3). A 64′ stop perhaps labeled "resultant bass" works similarly, sounding a $22\frac{1}{3}$′ rank with a 32′ rank.

The f_0 value of the middle C of a 32″ stop (C2) is 65.4 Hz and thus its bottom note is two octaves below this (C0) with an f_0 of $(65\frac{4}{4})$ or 16.35 Hz. The f_0 for the bottom note of a 64″ stop (C–1) is 8.175 Hz, which is below the human hearing range but within the frequency range of difference frequencies that are perceived as beats (see Figure 2.6). Harmonics that are within the human hearing range will contribute to a perception of pitch at these f_0 values which are themselves below the frequency range of the hearing system. Organists will sometimes play fifths in the pedals to imitate this effect, particularly on the last note of a piece. However, the effect is not as satisfactory as that obtained with a properly voiced resultant bass stop because the third harmonic (e.g., $10\frac{2}{3}$′) should be softer than the second harmonic (16′) for best effect.

Roederer (1975) describes an organ-based example to illustrate residue pitch, which constitutes a pitch illusion that is available as track 74 on the accompanying CD, while track 73 allows the chorale to be heard normally. The solo line of a chorale prelude, he suggests chorale number 40 from the

Orgelbuchlein by J. S. Bach, is played using a number of mutation stops (see Section 5.5) if available (e.g., 8′, 4′, 2⅔′, 2′, 1⅗′, 1⅓, 1) accompanied by 8′, 4′ in the left hand and 16′, 8′ in the pedal.

A musically trained audience should be asked to track the pitch of the melody and warned that timbre changes will occur. After playing a short snippet, play some more without the 8′, then without the 4′, then without the 2′, and finally without the 1′. What remains in the solo part is only mutation stops (i.e., those with a non-unison or non-octave pitch relationship to the fundamental). Roederer suggests making:

> *the audience aware of what was left in the upper voice and point out that the pitch of the written note was absent altogether (in any of its octaves)—they will find it hard to believe! A repetition of the experiment is likely to fail—because the audience will redirect their pitch processing strategies!*

Experience shows that such an experiment relies on pitch context being established when using such stimuli, usually through the use of a known or continuing musical melody.

A musical illusion only works by virtue of establishing a strong expectation in the mind's ear of the listener.

REFERENCES

ANSI, , 1960. American Standard Acoustical Terminology. American National Standards Institute, New York.

Bregman, A.S., 1990. Auditory Scene Analysis. MIT Press, Cambridge.

Brookes, T., Tyrrell, A.M., Howard, D.M., 2000. On the differences between conventional and auditory spectrograms of English consonants. Logoped. Phoniatr. Vocol. 25, 72–78.

Deutsch, D., 1974. An auditory illusion. Nature 251, 307–309.

Deutsch, D., 1975. Musical illusions. Sci. Am. 233, 92–104.

Deutsch, D., 1982. Grouping mechanisms in music. In: Deutsch, J. (ed.), The Psychology of Music. Academic Press, London, pp. 671–678.

Grey, J., 1977. Timbre discrimination in musical patterns. J. Acoust. Soc. Am. 64, 467–472.

Hall, D.E., 1991. Musical Acoustics: An introduction, second edn. Wadsworth Publishing Company, Belmont, CA.

Helmholtz, H.L.F. von., 1877. On the Sensations of Tone as a Physiological Basis for the Theory of Music, fourth edn. Dover, New York [trans. Ellis, A. J. (1954)].

Houtsma, A.J.M., Rossing, T.D., Wagenaars, W.M., 1987. Auditory Demonstrations. Acoustical Society of America, New York (Philips compact disc No.1126-061 and text).

Howard, D.M., 2005. Human hearing modeling real-time spectrography for visual feedback in singing training. Folia Phoniatr. Logo. 57 (5–6), 328–341.

Howard, D.M., Tyrrell, A.M., 1997. Psychoacoustically informed spectrography and timbre. Organ. Sound 2 (2), 65–76.

Howard, D.M., Hirson, A., Brookes, T., Tyrrell, A.M., 1995. Spectrography of disputed speech samples by peripheral human hearing modeling. Forensic Linguist. 2 (1), 28–38.

Howard, D.M., Disley, A., and Hunt, A.D. (2007). Timbral adjectives for the control of a music synthesizer. In: Proceedings of the 19th International Congress on Acoustics, ICA-07, paper MUS-07-004, 2–7 September, Madrid, Spain.

Katz, R., 2007. Mastering Audio: The Art and The Science, second edn. Focal Press, Oxford.

Maes, J., 1996. The MiniDisk. Focal Press, Oxford.

Moore, B.C.J., 1996. Masking and the human auditory system. In: Gilchrist, N., Grewin, C. (Eds), Collected Papers on Digital Audio Bit-rate Reduction. The Audio Engineering Society, pp. 9–22.

Padgham, C.A., 1986. The Well-Tempered Organ. Positif Press, Oxford.

Pierce, J.R., 1992. The Science of Musical Sound (Scientific American Books), second edn. W. H. Freeman and Company, New York.

Pollard, H.F., Jansson, E.V., 1982. A tristimulus method for the specification of musical timbre. Acustica 51, 162–171.

Rasch, R.A., Plomp, R., 1982. The perception of musical tones. In: Deutsch, D. (Ed.), The Psychology of Music. Academic Press, London, pp. 89–112.

Roederer, J.G., 1975. Introduction to the Physics and Psychophysics of Music. Springer-Verlag, New York.

Scholes, P.A., 1970. The Oxford Companion to Music. Oxford University Press, London.

Shepherd, R.N., 1964. Circularity in pitch judgement. J. Acoust. Soc. Am. 36, 2346–2353.

Sumner, W.L., 1975. The Organ. MacDonald and Company, London.

Sundberg, J., 1991. The Science of Musical Sounds. Academic Press, San Diego.

Hearing Music in Different Environments

In this chapter we will examine the behavior of the sound in a room with particular reference to how the room's characteristics affect the quality of the perceived sound. We will also examine strategies for analyzing and improving the acoustic quality of a room. Finally we will look at how rooms affect the output of loudspeakers.

6.1 ACOUSTICS OF ENCLOSED SPACES

In Chapter 1 the concept of a wave propagating without considering any boundaries was discussed. However, most music is listened to within a room, and is therefore influenced by the presence of boundaries, and so it is important to understand how sound propagates in such an enclosed space. Figure 6.1 shows an idealized room with a starting pistol and a listener; assume that at some time $(t = 0)$ the gun is fired. There are three main aspects to how the sound of a gun behaves in the room, which are as follows.

6.1.1 The direct sound

After a short delay the listener in the space will hear the sound of the starting pistol, which will have traveled the shortest distance between it and the listener. The delay will be a function of the distance, as sound travels 344 meters (1129 feet) per second or approximately 1 foot per millisecond. The shortest path between the starting pistol and the listener is the direct path, and therefore this is the first thing the listener hears. This component of the sound is called the direct sound, and its propagation path and associated time response are shown in Figure 6.2.

The direct component is important because it carries the information in the signal in an uncontaminated form. Therefore a high level of direct sound is required for a clear sound and good intelligibility of speech. The direct sound also behaves in the same way as sound in free space, because it has not yet interacted with any boundaries. This means that we can use the equation for the intensity of a free space wave some distance from the source

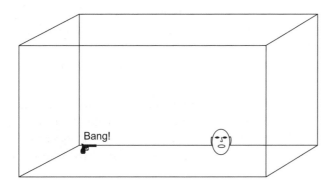

Bang!

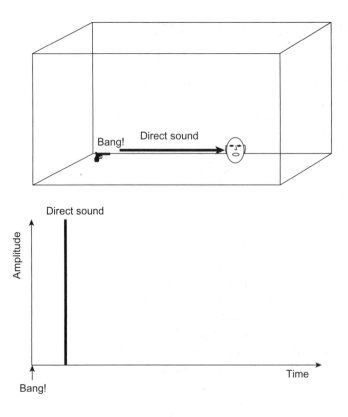

FIGURE 6.2

The direct sound in a room.

to calculate the intensity of the direct sound. The intensity of the direct sound is therefore given, from Chapter 1, by:

$$I_{\text{direct sound}} = \frac{QW_{\text{source}}}{4\pi r^2} \tag{6.1}$$

where $I_{\text{direct sound}}$ = the sound intensity (in W m^{-2})
Q = the directivity of the source (compared to a sphere)
W_{Source} = the power of the source (in W)
and r = the distance from the source (in m)

Equation 6.1 shows that the intensity of the direct sound reduces as the square of the distance from the source, in the same way as a sound in free space. This has important consequences for listening to sound in real spaces. Let us calculate the sound intensity of the direct sound from a loudspeaker.

EXAMPLE 6.1

A loudspeaker radiates a sound intensity level of 102 dB at 1 m. What is the sound intensity level (I_{direct}) of the direct sound at a distance of 4 m from the loudspeaker?

The sound intensity of the direct sound at a given distance can be calculated, using Equation 1.18 from Chapter 1, as:

$$IL = 10 \log_{10}\left(\frac{W_{source}}{W_{ref}}\right) - 20 \log_{10}(r) - 11 dB$$

As we already know the intensity level at 1 m this equation becomes:

$$I_{direct\ sound} = I_{1m} - 20 \log_{10}(r)$$

which can be used to calculate the direct sound intensity as:

$$I_{direct\ sound} = 102\ dB - 20 \log_{10}(4) = 102\ dB - 12 dB = 90 dB$$

Example 6.1 shows that the effect of distance on the direct sound intensity can be quite severe.

6.1.2 Early reflections

A little time later the listener will then hear sounds which have been reflected off one or more surfaces (walls, floor, etc.), as shown in Figure 6.3. These sounds are called early reflections and they are separated in both time and direction from the direct sound. These sounds will vary as the source or the listener moves within the space. We use these changes to give us information about both the size of the space and the position of the source in the space. If any of these reflections are very delayed, i.e., total path length difference longer than about 30 milliseconds (33 feet), then they will be perceived as echoes. Early reflections can cause interference effects, as discussed in Chapter 1, and these can both reduce the intelligibility of speech and cause unwanted timbre changes in music in the space.

The intensity levels of the early reflections are affected by both the distance and the surface from which they are reflected. In general most surfaces absorb some of the sound energy and so the reflection is weakened by the absorption. However, it is possible to have surfaces which "focus" the sound, as shown in Figure 6.4, and in these circumstances the intensity level at the listener will be enhanced.

FIGURE 6.3

FIGURE 6.3

The early reflections in a room.

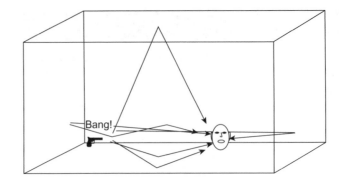

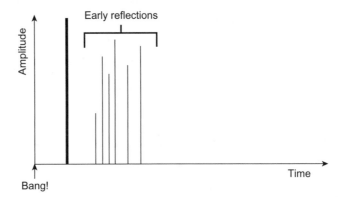

FIGURE 6.4

A focusing surface.

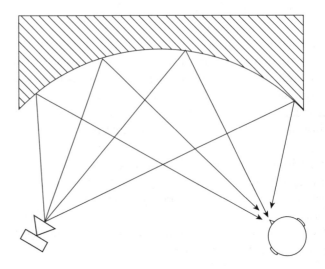

It is important to note, however, that the total power in the sound will have been reduced by the interaction with the surface. This means that there will be less sound intensity at other positions in the room. Also any focusing structure must be large when measured with respect to the sound wavelength, which tends to mean that these effects are more likely to happen for high-, rather than low-frequency components. In general therefore the level of direct reflections will be less than that which would be predicted by the inverse square law, due to surface absorption. Let us calculate the amplitude of an early reflection from a loudspeaker.

EXAMPLE 6.2

A loudspeaker radiates a peak sound intensity of 102 dB at 1 m. What is the sound intensity level ($I_{\text{reflection}}$), and delay relative to the direct sound, of an early reflection when the speaker is 1.5 m away from a hard reflecting wall and the listener is at a distance of 4 m in front of the loudspeaker?

The geometry of this arrangement is shown in Figure 6.5 and we can calculate the extra path length due to the reflection by considering the "image" of the loudspeaker, also shown in Figure 6.6, and by using Pythagoras' theorem. This gives the path length as 5 m.

Given the intensity level at 1 m, the intensity of the early reflection can be calculated because the reflected wave will also suffer from an inverse square law reduction in amplitude:

$$I_{\text{early reflection}} = I_{1m} - 20 \log_{10} (\text{Path length}) \qquad (6.2)$$

which can be used to calculate the direct sound intensity as:

$$I_{\text{early reflection}} = 102\,\text{dB} - 20 \log_{10} (5) = 102\,\text{dB} - 14\,\text{dB} = 88\,\text{dB}$$

Comparing this with the earlier example we can see that the early reflection is 2 dB lower in intensity compared with the direct sound. The delay is simply calculated from the path length as:

$$\text{Delay}_{\text{early reflection}} = \frac{\text{Path length}}{344\ \text{ms}^{-1}} = \frac{5\ \text{m}}{344\ \text{ms}^{-1}} = 14.5\ \text{ms}$$

Similarly the delay of the direct sound is:

$$\text{Delay}_{\text{direct}} = \frac{r}{344\ \text{ms}^{-1}} = \frac{4\ \text{m}}{344\ \text{ms}^{-1}} = 11.6\ \text{ms}$$

So the early reflection arrives at the listener 14.5 ms − 11.6 ms = 2.9 ms after the direct sound.

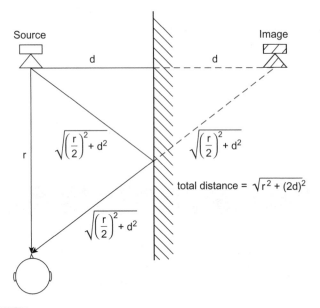

FIGURE 6.5 *A geometry for calculating the intensity of an early reflection.*

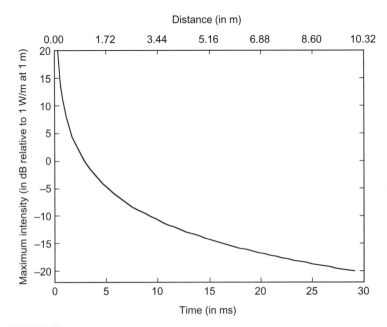

FIGURE 6.6 *The maximum bounds for early reflections assuming no absorption or focusing.*

Because there is a direct correspondence between delay, distance from the source, and the reduction in intensity due to the inverse square law, we can plot all this on a common graph (see Figure 6.6), which shows the maximum bounds of the intensity level of reflections, provided there are no focusing effects.

6.1.3 The effect of absorption on early reflections

How does the absorption of sound affect the level of early reflections heard by the listener? The absorption coefficient of a material defines the amount of energy, or power, that is removed from the sound when it strikes it. In general the absorption coefficient of real materials will vary with frequency, but for the moment we shall assume they do not. The amount of energy, or power, removed by a given area of absorbing material will depend on the energy, or power, per unit area striking it. As the sound intensity is a measure of the power per unit area this means that the intensity of the sound reflected is reduced in proportion to the absorption coefficient. That is:

$$\text{Intensity}_{\text{reflected}} = \text{Intensity}_{\text{incident}} \times (1 - \alpha) \qquad (6.3)$$

where $\text{Intensity}_{\text{reflected}}$ = the sound intensity reflected after absorption (in W m^{-2})
$\text{Intensity}_{\text{incident}}$ = the sound intensity before absorption (in W m^{-2})
and α = the absorption coefficient

Because a multiplication of sound levels is equivalent to adding the decibels together, as shown in Chapter 1, Equation 6.3 can be expressed directly in terms of the decibels as:

$$I_{\text{absorbed}} = I_{\text{incident}} + 10 \, \log(1 - \alpha) \qquad (6.4)$$

which can be combined with Equation 6.2 to give a means of calculating the intensity of an early reflection from an absorbing surface:

$$I_{\text{early reflection}} = I_{1 \, \text{m}} - 20 \, \log_{10}(\text{Path length}) + 10 \, \log(1 - \alpha) \qquad (6.5)$$

As an example consider the effect of an absorbing surface on the level of the early reflection level calculated earlier.

EXAMPLE 6.3

A loudspeaker radiates a peak sound intensity of 102 dB at 1 m. What is the sound intensity level ($I_{\text{early reflection}}$) of an early reflection, when the speaker is 1.5 m away from a reflecting wall and the listener is at a distance of 4 m in front of the loudspeaker, and the wall has an absorption of 0.9, 0.69, 0.5?

As we already know the intensity level at 1 m, the intensity of the early reflection can be calculated using Equation 6.5 because the reflected wave also suffers from an inverse square law reduction in amplitude:

$$I_{\text{early reflection}} = I_{1\,m} - 20\ \log_{10}(\text{Path length}) + 10\ \log(1 - \alpha)$$

The path length, from the earlier calculation, is 5 m so the sound intensity at the listener for the three different absorption coefficients is:

$$I_{\text{early reflection }(\alpha=0.9)} = 102\,dB - 20\ \log_{10}(5m) + 10\ \log(1 - 0.9)$$
$$= 102\,dB - 14\,dB - 10\,dB = 78\,dB$$
$$I_{\text{early reflection }(\alpha=0.69)} = 88\,dB + 10\ \log(1 - 0.69) = 88\,dB - 5\,dB = 83\,dB$$
$$I_{\text{early reflection }(\alpha=0.5)} = 88\,dB + 10\ \log(1 - 0.5) = 88\,dB - 3\,dB = 85\,dB$$

6.1.4 The reverberant sound

At an even later time the sound has been reflected many times and is arriving at the listener from all directions, as shown in Figure 6.7. Because there

FIGURE 6.7

The reverberant sound in a room.

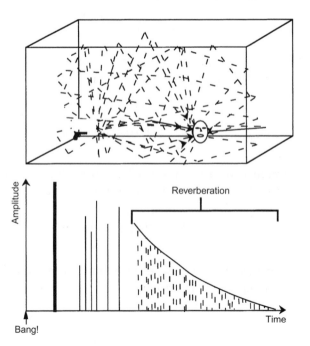

are so many possible reflection paths, each individual reflection is very close in time to its neighbors and thus there is a dense set of reflections arriving at the listener. This part of the sound is called reverberation and is desirable as it adds richness to, and supports, musical sounds. Reverberation also helps integrate all the sounds from an instrument so that a listener hears a sound which incorporates all the instruments' sounds, including the directional parts. In fact we find rooms which have very little reverberation uncomfortable and generally do not like performing music in them; it is much more fun to sing in the bathroom compared with the living room (consider tracks 75–77 on the accompanying CD).

The time taken for reverberation to occur is a function of the size of the room and will be shorter for smaller rooms, due to the shorter time between reflections and the losses incurred on each impact with a surface. In fact the time gap between the direct sound and reverberation is an important cue to the size of the space that the music is being performed in. Because some of the sound is absorbed at each reflection it dies away eventually. The time that it takes for the sound to die away is called the reverberation time and is dependent on both the size of the space and the amount of sound absorbed at each reflection. In fact there are three aspects of the reverberant field that the space affects (see Figure 6.8).

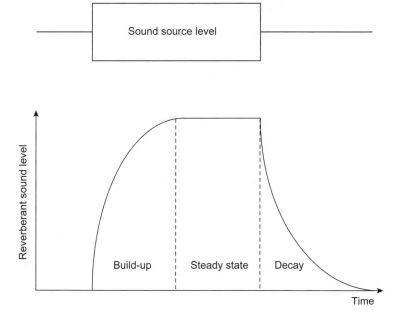

FIGURE 6.8

The time and amplitude evolution of the reverberant sound in a room.

- *The increase of the reverberant field level*: This is the initial portion of the reverberant field and is affected by the room size, which affects the time between reflections and therefore the time it takes the reverberant field to build-up. The amount of absorption in the room also affects the time that it takes the sound to get to its steady-state level. This is because, as shall be shown later, the steady-state level is inversely proportional to the amount of absorption in the room. As the rate at which sound builds up depends on the time between reflections and the absorption, the reverberant sound level will take more time to reach a louder level than reach a smaller one.

- *The steady-state level of the reverberant field*: If a steady tone, such as an organ note, is played in the space then after a period of time the reverberant sound will reach a constant level because at that point the sound power input balances the power lost by absorption in the space. This means that the steady-state level will be higher in rooms that have a small amount of absorption, compared with rooms that have a lot of absorption. Note that a transient sound in the space will not reach a steady-state level.

- *The decay of the reverberant field level*: When a tone in the space stops, or after a transient, the reverberant sound level will not reduce immediately but will instead decay at a rate determined by the amount of sound energy that is absorbed at each reflection. Thus in spaces with a small amount of absorption the reverberant field will take longer to decay.

Bigger spaces tend to have longer reverberation times and well-furnished spaces tend to have shorter reverberation times. Reverberation time can vary from about 0.2 of a second for a small well-furnished living room to about 10 seconds for a large glass and stone cathedral.

6.1.5 The behavior of the reverberant sound field

The reverberation part of the sound in a room behaves differently compared with the direct sound and early reflections from the perspective of the listener. The direct sound and early reflections follow the inverse square law, with the addition of absorption effects in the case of early reflections, and so their amplitude varies with position. However, the reverberant part of the sound largely remains constant with the position of the listener in the room. This is not due to the sound waves behaving differently from normal waves; instead it is due to the fact that the reverberant sound waves arrive at the listener from all directions.

The result is that at any point in the room there are a large number of sound waves whose intensities are being added together. These sound waves have many different arrival times, directions and amplitudes because the sound waves are reflected back into the room, and so shuttle forward, backward and sideways around the room as they decay. The steady-state sound level, at a given point in the room, therefore is an integrated sum of all the sound intensities in the reverberant part of the sound, as shown in Figure 6.9. Because of this behavior the reverberant part of the sound in a room is often referred to as the "reverberant field."

6.1.6 The balance of reverberant to direct sound

This behavior of the reverberant field has two consequences. Firstly, the balance between the direct and reverberant sounds will alter depending on the position of the listener relative to the source. This is due to the fact that the level of the reverberant field is independent of the position of the

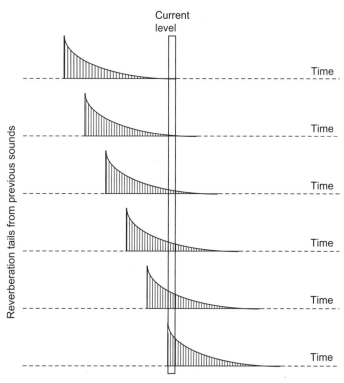

FIGURE 6.9

The source of the steady-state sound level of the reverberant field.

Current reverberation level is the sum of the previous reverberation tails listener with respect to the source, whereas the direct sound level is dependent on the distance between the listener and the sound source. These effects are summarized in Figure 6.10, which shows the relative levels of direct to reverberant field as a function of distance from the source. This figure shows that there is a distance from the source at which the reverberant field will begin to dominate the direct field from the source. The transition occurs when the two are equal and this point is known as the "critical distance."

6.1.7 The level of the reverberant sound in the steady state

Secondly, because in the steady state the reverberant sound at any time instant is the sum of all the energy in the reverberation tail, the overall sound level is increased by reverberation. The level of the reverberation will depend on how fast the sound is absorbed in the room. A low level of absorption will result in sound that stays around in the room for longer and so will give a higher level of reverberant field. In fact, if the average level of absorption coefficient for the room is given by α, the power level in the reverberant sound in a room can be calculated using the following equation:

$$W_{\text{reverberant}} = W_{\text{Source}} 4 \left(\frac{1 - \alpha}{S\alpha} \right) \tag{6.6}$$

FIGURE 6.10

The composite effect of direct sound and reverberant field on the sound intensity, as a function of the distance from the source.

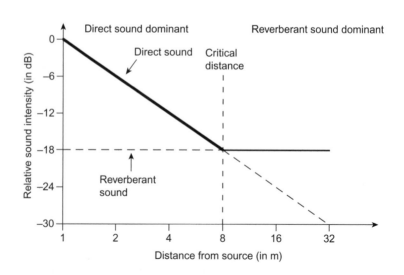

where $W_{\text{reverberant}}$ = the reverberant sound power (in W)

S = the total surface area in the room (in m^2)

W_{Source} = the power of the source (in W)

and α = the average absorption coefficient in the room

Equation 6.6 is based on the fact that, at equilibrium, the rate of energy removal from the room will equal the energy put into its reverberant sound field. As the sound is absorbed when it hits the surface, it is absorbed at a rate which is proportional to the surface area times the average absorption, or $S\alpha$. This is similar to a leaky bucket being filled with water where the ultimate water level will be that at which the water runs out at the same rate as it flows in (see Figure 6.11.)

The amount of sound energy available for contribution to the reverberant field is also a function of the absorption because if there is a large amount of absorption then there will be less direct sound reflected off a

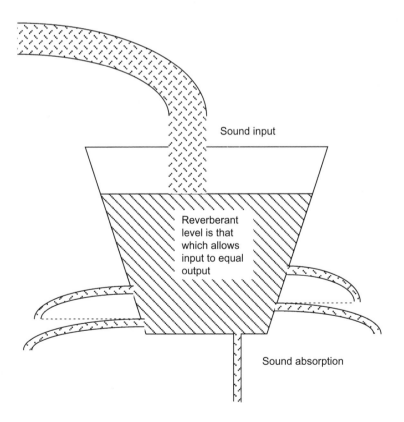

Sound input

FIGURE 6.11

The leaky bucket model of reverberant field intensity level.

Reverberant level is that which allows input to equal output

Sound absorption

surface to contribute to the reverberant field—remember that before the first reflection the sound is direct sound. The amount of sound energy available to contribute to the reverberant field is therefore proportional to the residual energy left after the first reflection, or $(1 - \alpha)$ because α is absorbed at the first surface. The combination of these two effects gives $(1 - \alpha)/S\alpha$— the term in Equation 6.6. The factor of four in Equation 6.6 arises from the fact that sound is approaching the surfaces in the room from all possible directions.

An interesting result from Equation 6.6 is that it appears that the level of the reverberant field depends only on the total absorbing surface area. In other words it is independent of the volume of the room. However, in practice the surface area and volume are related because one encloses the other. In fact, because the surface area in a room becomes less as its volume decreases, the reverberant sound level becomes higher for a given average absorption coefficient in smaller rooms although the reverberation time in the smaller room will always be shorter. Another way of visualizing this is to realize that in a smaller room there is less volume for a given amount of sound energy to spread out in, like a pat of butter on a smaller piece of toast. Therefore the energy density, and thus the sound level, must be higher in smaller rooms. However, there are more impacts per second with the surface of a smaller room, which gives rise to the more rapid decay than in a larger room.

The term $(1 - \alpha)/S\alpha$ in Equation 6.6 is often inverted to give a quantity known as the room constant, R, which is given by:

$$R = \frac{S\alpha}{(1 - \alpha)} \qquad (6.7)$$

where R = the room constant (in m^2)
and α = the average absorption coefficient in the room

Using the room constant Equation 6.6 simply becomes:

$$W_{\text{reverberant}} = W_{\text{Source}}\left(\frac{4}{R}\right) \qquad (6.8)$$

In terms of the sound power level this can be expressed as:

$$SWL_{\text{reverberant}} = 10 \, \log_{10}\left(\frac{W_{\text{Source}}}{W_{\text{ref}}}\right) + 10 \, \log_{10}\left(\frac{4}{R}\right) \qquad (6.9)$$

As α is a number between 0 and 1 this also means that the level of the reverberant field will be greater in a room with a small surface area, compared with a larger room, for a given level of absorption coefficient. However, one must be careful in taking this result to extremes. A long and very thin cylinder will have a large surface area, but Equation 6.6 will not predict the reverberation level correctly because in this case the sound will not visit all the surfaces with equal probability. This will have the effect of modifying the average absorption coefficient and so will alter the prediction of Equation 6.6.

Therefore one must take note of an important assumption behind Equation 6.6 which is that *the reverberant sound visits all surfaces with equal probability and from all possible directions*. This is known as the "diffuse field assumption." It can also be looked at as a definition of a diffuse field. In general the assumption of a diffuse field is reasonable and it is usually a design goal for most acoustics. However, it is important to recognize that there are situations in which it breaks down, for example at low frequencies.

As an example consider the effect of different levels of absorption and surface area on the level of the reverberant field that might arise from the loudspeaker described earlier.

EXAMPLE 6.4

A loudspeaker radiates a peak sound intensity of 102 dB at 1 m. What is the sound pressure level of the reverberant field if the surface area of the room is 75 m², and the average absorption coefficient is (a) 0.9 and (b) 0.2? What would be the effect of doubling the surface area in the room while keeping the average absorption the same?

From Equation 1.18 we can say:

$$SIL = 10 \log_{10}\left(\frac{W_{Source}}{W_{ref}}\right) - 20 \log_{10}(r) - 11dB$$

Thus the sound power level (SWL) radiated by the loudspeaker is:

$$SWL = 10 \log_{10}\left(\frac{W_{Source}}{W_{ref}}\right) = SIL + 11dB = 102dB + 11dB = 113dB$$

The power in the reverberant field is given by:

$$SWL_{reverberant} = 10 \log_{10}\left(\frac{W_{Source}}{W_{ref}}\right) + 10 \log_{10}\left(\frac{4}{R}\right)$$

The room constant "R" for the two cases is:

$$R_{(\alpha=0.9)} = \frac{S\alpha}{(1-\alpha)} = \frac{75 \text{ m}^2 \times 0.9}{(1-0.9)} = 675 \text{ m}^2$$

$$R_{(\alpha=0.2)} = \frac{S\alpha}{(1-\alpha)} = \frac{75 \text{ m}^2 \times 0.2}{(1-0.2)} = 18.75 \text{ m}^2$$

The level of the reverberant field can therefore be calculated from:

$$SWL_{\text{reverberant}} = 10 \, \log_{10}\left(\frac{W_{\text{Source}}}{W_{\text{ref}}}\right) + 10 \, \log_{10}\left(\frac{4}{R}\right)$$

$$= 113 \text{dB} + 10 \, \log_{10}\left(\frac{4}{R}\right)$$

which gives:

$$SWL_{\text{reverberant} \, (\alpha=0.9)} = 113 \text{dB} + 10 \, \log_{10}\left(\frac{4}{675}\right)$$

$$= 113 \text{dB} - 22.3 \text{dB} = 90.7 \text{dB}$$

and:

$$SWL_{\text{reverberant} \, (\alpha=0.2)} = 113 \text{dB} + 10 \, \log_{10}\left(\frac{4}{18.75}\right)$$

$$= 113 \text{dB} - 6.7 \text{dB} = 106 \text{dB}$$

The effect of doubling the surface area is to increase the room constant by the same proportion, so we can say that:

$$SWL_{\text{reverberant (S doubled)}} = 10 \, \log_{10}\left(\frac{W_{\text{Source}}}{W_{\text{ref}}}\right) + 10 \, \log_{10}\left(\frac{4}{2R}\right)$$

$$= 113 \text{dB} + 10 \, \log_{10}\left(\frac{4}{R}\right) + 10 \, \log_{10}\left(\frac{1}{2}\right)$$

which gives:

$$SWL_{\text{reverberant (S doubled)}} = 113 \text{dB} + 10 \, \log_{10}\left(\frac{4}{R}\right) - 3 \text{dB}$$

Thus the effect of doubling the surface area is to reduce the level of the reverberant field by 3 dB in both cases.

Clearly the level of the reverberant field is strongly affected by the level of average absorption. The first example would be typical of an extremely "dead" acoustic environment, as found in some studios, whereas the second is typical of an average living room. The amount of loudspeaker energy required to produce a given volume in the room is clearly much greater, about 15 dB, in the first room compared with the second. If there is a musician in the room then they will experience a "lift" in output due to the reverberant field in the room. Because of this musicians feel uncomfortable playing in rooms with a low level of reverberant field and prefer performing in rooms which help them in producing more output. This is also one of the reasons we prefer singing in the bathroom. However, where the quality control of a recording is the goal, many recording engineers need much shorter room decay times because room reverberation can mask low-level detail.

6.1.8 Calculating the critical distance

The reverberant field is, in most cases, diffuse, and therefore visits all parts of the room with equal probability. Also at any point, and at any instant, we hear the total power in the reverberant field, as discussed earlier. Because of this it is possible to equate the power in the reverberant field to the sound pressure level. Thus we can say:

$$SPL_{reverberant} \approx SWVL_{reverberant} = 10 \ \log_{10}\left(\frac{W_{Source}}{W_{ref}}\right) + 10 \ \log_{10}\left(\frac{4}{R}\right) \quad (6.10)$$

The distance at which the reverberant level equals the direct sound—the critical distance—can also be calculated using the above equations. At the critical distance the intensity due to the direct field and the power in the reverberant field at a given point are equal so we can equate Equation 6.1 and Equation 6.8 to give:

$$\frac{QW_{source}}{4\pi r^2_{critical \ distance}} = W_{Source}\left(\frac{4}{R}\right)$$

Which can be rearranged to give:

$$r^2_{critical \ distance} = \left(\frac{R}{4}\right)\frac{Q}{4\pi}$$

Thus the critical distance is given by:

$$r_{critical \ distance} = \sqrt{\left(\frac{1}{16\pi}\right)}\sqrt{RQ} = 0.141\sqrt{RQ} \quad (6.11)$$

Equation 6.11 shows that the critical distance is determined only by the room constant and the directivity of the sound source. Because the room constant is a function of the surface area of the room, the critical distance will tend to increase with larger rooms. However, many of us listen to music in our living rooms so let us calculate the critical distance for a hi-fi loudspeaker in a living room.

EXAMPLE 6.5

What is the critical distance for a free-standing, omnidirectional, loudspeaker radiating into a room whose surface area is 75 m², and whose average absorption coefficient is 0.2? What would be the effect of mounting the speaker into a wall?

The speaker is omnidirectional so the "Q" is equal to 1. The room constant "R" is the same as was found in the earlier example, 18.75 m². Substituting both these values into Equation 6.11 gives:

$$r_{\text{critical distance}} = 0.141\sqrt{RQ} = 0.141\sqrt{18.75 \times 1} = 0.61 \text{ m (61cm)}$$

This is a very short distance! If the speaker is mounted in the wall the "Q" increases to 2, because the speaker can only radiate into 2 π steradians; so the critical distance increases to:

$$r_{\text{critical distance}} = 0.141\sqrt{RQ} = 0.141\sqrt{18.75 \times 2} = 0.86 \text{m (86cm)}$$

Which is still quite small!

As most people would be about 2 m away from their loudspeakers when they are listening to them this means that in a normal domestic setting the reverberant field is the most dominant source of sound energy from the hi-fi, and not the direct sound. Therefore the quality of the reverberant field is an important aspect of the performance of any system which reproduces recorded music in the home. There is also an effect on speech intelligibility in a reverberant space as the direct sound is the major component of the sound which provides this.

The level of the reverberant field is a function of the average absorption coefficient in the room. Most real materials, such as carpets, curtains, sofas and wood paneling have an absorption coefficient which changes with frequency. This means that the reverberant field level will also vary with frequency, in some cases quite strongly. Therefore in order to hear music, recorded or otherwise, with good fidelity, it is important to have a reverberant field which has an appropriate frequency response. As seen in the previous chapter, one of the cues for sound timbre is the spectral content of the sound which is being heard, and this means that when the reverberant field is dominant, as it is beyond the critical distance, it will determine the perceived timbre of the sound. This subject will be considered in more detail later in the chapter.

6.1.9 The effect of source directivity on the reverberant sound

There is an additional effect on the reverberation field, and that is the directivity of the source of sound in the room. Most hi-fi loudspeakers, and musical instruments, are omnidirectional at low frequencies but are not necessarily so at higher ones. As the level of the reverberant field is a function of both the average absorption and the directivity of the source, the variation in directivity of real musical sources will also have an effect on the reverberant sound field and hence the perception of the timbre of the sound. Consider the following example of a typical domestic hi-fi speaker in the living room considered earlier.

EXAMPLE 6.6

A hi-fi loudspeaker, with a flat-on axis, direct field, response, has a "Q" which varies from 1 to 25, and radiates a peak on axis sound intensity of 102 dB at 1 m. The surface area of the room is 75 m², and the average absorption coefficient is 0.2. Over what range does the sound pressure level of the reverberant field vary?

As the speaker has a flat-on axis response the intensity of the direct field given by Equation 6.1 should be constant. That is:

$$I_{\text{directive source}} = \frac{QW_{\text{Source}}}{4\pi r^2} \tag{6.12}$$

where $I_{\text{directive source}}$ = the sound intensity (in W m^{-2})
Q = the directivity of the source (compared to a sphere)
W_{Source} = the power of the source (in W)
and r = the distance from the source (in m)

should be constant. Therefore the sound power radiated by the loudspeaker can be calculated by rearranging Equation 6.12 to give:

$$W_{\text{Source}} = \left(\frac{4\pi}{Q}\right) I_{\text{directive source}} \tag{6.13}$$

Equation 6.13 shows that in order to achieve a constant direct sound response the power radiated by the source must reduce as the "Q" increases. The power in the reverberant field is given by:

$$W_{\text{reverberant}} = W_{\text{Source}}\left(\frac{4}{R}\right) \tag{6.14}$$

By combining Equations 6.13 and 6.14 the reverberant field due to the loudspeaker can be calculated as:

$$W_{\text{reverberant}} = I_{\text{directive source}}\left(\frac{4\pi}{Q}\right)\left(\frac{4}{R}\right)$$

which gives a level for the reverberant field as:

$$SWL_{\text{reverberant}} = 10\ \log_{10}\left(\frac{I_{\text{directive source}}}{I_{\text{ref}}}\right) + 10\ \log_{10}(4\pi) - 10\ \log_{10}(Q)$$
$$+ 10\ \log_{10}\left(\frac{4}{R}\right)$$

The room constant "R" is $18.75\,\text{m}^2$, as calculated in Example 6.4. The level of the reverberant field can therefore be calculated as:

$$SWL_{\text{reverberant}} = 102\,\text{dB} + 11\,\text{dB} - 10\ \log_{10}(Q)$$
$$+ 10\ \log_{10}\left(\frac{4}{18.75\text{m}^2}\right)$$

which gives:

$$SWL_{\text{reverberant (Q=1)}} = 102\,\text{dB} + 11\,\text{dB} - 10\ \log_{10}(1) - 6.7\,\text{dB}$$
$$= 106.3\,\text{dB}$$

for the level of the reverberant field when the "Q" is equal to 1, and:

$$SWL_{\text{reverberant (Q=25)}} = 102\,\text{dB} + 11\,\text{dB} - 10\ \log_{10}(25) - 6.7\,\text{dB}$$
$$= 92.3\,\text{dB}$$

when the "Q" is equal to 25.
Thus the reverberant field varies by $106.3 - 92.3 = 14\,\text{dB}$ over the frequency range.

The effect therefore of a directive source with constant on-axis response is to reduce the reverberant field as the "Q" gets higher. The subjective effect of this would be similar to reducing the high "Q" regions via the use of a tone control, which would not normally be acceptable as a sound quality.

A typical reverberant response of a typical domestic hi-fi speaker is shown in Figure 6.12. Note that the reverberant response tends to drop in both the midrange and high frequencies. This is due to the bass and treble speakers becoming more directive at the high ends of their frequency range. The dip in reverberant energy will make the speaker less "present" and may make sounds in this region harder to hear in the mix. The drop in reverberant field at the top end will make the speaker sound "duller." Some manufacturers try to compensate for these effects by allowing the on-axis

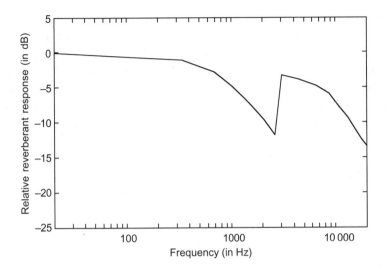

FIGURE 6.12

The reverberant response of a domestic two-way high-fidelity loudspeaker.

response to rise in these regions; however, this brings other problems. The reduction in reverberant field with increasing "Q" is used to advantage in speech systems to raise the level of direct sound above the reverberant field and so improve the intelligibility.

However, in many professional recording studios' control rooms, acoustic means are used to control the off-axis response irregularities and to cause a generally flat on-axis response, but such methods are often beyond the means of domestic listeners.

6.1.10 Reverberation time

Another aspect of the reverberant field is that sound energy which enters it at a particular time dies away. This is because each time the sound interacts with a surface in the room it loses some of its energy due to absorption. The time that it takes for sound at a given time to die away in a room is called the reverberation time. Reverberation time is an important aspect of sound behavior in a room. If the sound dies away very quickly we perceive the room as being "dead" and we find that listening to, or producing, music within such a space is unrewarding. On the other hand when the sound dies away very slowly we perceive the room as being "live." A live room is preferred to a dead room when it comes to listening to, or producing, live music. On the other hand when listening to recorded music, which already has reverberation as part of the recording, a dead room is often preferred.

However, as in many pleasurable aspects of life, reverberation must be taken in moderation. In fact the most appropriate length of reverberation time depends on the nature of the music being played. For example, fast pieces of contrapuntal music, like that of Scarlatti or Mozart, require a shorter reverberation time compared with large romantic works, like that of Wagner or Berlioz, to be enjoyed at their best. The most extreme reverberation times are often found in cathedrals, ice rinks, and railway stations and these acoustics can convert many musical events to "mush" yet to hear slow vocal polyphony, for example works by Palestrina, in a cathedral acoustic can be ravishing! This is because the composer has made use of the likely performance acoustic as part of the composition.

Because of the importance of reverberation time in the perception of music in a room, and because of the differing requirements for speech and different types of music, much effort is focused on it. In fact a major step in room acoustics occurred when Wallace Clement Sabine enumerated a means of calculating, and so predicting, the reverberation time of a room in 1898. Much design work on auditoria in the first half of this century focused almost exclusively on this one parameter, with some successes and some spectacular failures. Nowadays other acoustical and psychoacoustical factors are also taken into consideration.

6.1.11 Calculating and predicting reverberation time

Clearly the length of time that it takes for sound to die is a function not only of the absorption of the surfaces in a room, but also of the length of time between interactions with the surfaces of the room. We can use these facts to derive an equation for the reverberation time in a room. The first thing to determine is the average length of time that a sound wave will travel between interactions with the surfaces of the room. This can be found from the mean free path of the room which is a measure of the average distances between surfaces, assuming all possible angles of incidence and position. For an approximately rectangular box the mean free path is given by the following equation:

$$MFP = \frac{4V}{S} \tag{6.15}$$

where MFP = the mean free path (in m)
V = the volume (in m^3)
and S = the surface area (in m^2)

The time between surface interactions may be simply calculated from Equation 6.15 by dividing it by the speed of sound to give:

$$\tau = \frac{4V}{Sc} \qquad (6.16)$$

where τ = the time between reflections (in s)
and c = the speed of sound (in ms^{-1})

Equation 6.16 gives us the time between surface interactions and at each of these interactions α is the proportion of the energy absorbed, where α is the average absorption coefficient discussed earlier. If α of the energy absorbed at the surface then $(1 - \alpha)$ is the proportion of the energy reflected back to interact with further surfaces. At each surface a further proportion, α, of energy will be removed so the proportion of the original sound energy that is reflected will reduce exponentially. The combination of the time between reflections and the exponential decay of the sound energy, through progressive interactions with the surfaces of the room, can be used to derive an expression for the length of time that it would take for the initial energy to decay by a given ratio. (See Appendix 3 for details.)

There is an infinite number of possible ratios that could be used. However, the most commonly used ratio is that which corresponds to a decrease in sound energy of 60 dB, or 10^6. This gives an equation for the 60 dB reverberation time, known as T_{60}, which is, from Appendix 4:

$$T_{60} = \frac{-0.161V}{S \ln(1 - \alpha)} \qquad (6.17)$$

where T_{60} = the 60 dB reverberation time (in s)

Equation 6.17 is known as the "Norris–Eyring reverberation formula;" the negative sign in the top compensates for the negative sign that results from the natural logarithm resulting in a reverberation time which is positive. Note that it is possible to calculate the reverberation time for other ratios of decay and that the only difference between these and Equation 6.17 would be the value of the constant. The argument behind the derivation of reverberation time is a statistical one and so there are some important assumptions behind Equation 6.17. These assumptions are:

- that the sound visits all surfaces with equal probability, and at all possible angles of incidence. That is, the sound field is diffuse. This is required in order to invoke the concept of an average absorption coefficient for the room. Note that this is a desirable acoustic goal for subjective reasons as well; we prefer to listen to and perform music in rooms with a diffuse field.

■ that the concept of a mean free path is valid. Again this is required in order to have an average absorption coefficient, but in addition it means that the room's shape must not be too extreme. This means that this analysis is not valid for rooms which resemble long tunnels. However, most real rooms are not too deviant and the mean free path equation is applicable.

6.1.12 The effect of room size on reverberation time

The result in Equation 6.17 also allows some broad generalizations to be made about the effect of the size of the room on the reverberation time, irrespective of the quantity of absorption present. Equation 6.17 shows that the reverberation time is a function of the surface area, which determines the total amount of absorption, and the volume, which determines the mean time between reflections in conjunction with the surface area. Consider the effect of altering the linear dimensions of the room on its volume and surface area. These clearly vary in the following way:

$$V \propto (\text{Linear dimension})^3$$

and

$$S \propto (\text{Linear dimension})^2$$

However, both the mean time between reflections, and hence the reverberation time, vary as:

$$\frac{V}{S} \propto \frac{(\text{Linear dimension})^3}{(\text{Linear dimension})^2} \propto \text{Linear dimension}$$

Hence as the room size increases the reverberation time increases proportionally, if the average absorption remains unaltered. In typical rooms the absorption is due to architectural features such as carpets, curtains and people, and so tends to be a constant fraction of the surface area. The net result is that, in general, large rooms have a longer reverberation time than smaller ones and this is one of the cues we use to ascertain the size of a space, in addition to the initial time delay gap. Thus one often hears people referring to the sound of a "big" or "large" acoustic as opposed to a "small" one when they are really referring to the reverberation time. Interestingly, now that it is possible to provide a long reverberation time in a small room, via electronic reverberation enhancement systems, with good quality, people have found that long reverberation times in a small room sound "wrong" because the visual cues contradict the audio ones. That is, the listener, on the basis of the apparent size of the space and their experience, expects a shorter reverberation time than they are hearing. Apparently closing one's eyes restores the illusion by removing the distracting visual cue!

Let us use Equation 6.17 to calculate some reverberation times.

EXAMPLE 6.7

What is the reverberation time of a room whose surface area is 75 m², whose volume is 42 m³, and whose average absorption coefficient is 0.9, 0.2? What would be the effect of doubling all the dimensions of the room while keeping the average absorption coefficients the same?

Using Equation 6.17 and substituting in the above values gives, for $\alpha = 0.9$:

$$T_{60} = \frac{-0.161V}{S \ln(1 - \alpha)} = \frac{-0.161 \times 42 m^2}{75 \, m^2 \times \ln(1 - 0.09)} = 0.042 \text{ s } (42 \times 10^{-3} s)$$

which is very small! For $\alpha = 0.2$ we get:

$$T_{60} = \frac{-0.161V}{S \ln(1 - \alpha)} = \frac{-0.161 \times 42 \, m^3}{75 \, m^2 \times \ln(1 - 0.2)} = 0.43 \text{ s}$$

which would correspond well with the typical T_{60} of a living room, which is in fact what it is.

If the room dimensions are doubled then the ratio of volume with respect to the surface area also doubles so the new reverberation times are given by:

$$\frac{V_{doubled}}{S_{doubled}} = (\text{Linear dimension})_{doubled} = 2$$

so the old reverberation times are increased by a factor of 2:

$$T_{60 \, doubled} = T_{60} \times 2$$

which gives a reverberation time of:

$$T_{60 \, doubled} = T_{60} \times 2 = 0.042 \times 2 = 0.084 \text{ s}$$

when $\alpha = 0.9$ and:

$$T_{60 \, doubled} = T_{60} - 2 = 0.43 \times 2 = 0.86 \text{ s}$$

when $\alpha = 0.2$

6.1.13 The problem of short reverberation times

The very short reverberation times that occur when the absorption is high pose an interesting problem. Remember that one of the assumptions behind the derivation of the reverberation time calculation was that the sound energy visited all the surfaces in the room with equal probability. For our example room the mean time between reflections, using Equation 6.16, is given by:

$$\tau = \frac{4V}{Sc} = \frac{4 \times 42 \text{ m}^3}{75 \text{ m}^2 \times c} = \frac{2.24 \text{ m}}{344 \text{ ms}^{-1}} = 6.51 \text{ ms } (0.00651 \text{ s})$$

If the reverberation time calculated in Example 6.7, when $\alpha = 0.9$, is divided by the mean time between reflections then the average number of reflections that have occurred during the reverberation time can be calculated to be:

$$N_{\text{reflections}} = \frac{T_{60}}{\tau} = \frac{42 \times 10^{-3}}{6.51 \times 10^{-3} \text{s}} = 6.45 \text{ reflections}$$

These are barely enough reflections to have hit each surface once! In this situation the reverberant field does not really exist; instead the decay of sound in the room is really a series of early reflections to which the concept of reverberant field or reverberation does not really apply. In order to have a reverberant field there must be much more than six reflections. A suitable number of reflections, in order to have a reverberant field, might be nearer 20, although this is clearly a hard boundary to accurately define. Many studios and control rooms have been treated so that they are very "dead" and so do not support a reverberant field.

6.1.14 A simpler reverberation time equation

Although the Norris–Eyring reverberation formula is often used to calculate reverberation times there is a simpler formula known as the "Sabine formula," named after its developer Wallace Clement Sabine, which is also often used. Although it was originally developed from considerations of average energy loss from a volume, a derivation which involves solving a simple differential equation, it is possible to derive it from the Norris–Eyring reverberation formula. This also gives a useful insight into the contexts in which the Sabine formula can be reasonably applied. Consider the Norris–Eyring reverberation formula below:

$$T_{60} = \frac{-0.161V}{S \ln(1 - \alpha)}$$

The main difficulty in applying this formula is due to the need to take the natural logarithm of $(1 - \alpha)$. However, the natural logarithm can be expanded as an infinite series to give:

$$T_{60} = \frac{-0.161V}{S\left(\alpha - \dfrac{\alpha^2}{2} - \dfrac{\alpha^3}{3} - \ldots - \dfrac{\alpha^n}{n} - \ldots - \dfrac{\alpha^\infty}{\infty}\right)} \tag{6.18}$$

Because $\alpha < 1$ the sequence always converges. However, if $\alpha < 0.3$ then the error due to all the terms greater than $-\alpha$ is less than 5.7%. This means that Equation 6.18 can be approximated as:

$$T_{60(\alpha<0.3)} \approx \frac{-0.161V}{S(-\alpha)} = \frac{0.161V}{S\alpha} \tag{6.19}$$

Equation 6.19 is known as the "Sabine reverberation formula" and, apart from being useful, was the first reverberation formula. It was developed on the basis of experimental measurements made by W. C. Sabine, thus initiating the whole science of architectural acoustics. Equation 6.19 is much easier to use and gives accurate enough results provided the absorption, α, is less than about 0.3. In many real rooms this is a reasonable assumption. However, it becomes increasingly inaccurate as the average absorption increases and in the limit predicts a reverberation time when $\alpha = 1$, that is reverberation without walls!

6.1.15 Reverberation faults

As stated previously, the basic assumption behind these equations is that the reverberant field is statistically random, that is, a diffuse field. There are, however, acoustic situations in which this is not the case. Figure 6.13 shows the decay of energy, in dB, as a function of time for an ideal diffuse field reverberation. In this case the decay is a smooth straight line representing an exponential decay of an equal number of dBs per second.

Figure 6.14 on the other hand shows two situations in which the reverberant field is no longer diffuse. In the first situation all the absorption is only on two surfaces, for example an office with acoustic tiles on the ceiling, carpets on the floor, and nothing on the walls. Here the sound between the absorbing surfaces decays quickly whereas the sound between the walls decays much more slowly, due to the lower absorption. In the second case there are two connected spaces, such as the transept and nave in a church, or under the balconies in a concert hall. In this case the sound energy does not couple entirely between the two spaces and so they will decay at

FIGURE 6.13

The ideal decay versus time curve for diffuse field reverberation.

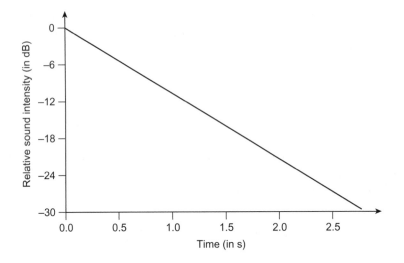

FIGURE 6.14

Two situations which give poor reverberation decay curves (see text).

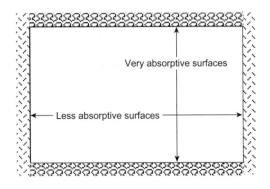

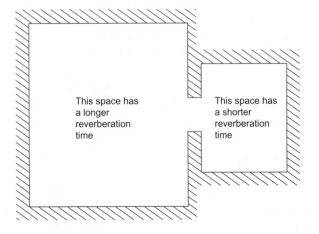

different rates that depend on the level of absorption in them. In both of these cases the result is a sound energy curve as a function of time which has two or more slopes, as shown in Figure 6.15. This curve arises because the faster decaying waves die away before the more slowly decaying ones and so allow them to dominate in the end.

The second major acoustical defect in reverberant decay occurs when there are two precisely parallel and smooth surfaces, as shown in Figure 6.16. This results in a series of rapidly spaced echoes, onomatopoeically called flutter echoes, which result as the energy shuttles backward and forward between the two surfaces. These are most easily detected by clapping one's hands between the parallel surfaces to provide the packet of sound energy to excite

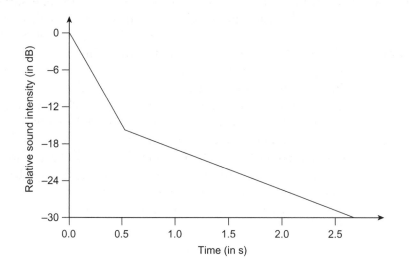

FIGURE 6.15

A double slope reverberation decay curve.

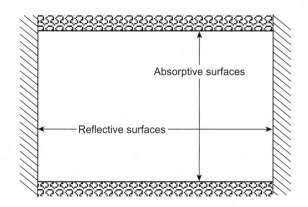

FIGURE 6.16

A situation which can cause flutter.

the flutter echo. The decay of energy versus time in this situation is shown in Figure 6.17 and the presence of the flutter echo manifests itself as a series of peaks in the decay curve. Note that this behavior is also often associated with the double-slope decay characteristic shown in Figure 6.15 because the energy shuttling between the parallel surfaces suffers less absorption compared with a diffuse sound.

6.1.16 Reverberation time variation with frequency

Equations 6.17 and 6.18 show that the reverberation time depends on the volume, surface area, and the average absorption coefficient in the room. However, the absorption coefficients of real materials are not constant with frequency. This means that, assuming that the room's volume and surface area are constant with frequency, which is not an unreasonable assumption, the reverberation time in the room will also vary with frequency. This will subjectively alter the timbre of the sound in the room due to both the effect on the level of the reverberant field discussed earlier and the change in timbre as the sound in the room decays away.

As an extreme example, if a particular frequency has a much slower rate of decay compared with other frequencies, then, as the sound decays away, this frequency will ultimately dominate and the room will "ring" at that particular frequency. The sound power for steady-state sounds will also have a strong peak at that frequency because of the effect on the reverberant field level.

Table 6.1 shows some typical absorption coefficients for some typical materials which are used in rooms as a function of frequency. Note that

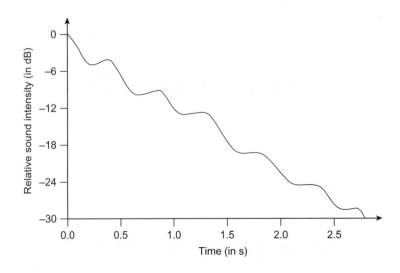

FIGURE 6.17

The decay versus time curve for flutter.

they are measured over octave bands. One could argue that third octave band measurements would be more appropriate psychoacoustically as the octave measurement will tend to blur variations within the octave, which might be perceptually noticeable. In many cases, because the absorption coefficient varies smoothly with frequency, octave measurements are sufficient. However, especially when considering resonant structures, more resolution would be helpful. Note also that there are often no measurements of the absorption coefficients below 125 Hz. This is due to both the difficulty in making such measurements and the fact that below 125 Hz other factors in the room become more important, as we shall see later.

In order to take account of the frequency variation of the absorption coefficients we must modify the equations used to calculate the reverberation time as follows:

$$T_{60} = \frac{-0.161V}{S \ln(1 - \alpha(f))}$$

where $\alpha(f)$ = frequency-dependent absorption coefficient for the Norris–Eyring reverberation time equation and:

$$T_{60(\alpha < 0.3)} = \frac{0.161V}{S\alpha(f)}$$

for the Sabine reverberation time equation.

Table 6.1   Typical absorption coefficients as a function of frequency for various materials

Material	Frequency					
	125 Hz	250 Hz	500 Hz	1 kHz	2 kHz	4 kHz
Plaster on lath	0.14	0.10	0.06	0.05	0.04	0.03
Carpet on concrete	0.02	0.06	0.14	0.37	0.60	0.65
Floor (wood joist)	0.15	0.11	0.10	0.07	0.06	0.07
Painted plaster	0.01	0.01	0.02	0.02	0.02	0.02
Walls (½ inch plasterboard)	0.29	0.10	0.05	0.04	0.07	0.09
Windows (float glass)	0.35	0.25	0.18	0.12	0.07	0.04
Wood paneling	0.30	0.25	0.20	0.17	0.15	0.10
Curtains (cotton draped to half area)	0.07	0.31	0.49	0.81	0.66	0.54
Air absorption (per m³ @ 20°C and 30% RH)	—	—	—	—	0.012	0.038

6.1.17 Reverberation time calculation with mixed surfaces

In real rooms we must also allow for the presence of a variety of different materials, as well as accounting for their variation of absorption as a function of frequency. This is complicated by the fact that there will be different areas of material, with different absorption coefficients, and these will have to be combined in a way that accurately reflects their relative contribution. For example, a large area of a material with a low value of absorption coefficient may well have more influence than a small area of material with more absorption.

In the Sabine equation this is easily done by multiplying the absorption coefficient of the material by its total area and then adding up the contributions from all the surfaces in the room. These resulted in a figure which Sabine called the "equivalent open window area", as he assumed, and experimentally verified, that the absorption coefficient of an open window was equal to 1.

The denominator in the Sabine reverberation equation, Equation 6.19, is also equivalent to the open window area of the room, but has been calculated using the average absorption coefficient in the room. It is therefore easy to incorporate the effects of different materials by simply calculating the total open window area for different materials, using the method described above, and substituting it for $S\alpha$ in Equation 6.19. This gives a modified equation which allows for a variety of frequency-dependent materials in the room as:

$$T_{60(\alpha<0.3)} = \frac{0.161V}{\displaystyle\sum_{\text{All surfaces } S_i} S_i \alpha_i(f)} \tag{6.20}$$

where $\alpha_i(f)$ = absorption coefficient for a given material
and S_i = its area

For the Norris–Eyring reverberation time equation the situation is a little more complicated because the equation does not use the open window area directly. However, the Norris–Eyring reverberation time equation can be rewritten in a modified form, as shown in Appendix 4, which allows for the variation in material absorption due to both nature and frequency, as:

$$T_{60} = \frac{-0.161V}{\displaystyle\sum_{\text{All surfaces } S_i} S_i \ln(1 - \alpha_i(f))} \tag{6.21}$$

Equation 6.21 is also known as the "Millington–Sette equation." Although Equation 6.21 can be used irrespective of the absorption level it is still more complicated than the Sabine equation and, if the average absorption coefficient is less than 0.3, it can be approximated very effectively by it, as discussed previously. Thus in many contexts the Sabine equation, Equation 6.20, is preferred.

Equation 6.20 is readily used in conjunction with tables of absorption coefficients to calculate the reverberation time and can be easily programmed into a spreadsheet. As an example, consider the reverberation time calculation for a living room outlined in Example 6.8.

EXAMPLE 6.8

What is the 60 dB reverberation time (T_{60}) of a living room as a function of frequency whose surface area is 75 m^2 and whose volume is 42 m^3? The floor is carpet on concrete, the ceiling is plaster on lath, and both have an area of 16.8 m^2. There are 6 m^2 of windows and the rest of the surfaces are painted plaster on brick; ignore the effect of the door.

Using the data in Table 6.1, set up a spreadsheet or table, as shown in Table 6.2, and calculate the equivalent open window area for each surface as a function of frequency. Having done that add up the individual surface contributions for each frequency band and apply Equation 6.20 to the result in order to calculate the reverberation time.

From the results shown in Table 6.2, which are also plotted in Figure 6.18, one can see that the reverberation varies from 1.49 seconds at low frequencies to 0.55 seconds at high frequencies. This is a normal result for such a structure and would tend to sound a bit "woolly" or "boomy." The relative level of reverberant field for this room is also shown in Figure 6.19 and this shows approximately a 5 dB increase in the reverberant field at low frequencies.

Table 6.2 Absorption and reverberation time calculations for an untreated living room

Surface (material)	Area (m²)	Frequency					
		125 Hz	250 Hz	500 Hz	1 kHz	2 kHz	4 kHz
Ceiling (plaster on lath)	16.8	2.35	1.68	1.01	0.84	0.67	0.50
Floor (carpet on concrete)	16.8	0.34	1.01	2.35	6.22	10.08	10.92
Walls (painted plaster)	35.4	0.35	0.35	0.71	0.71	0.71	0.71
Windows (float glass)	6.0	2.10	1.50	1.08	0.72	0.42	0.24
Total open window area		5.14	4.54	5.15	8.48	11.88	12.37
Room volume (m³)	42						
Reverberation time (s)		1.32	1.49	1.31	0.80	0.57	0.55

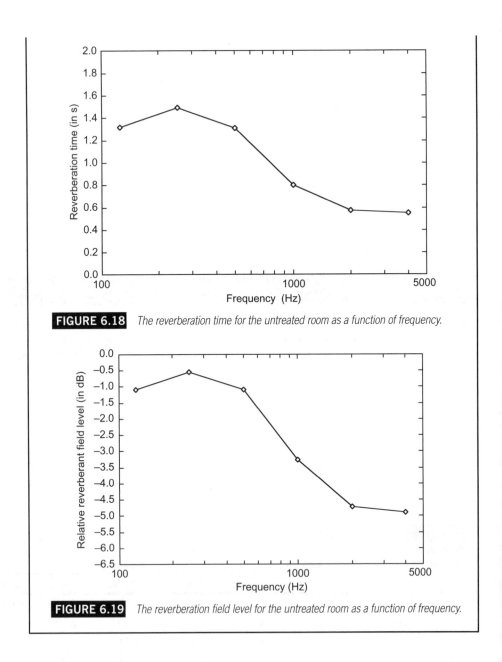

FIGURE 6.18 *The reverberation time for the untreated room as a function of frequency.*

FIGURE 6.19 *The reverberation field level for the untreated room as a function of frequency.*

6.1.18 Reverberation time design

The results of Example 6.8 beg the question: "How can we improve the evenness of the reverberation time?" The answer is to either add, or remove, additional absorbing materials into or from the room in order to achieve the desired reverberation characteristic. Here the concept of an open window area budget is useful. The idea is that, given the volume of the room, and the desired reverberation time, the necessary open window area required is calculated. The open window area already present in the room is then examined and, depending on whether the room is over or under budget, appropriate materials are added or removed. Consider Example 6.9 which tries to improve the reverberation of the previous room.

EXAMPLE 6.9

Which single material could be added to the room in Example 6.8 which would result in an improved reverberation time, and what amount would be required to effect the improvement?

A material which has a high absorption at low frequencies, such as wood paneling, needs to be added to the room. If the absorption budget is set as being equivalent to the open window area at 4 kHz then we must achieve an open window area of 12.5 m over the whole frequency range. The worst frequency in the previous example is 250 Hz, which only has 4.5 m of open window area at that frequency. This means that any additional absorber must add 12.5 − 4.5 = 8 m of open window area at that frequency. The absorption of wood paneling, from Table 6.1, at 250 Hz is 0.25. Therefore the amount of wood paneling required is:

$$\text{Area}_{\text{Wood paneling}} = \frac{\text{Required open window area}}{\text{Absorption coefficient}} = \frac{8 \text{ m}}{0.25} = 32\text{m}$$

Table 6.3, Figure 6.20 and Figure 6.21 show the effect of applying the treatment, which dramatically improves the reverberation time characteristics. The reverberation time now only varies from 0.59 to 0.41 s, which is a much smaller variation than before. The peak-to-peak variation in the level of the reverberant field has also been reduced to less than 2 dB.

Table 6.3 Absorption and reverberation time calculations for a treated living room

Surface (material)	Area (m²)	Frequency					
		125 Hz	250 Hz	500 Hz	1 kHz	2 kHz	4 kHz
Ceiling (plaster on lath)	16.8	2.35	1.68	1.01	0.84	0.67	0.50
Floor (carpet on concrete)	16.8	0.34	1.01	2.35	6.22	10.08	10.92
Walls (painted plaster)	35.4	0.35	0.35	0.71	0.71	0.71	0.71
Windows (float glass)	6.0	2.10	1.50	1.08	0.72	0.42	0.24
Wood paneling	32.0	9.60	8.00	6.40	5.44	4.80	3.20
Total open window area		14.74	12.54	11.55	13.92	16.68	15.57
Room volume (m³)	42.0						
Reverberation time (s)		0.46	0.54	0.59	0.49	0.41	0.43

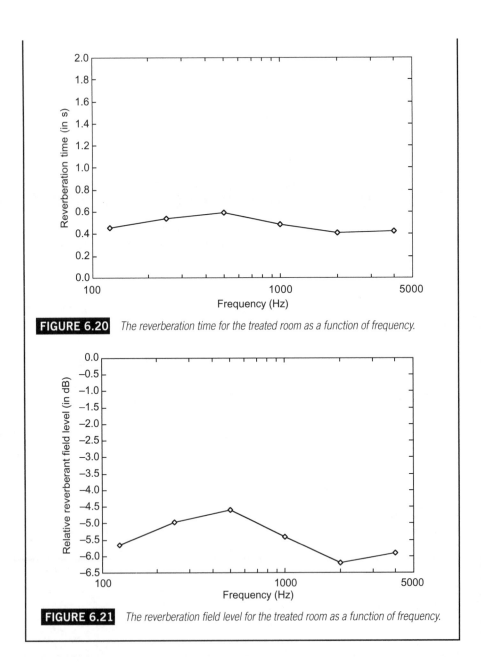

FIGURE 6.20 *The reverberation time for the treated room as a function of frequency.*

FIGURE 6.21 *The reverberation field level for the treated room as a function of frequency.*

However, the overall reverberation time has gone down, especially at the lowest frequencies, because of the effect of the wood paneling at frequencies other than the one being concentrated on. Thus in practice an iterative approach to deciding on the most suitable treatment for a room is often required. Another point to consider is that the treatment proposed only just fits in the room, and sometimes it proves impossible to achieve a desired reverberation characteristic due to physical limitations.

6.1.19 Ideal reverberation time characteristics

What is an ideal reverberation characteristic? We have seen that the decay should be a smooth exponential of a constant number of decibels of decay per unit time. We also know that different sorts of music require different reverberation times. In many cases the answer is: "it depends on the situation." However, there are a few general rules which seem to be broadly accepted.

Firstly, there is a range of reverberation times which are a function of the type of music being played; music with a high degree of articulation needs a drier acoustic than music which is slower and more harmonic. Secondly, as the performance space gets larger the reverberation time required for all types of music becomes longer. This result is summarized in Figure 6.22 which shows the "ideal" reverberation time as a function of both music and room volume. Thirdly, in general, listeners prefer a rise in reverberation time in the bass (125 Hz) of about 40% relative to the midrange (1 kHz) value as shown in Figure 6.23. This rise in bass reverberation adds "warmth" and it also helps increase the sound level of bass instruments, which often have weak fundamentals, by raising the level of the reverberant field at low frequencies. However, when recording musical

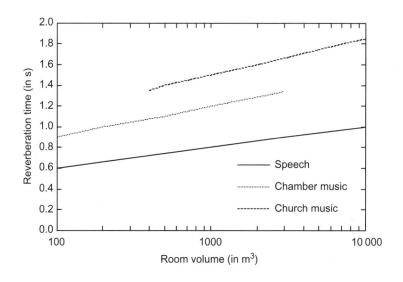

FIGURE 6.22

Ideal reverberation times as a function of room volume and musical style.

FIGURE 6.23

*The ideal reverberation
time versus frequency
curves.*

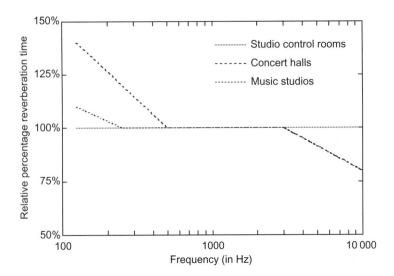

FIGURE 6.23

The ideal reverberation time versus frequency curves.

instruments, or when listening to recorded music, this bass lift due to the reverberant field may be undesirable and therefore a flat reverberation characteristic is preferred.

There are many other aspects of reverberation, too numerous to mention here, which must be considered when designing acoustic spaces. However, there are four aspects that are worthy of mention as they have proved to be the downfall of more than one acoustic designer, or manufacturer, of reverberation units.

6.1.20 Early decay time

The first aspect is that the measure of reverberation time as being the time it takes the sound to fall by 60 dB is not particularly relevant psychoacoustically; it is also very difficult to measure in situ. This is due to the presence of background noise, either unwanted or the music being played, which often results in less than 60 dB of energy decay before the decay sound becomes less than the residual noise in the environment. Even in the quieter environment of a Victorian town in the days before road traffic, Sabine had to do measurements, using his ears, at night to avoid the results being affected by the level of background noise. Because we rarely hear a full reverberant decay, our ears and brains have adapted, quite logically, to focus on what can be heard. Thus we are more sensitive to the effects of the first 20 to 30 dB of the reverberant decay curve.

In principle, provided we have an even exponential decay curve, the 60 dB reverberation is directly proportional to the earlier curves and so this should not cause any problems. However, if the curve is of the double-slope form shown in Figure 6.15 then this simple relationship is broken. The net result is that, although the T_{60} reverberation time may be an appropriate value, because of the faster early decay to below 30 dB we perceive the reverberation as being shorter than it really is. The psychoacoustic effect of this is that the space sounds "drier" than one would expect from a simple measurement of T_{60}. Modern acoustic designers therefore worry much more about the early decay time (EDT) than they used to when designing concert halls.

6.1.21 Lateral reflections

The second factor which has been found to be important for the listener is the presence of dense diffuse reflections from the side walls of a concert hall, called lateral reflections, as shown in Figure 6.24. The effect of these is to envelop or bathe the listener in sound and this has been found to be necessary for the listener to experience maximum enjoyment from the sound. It is important that these reflections be diffuse, as specular reflections will result in disturbing comb filter effects, as discussed in Chapter 1, and distracting images of the sound sources in unwanted and unusual directions. Providing diffuse reflections is thus important and this has been recognized for some time. Traditionally, plaster mouldings, niches and other decorative surface irregularities have been used to provide diffusion in an ad hoc manner. More recently diffusion structures based on patterns of wells whose depths are formally defined by an appropriate mathematical sequence have been proposed and used.

However, it is not just the provision of diffusion on the side walls that must be considered. The traditional concert hall is called a shoebox hall, because of its shape, as shown in Figure 6.25, and this naturally provides a large number of lateral reflections to the audience. This shape, combined with the Victorian penchant for florid plaster decoration, resulted in some excellent sounding spaces. Unfortunately shoebox halls are harder to make a profit with because they cannot seat as many people as some other structures.

Another popular structure, which has a different behavior as regards to lateral reflections, is the fan-shaped hall shown in Figure 6.26. This structure has the advantage of being able to seat more people with good sightlines but unfortunately it directs the lateral reflections away from the audience and those few that do arrive are very weak at the wider part of the fan.

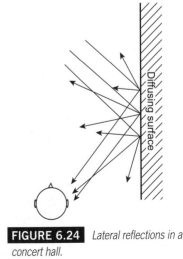

FIGURE 6.24 *Lateral reflections in a concert hall.*

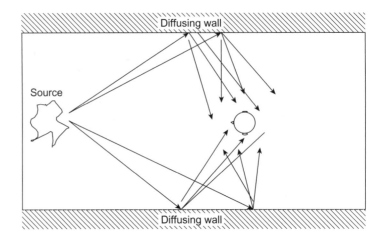

FIGURE 6.25

Lateral reflections in a shoebox concert hall.

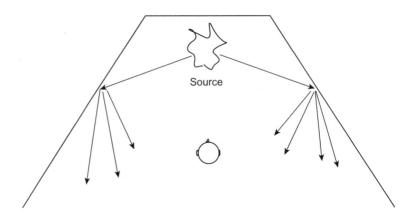

FIGURE 6.26

Lateral reflections in a fan-shaped concert hall.

The situation can be improved via the use of explicit diffusion structures on the walls, ceilings, and mid-air as floating shapes, as shown in Figure 6.27. However, it has been found that the pseudo-lateral diffuse reflections from the ceiling are not quite comparable in effect to reflections from the side walls, and so the provision of a good listening environment within the realities of economics is still a challenge.

6.1.22 Early reflections and performer support

A third factor, which is often ignored, is the acoustics that the performers experience. Pop groups have known about this for years and take elaborate precautions to provide each performer on stage with their own individual balance of acoustic sounds via a technique known as "foldback." In fact

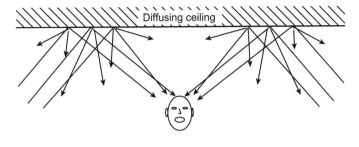

FIGURE 6.27
Lateral reflections from ceiling diffusion in a concert hall.

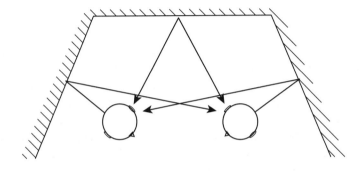

FIGURE 6.28
Early reflections to provide acoustic foldback for the performer.

some performers now receive their foldback directly into their ears via a technique known as "in-ear monitoring" and in many large gigs the equipment providing foldback to the performer can rival, or even exceed, that which provides the sound for the audience. The classical musician, however, only has the acoustics of the hall to provide them with "foldback." Thus the musicians on the stage must rely on reflections from the nearby surfaces to provide them with the necessary sounds to enable them to hear themselves and each other.

There are two requirements for the sound reaching the performer on stage. It must be at a sufficient level, and arrive soon enough to be useful. To begin with it is important that the surfaces surrounding the performers direct some sound back to them. Note that there is a conflict between this and providing a maximum amount of sound to the audience so some compromise must be reached. The usual compromise is to make use of the sound which radiates behind the performers and direct it out to the audience via the performers, as shown in Figure 6.28. This has the twofold advantage of providing the performers with acoustic foldback and redirecting sound energy that might have been lost back to the audience. Ideally the sound that is redirected back to the performers should be diffuse as this will blend the sounds of the different instruments together for all the

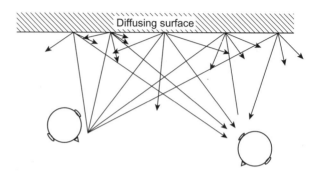

performers, as shown in Figure 6.29, whereas specular reflectors can have hot and cold spots for a given instrument on the stage.

An important aspect of acoustic foldback, however, is the time that it takes to arrive back at the performers. Ideally it should arrive immediately, and some does via the floor and direct sound from the instrument. However, the majority will have to travel to a reflecting or diffusing surface and back to the performers. There is evidence to show that, in order to maintain good ensemble comfortably, the musicians should receive the sound from other musicians within about 20 ms of the sound being produced. This means that ideally there should be a reflecting or diffusing surface within 10 ms (3.44 m or 11.5 ft) of the performer; the time is divided by 2 to allow for going to the reflecting surface and back. In practice some of the surfaces may have to be further away when large orchestral forces are being mustered, although the staging used can assist the provision of acoustic foldback. Sometimes, however, the orchestra enclosure is so large that the reflections arrive later than this. If they arrive later than about 50 ms the musicians will perceive them as echoes and ignore them. On the other hand if these reflections arrive at the boundary between perceiving it as part of the sound or an echo of a previous sound it can cause severe disruption of the performers' perception of it.

The net effect of these "late early reflections" is to damage the performers' ability to hear other instruments close to them, and this further reduces their ability to maintain ensemble. In one prestigious hall, the reason musicians used to complain that they couldn't hear each other, and so hated playing there, was traced to the problem of late early reflections. As a postscript it is interesting to note that the orchestra enclosure in shoebox halls often did the right things. However, in modern multipurpose facilities it is often a challenge to provide the necessary acoustic foldback while allowing space for scenery and machinery, etc.

6.1.23 The effect of air absorption

The fourth aspect of reverberation, which caught early reverberation unit designers by surprise, is the observation that, as well as suffering many reflections, the sound energy in a reverberant decay will have traveled through a lot of air. In fact the distance that the sound will have traveled will be directly proportional to the reverberation time, so a one second reverberation time implies that the sound will have traveled 344 m by the end of the decay. Although for low frequencies air absorbs a minimal amount of sound energy, at high frequencies this is not the case. In particular, humidity, smoke particles and other impurities will absorb high-frequency energy and so reduce the level of high frequencies in the sound. This is one of the reasons why people sound duller when they are speaking at a distance.

In terms of reverberation time, and also the level of the reverberant field, the effect of this extra absorption is to reduce the reverberation time, and the level of the reverberant field, at high frequencies. Fortunately this effect only becomes dominant at higher frequencies, above 2 kHz. Unfortunately, though, it is dependent on the level of humidity and smoke in the venue and so the high-frequency reverberation time, and the reverberant field level, will change as the audience stays in the space. Note this is an additional dynamic effect over and above the static absorption simply due to the presence of a clothed person in a space and is due to the fact that people exhale water vapor and perspire. Clearly then the degree of change will be a function of both the physical exertions of the audience and the quality of the ventilation system!

As the effect of air absorption is determined by the distance the sound has traveled, rather than its interaction with a surface, it is difficult to incorporate the effect into the reverberation time equations discussed earlier. An approximation that seems to work is to convert the effect of the air absorption into an equivalent absorption area by scaling an air absorption coefficient by the volume of the space. This is reasonable because as the volume of the room increases, the more air the wave must travel through and the longer the distance that it travels. This coefficient is shown at the bottom of Table 6.1 and from it one can see that for small rooms the effect can be ignored because until the volume becomes greater than $40\,m^3$ the equivalent absorbing area is less than $1\,m^2$. However, the effect does become significant if one is designing artificial reverberation units because, if it is not allowed for, the result will be an overbright reverberation, which sounds unnatural.

In this section the concept of reverberation time and reverberant field has been discussed. The assumption behind the equations has been that the sound field is diffuse. However, if this is not the case then the equations are invalid. Although at mid and high audio frequencies a diffuse field might be possible, either by accident or design, at low frequencies this is not the case due to the effect of the room's boundaries causing standing waves.

6.2 ROOM MODES AND STANDING WAVES

When a room is excited by an impulse, the sound energy is reflected from its surfaces. At each reflection some of the sound is absorbed and therefore the sound energy decays exponentially. Ideally the sound should be reflected from each surface with equal probability, forming a diffuse field. This results in a single exponential decay with a time constant proportional to the average absorption in the room. However, in practice not all the energy is reflected in a random fashion. Instead some energy is reflected in cyclic paths, as shown in Figure 6.30. If the length of the path is a precise number of half wavelengths then they will form standing waves in the room. These standing waves (resonant modes) have pressure and velocity distributions which are spatially static and so behave differently from the rest of the sound in the room in the following ways:

- They do not visit each surface with equal probability. Instead a subset of the surfaces is involved.

- They do not strike these surfaces with random incidence. Instead a particular angle of incidence is involved in the reflection of the standing wave.

- They require a coherent return of energy back to an original surface: a cyclic path. This is of necessity strongly frequency dependent and so these paths only exist for discrete frequencies which are determined by the room geometry.

Another name for these standing waves in a room are resonant modes and the frequencies at which they occur are known as "modal frequencies." Because the modes are spatially static there will be a strong variation of sound pressure level as one moves around the room, which is undesirable. There are three basic types of room mode, which are outlined in Sections 6.2.1 to 6.2.3.

6.2.1 Axial modes

These modes occur between two opposing surfaces, as shown in Figure 6.31, and so are a function of the linear dimensions of the room. The frequencies of an axial mode are given by the following equation:

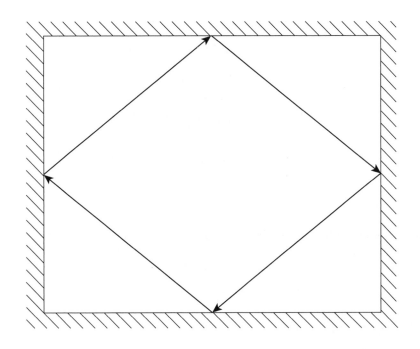

FIGURE 6.30

Cyclic reflection paths in a room.

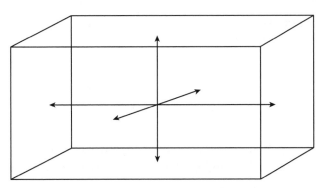

FIGURE 6.31

Axial modal paths in a room.

$$f_{x(\text{axial})} = \frac{c}{2}\left(\frac{x}{L}\right)$$

where $f_{x(\text{axial})}$ = the axial mode frequencies (in Hz)

   x = the number of half wavelengths that fit between the surfaces (1, 2, ..., ∞)

   L = the distance between the reflecting surfaces (in m)

   and c = the speed of sound (in ms^{-1})

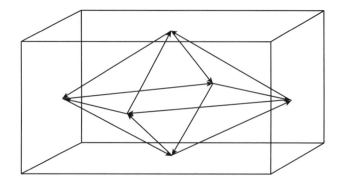

This equation shows that there are an infinite number of possible modal frequencies at which an integer number of wavelengths fit into the room, with lowest modal frequency occurring when just one half wavelength fits into the space between the reflecting surfaces.

6.2.2 Tangential modes

These modes occur between four surfaces, as shown in Figure 6.32, and so are a function of two of the dimensions of the room. The frequencies of the tangential modes are given by the following equation:

$$f_{xy(\text{tangential})} = \frac{c}{2}\sqrt{\left(\frac{x}{L}\right)^2 + \left(\frac{y}{W}\right)^2}$$

where $f_{xy(\text{tangential})}$ = the tangential modal frequencies (in Hz)

x = the number of half wavelengths between one set of two surfaces $(1, 2, ..., \infty)$

y = the number of half wavelengths between the other set of two surfaces $(1, 2, ..., \infty)$

and L, W = the distance between the reflecting surfaces (in m)

There is also an infinite number of tangential modes, but they must fit an integral number of half wavelengths in two dimensions. This has the interesting consequence that the lowest modal frequencies are higher than the axial modes, despite the fact that the apparent path length is greater. The reason is that the standing waves must fit between the opposing surfaces, that is, on the sides rather than the hypotenuse of the triangular path, and

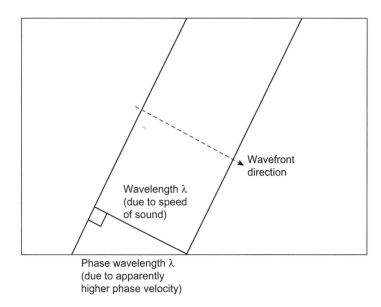

FIGURE 6.33

The phase velocity of tangential room modes.

as the propagating wave travels down the hypotenuse, the effective wavelength, or phase velocity, on the sides of the room is larger, as shown in Figure 6.33. The lowest modal frequency for a tangential mode occurs when precisely one half wavelength, at the phase velocity, fits into each dimension.

6.2.3 Oblique modes

These modes occur between all six surfaces, as shown in Figure 6.34, and so are a function of all three dimensions of the room. The frequencies of the oblique modes are given by the following equation:

$$f_{xy(\text{oblique})} = \frac{c}{2}\sqrt{\left(\frac{x}{L}\right)^2 + \left(\frac{y}{W}\right)^2 + \left(\frac{z}{H}\right)^2}$$

where $f_{xyz(\text{oblique})}$ = the oblique modal frequencies (in Hz)

x, y, z = the number of half wavelengths between the surfaces (1, 2,..., ∞)

and L, W, H = the distance between the reflecting surfaces (in m)

The lowest frequencies of these modes are also higher than the lowest axial modes, for the reasons discussed earlier.

FIGURE 6.34

An oblique modal path in a room.

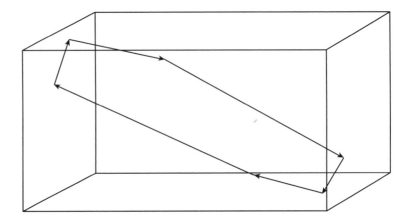

6.2.4 A universal modal frequency equation

The combination of these three types of mode forms a dense set of possible standing wave frequencies in the room and they can be combined into one equation by simply allowing x, y, and z in the oblique mode equation to range from 0, 1, 2 to infinity, giving the following equation which will give the frequencies of all possible modes in the room:

$$f_{xyz} = \frac{c}{2}\sqrt{\left(\frac{x}{L}\right)^2 + \left(\frac{y}{W}\right)^2 + \left(\frac{z}{H}\right)^2} \qquad (6.22)$$

where x, y, z = the number of half wavelengths between the surfaces (0, 1, 2, ..., ∞)

The above equation also shows that if any of the dimensions are integer multiples of each other then some of the modal frequencies will be the same, which can cause problems. It is therefore better to choose non-commensurate ratios for the wall dimensions to ensure that the modes are spread out as much as possible. Much work has been done on ideal room ratios and one set of favorable room dimensions is shown in Table 6.4. However, these dimensions are not necessarily the only optimum ones for all room sizes. It is also important to realize that room modes are inherent in any structure which encloses the sound sources. This means that changing the shape of the room, for example by angling the walls, does not remove the resonances—it merely changes their frequencies from values that are easily calculated to ones which are not. Both Walker (1996) and Cox et al. (2004) discuss more general and useful approaches to optimum room dimensions.

Table 6.4	Some favorable room dimensions		
	Height	**Width**	**Length**
A	1.00	1.14	1.39
B	1.00	1.28	1.54
C	1.00	1.60	2.33

6.2.5 The Bonello criteria

In general the number of resonances within a given frequency bandwidth increases with frequency. In fact it can be shown that they increase proportionally to the square of the frequency, and in large well-behaved acoustical spaces, which sound good, this increase in mode density with frequency is smooth. This is the rationale behind a method for assessing the modal behavior in a room known as the "Bonello criteria." These criteria try to ascertain how significant the modal behavior of a room is in perceptual terms. This is done by dividing the audio frequency spectrum into third octave bands, as an approximation of critical bands, and then counting the number of modes per band. If the number of modes per third octave band increases monotonically then there is a good chance that we will perceive the room as having a "smooth" frequency response despite the resonances. If the number of resonances per third octave drops as the frequency rises then there will be a perceptually noticeable peak in the frequency response.

Coincident modes are also another way of creating a perceptually noticeable frequency response peak and the Bonello criteria do further stipulate that there should be no modal coincidence within a third octave band unless there are at least three additional non-coincident resonances to balance the two that are coincident. As an example of the calculation of mode frequencies let us calculate some for a typical living room.

EXAMPLE 6.10

Calculate the lowest frequency mode in a room which measures 3.5 m × 5 m × 2.5 m. At what frequency would a tangential mode with one half wavelength along the 3.5 m dimension and three half wavelengths along the 5 m dimension occur, at what frequency would the (2 2 2) oblique mode occur, and at what frequency is the first coincident mode?

Using Equation 6.21 calculate the modes as follows. The lowest frequency mode is the first axial mode along the longest dimension of the room, which is the (0 1 0) or axial mode in this example, so the lowest modal frequency in the room is:

$$f_{10} = \frac{c}{2}\sqrt{\left(\frac{0}{3.5m}\right)^2 + \left(\frac{1}{5m}\right)^2 + \left(\frac{0}{2.5m}\right)^2} = \frac{344 \text{ ms}^{-1}}{2}\sqrt{\left(\frac{1}{5m}\right)^2}$$

$$= 34.4Hz$$

The mode with one half wavelength along the 3.5 m dimension and three half wavelengths along the 5 m dimension is the (1 3 0) or tangential mode so its frequency is:

$$f_{130} = \frac{c}{2}\sqrt{\left(\frac{1}{3.5m}\right)^2 + \left(\frac{3}{5m}\right)^2 + \left(\frac{0}{2.5m}\right)^2}$$

$$= 172ms^{-1}\sqrt{0.082 + 0.36} = 114.4Hz$$

The frequency of the (2 2 2) or oblique mode is:

$$f_{222} = \frac{c}{2}\sqrt{\left(\frac{2}{3.5m}\right)^2 + \left(\frac{2}{5m}\right)^2 + \left(\frac{2}{2.5m}\right)^2}$$

$$= 172 \text{ ms}^{-1}\sqrt{0.327 + 0.16 + 0.64} = 182.6 \text{ Hz}$$

The dimensions of 2.5 m and 5 m are related by a factor of 2 so the second axial mode along the 5 m dimension will be at the same frequency as the first axial mode along the 2.5 m dimension. That is:

$$f_{020} = f_{001}$$

The (0 2 0) mode has a frequency of:

$$f_{020} = \frac{c}{2}\sqrt{\left(\frac{0}{3.5m}\right)^2 + \left(\frac{2}{5m}\right)^2 + \left(\frac{0}{2.5m}\right)^2}$$

$$= 172 \text{ ms}^{-1}\sqrt{\left(\frac{2}{5m}\right)^2} = 68.8Hz$$

and the (0 0 1) mode has a frequency of:

$$f_{001} = \frac{c}{2}\sqrt{\left(\frac{0}{3.5m}\right)^2 + \left(\frac{0}{5m}\right)^2 + \left(\frac{1}{2.5m}\right)^2}$$

$$= 172 \text{ ms}^{-1}\sqrt{\left(\frac{1}{2.5 m}\right)^2} = 68.8Hz$$

which are both at the same frequency and are therefore coincident.

6.2.6 The behavior of modes

As has been already discussed, modes behave differently to diffuse sound and this has the following consequences:

■ The modes are not absorbed as strongly as sound which visits all surfaces. This is due to both the reduction in the number of surfaces visited and the change in absorption due to non-random incidence.

■ This reduction in absorption is strongly frequency dependent and results in less absorption and therefore a longer decay time at the frequencies at which standing waves occur.

■ The decay of sound energy in the room is no longer a single exponential decay with a time constant proportional to the average absorption in the room. Instead there are several decay times. The shortest one tends to be due to the diffuse sound field whereas the longer ones tend to be due to the resonant modes in the room. This results in excess energy at those frequencies with the attendant degradation of the sound in the room.

How does the energy in a mode decay as a function of time, how can it be related to the reverberation, and what is the effect of absorption in a mode on the frequency response?

6.2.7 The decay time of axial modes

The decay of sound energy in modes is in many respects identical to the decay of sound energy which is analyzed in Appendix 4. The main difference is that the absorption coefficient is sometimes smaller, because the modal sound wave does not have random incidence; it will also be specific to the surfaces involved instead of being an average value for the whole room. In addition the time between reflections will be dependent on the length of the modal path rather than the mean free path. This means that the decay time for a mode is likely to be different to the diffuse sound.

For example the length of an axial mode path is determined by the distance between the two reflecting surfaces that support the mode, which will be one of the room's dimensions. Thus for an axial mode the energy left after a given time period is given by modifying Equation A4.5 in Appendix 4 using the distance between the surfaces instead of the mean free path to give:

$$\text{Modal energy}_{\text{After } t \text{ seconds}} = \text{Modal energy}_{\text{Initial}} \left(1 - \alpha_{\text{mode}}\right)^{t(c/L_{\text{mode}})}$$

where L_{mode} = distance between the surfaces in the mode (in m)
and α_{mode} = absorption per reflection in the modal structure

The above equation can be manipulated to give a 60 dB decay time, analogous to Equation 6.17, which is:

$$T_{60(Modal)} = \left(\frac{L_{mode}}{c}\right)\frac{\ln(10^{-6})}{\ln(1-\alpha_{mode})} = \left(\frac{L_{mode}}{\ln(1-\alpha_{mode})}\right)\frac{(-13.82)}{344 \text{ ms}^{-1}}$$

$$= \frac{-0.04 L_{mode}}{\ln(1-\alpha_{mode})}$$

where $T_{60 \ (Modal)}$ = the 60 dB decay time of the mode (in s)

This expresses a similar result to Equation 6.17 except for the difference caused by the differing length of the modal structure with respect to the mean free path. If the length of the modal structure is longer than the mean free path then, assuming similar levels of absorption, the decay time for the mode will be longer than the diffuse field whereas if the length is smaller then the modal decay will be shorter than the diffuse field. The length between reflections is both a function of the surfaces that support the mode and the type of mode—axial, tangential, or oblique—that occurs. For axial modes the mode length, L_{mode}, is simply the relevant room dimension.

6.2.8 The decay time of other mode types

For the other types of mode the situation is more complicated, as shown in Figure 6.35 for a tangential mode. However, one could argue that the path length for this type of mode is given by the length of the hypotenuse of the triangle formed by half the length and half the width of the four walls that support the mode. That is, the modal length is given by:

$$L_{mode \ (tangential)} = \sqrt{\left(\frac{\text{Length}}{2}\right)^2 + \left(\frac{\text{Width}}{2}\right)^2}$$

$$= \frac{1}{2}\sqrt{(\text{Length}^2 + \text{Width}^2)}$$

This equation shows that the distance between reflections for a tangential mode is essentially the diagonal dimension between the four surfaces that support the mode divided by two to allow for the fact that the wave suffers

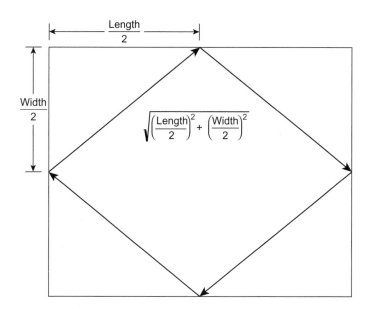

FIGURE 6.35

The path length for a tangential mode.

two reflections along this path. The length so derived can then be used as the modal length in the equation for the modal decay. A similar argument can be applied to the oblique modes, which visit all six surfaces. Because of this the modal supporting structure is a cuboid and the diagonal between opposing corners must be used. In addition the wave will suffer three reflections along this path. This gives a path length for the oblique mode as:

$$L_{\text{mode (oblique)}} = \sqrt{\left(\frac{\text{Length}}{3}\right)^2 + \left(\frac{\text{Width}}{3}\right)^2 + \left(\frac{\text{Height}}{3}\right)^2}$$

$$= \frac{1}{3}\sqrt{(\text{Length}^2 + \text{Width}^2 + \text{Height}^2)}$$

The absorption does more than cause the mode to decay; it reduces the total energy stored in the mode, in a similar manner to the effect of absorption on the reverberant field, and also causes the mode to have a finite bandwidth which is proportional to the amount of absorption, as shown in Figure 6.36. The absorption also reduces the peak to minimum variation in the standing wave pattern, and so reduces the spatial variation of the sound pressure, as shown in Figure 6.37. The bandwidth of a mode can be calculated from the 60 dB decay time using the following equation:

$$Bw_{\text{mode}} = \frac{2.2}{T_{60(\text{Modal})}}$$

FIGURE 6.36

The bandwidth of modes for a given value of absorption.

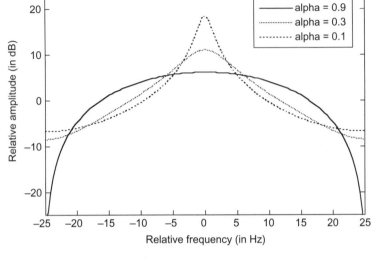

FIGURE 6.37

The spatial variation in the amplitude of modes for a given value of absorption.

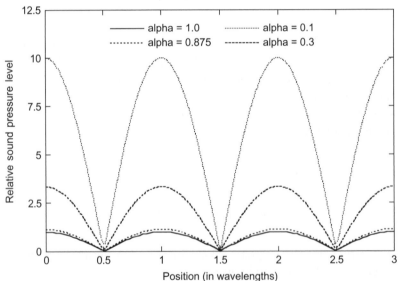

where Bw_{mode} = the $-3\,\text{dB}$ bandwidth of the mode (in Hz)

Because it is not always possible to calculate the true modal decay time this equation can be dangerously approximated using the reverberation time of the room as:

$$Bw_{\text{mode}} \approx \frac{2.2}{T_{60}}$$

This assumption is dangerous because the mode will not be diffuse whereas the reverberation time calculation assumes a diffuse sound field. In general the bandwidths and intensity levels of a mode are proportional to the number of reflections required to support them. This means that axial modes tend to be the strongest followed by tangential and then oblique modes in order of strength, as shown in Figure 6.38. However, this is not always the case as a tangential mode in a room with four reflecting surfaces could be stronger than the axial mode between the other two absorbing surfaces.

EXAMPLE 6.11

Calculate the approximate modal bandwidth in a room which has a reverberation time (T_{60}) of 0.44 seconds. What would be the modal bandwidth of axial modes along the 5 m dimension of the room if the absorption coefficients on the opposing walls were equal to the average room absorption coefficient of 0.2?

The approximate modal bandwidth can be calculated as:

$$Bw_{mode} \approx \frac{2.2}{T_{60}} = \frac{2.2}{0.44 \text{ s}} = 5 \text{Hz}$$

To answer the second problem one must calculate the modal decay time, $T_{60 \text{ (Modal)}}$, which is given by:

$$T_{60 \text{ (Modal)}} = \frac{-0.04L}{\ln(1 - \alpha_{mode})} = \frac{-0.04 \times 5\text{m}}{\ln(1 - 0.2)} = 0.9 \text{ s}$$

This value of decay can be used to calculate the actual modal bandwidth as:

$$Bw_{mode} = \frac{2.2}{T_{60 \text{ (Modal)}}} = \frac{2.2}{0.9 \text{ s}} = 2.4 \text{ Hz}$$

Clearly care must be taken when calculating modal bandwidths due to the fact that the diffuse field assumptions no longer apply. In the case above, even though an even distribution of absorption was assumed, the decay time and bandwidth were radically different from that predicted by the diffuse field assumptions simply because the path traveled by the sound wave was longer than the mean free path. In practice the absorption coefficient

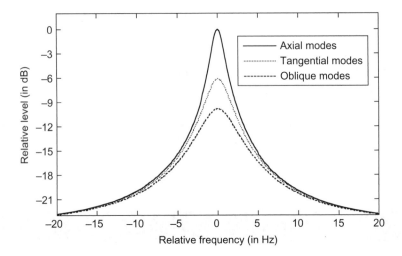

FIGURE 6.38

Typical variation in the amplitudes and bandwidths for the different mode types, assuming an even distribution of absorption.

is likely to be different as well, making prediction even more difficult. Note that, if the absorption remains constant with frequency, the bandwidths of a mode are independent of their frequency—they are simply a function of the modal decay time.

6.2.9 Critical frequency

Because all rooms have modes in their lower frequency ranges, there will always be a frequency below which the modal effects dominate and the room can no longer be treated as diffuse. Even anechoic rooms have lower frequency limits to their operation. One of the effects of room modes is to cause variations in the frequency response of the room, via its effect on the reverberant field. The frequency response due to modal behavior will also be room position dependent, due to the spatial variation of standing waves. An important consequence of this is that the room no longer supports a diffuse field in the modal region and so the reverberation time concept is invalid in this frequency region. Instead an approach based on modal decay should be used. But at what frequency does the transition occur, and can it even be calculated? Consider the typical frequency response of the room shown in Figure 6.39. In it, three different frequency regions can be identified.

- *The cut-off region*: This is the region below the lowest resonance, sometimes called the room cut-off region or the pressure zone frequency. In this region the room is smaller than a half wavelength in all dimensions. This does not mean that the room does not support sound propagation, in fact it behaves more like

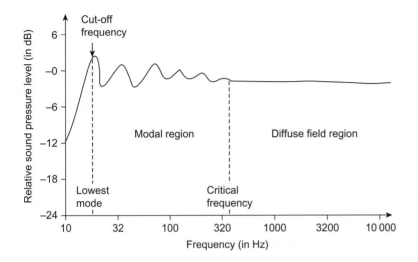

FIGURE 6.39

The frequency response of a typical room.

the air in a bicycle pump when the end is blocked. This means that the environment "loads" any sources of sound in the room differently (such as loudspeakers or musical instruments), and often, but not always, the effect of this loading is to reduce the ability of the source to radiate sound into the room, and so results in reduced sound levels at these frequencies. This is because the acoustic impedance at the source is altered, making it harder for it to radiate sound. The low-frequency cut-off can be calculated simply from:

$$f_{\text{cut-off}} = \frac{c}{2 \times \text{Longest dimension}}$$

$$= \frac{344 \text{ ms}^{-1}}{2 \times \text{Longest dimension (m)}}$$

- *The modal region*: The next region is the modal region in which the modal behavior of the room dominates its acoustic performance. In this region the analysis based on the assumption of a diffuse field is doomed to fail.

- *The diffuse field region*: The final region is the region in which a diffuse field can exist and therefore the concept of reverberation time is valid. In general this region of the frequency range is the one that will sound the best, provided the reverberation characteristics are good, because the effects of room modes are minimal and so the listener experiences an even reverberant sound level throughout the room.

The transition boundary between the region of modal behavior and the region of diffuse behavior is known as the "critical frequency." As is usual in these situations, although the critical frequency is a single frequency it is not a sharp boundary: it represents some defined point in a transition region between the two regions.

6.2.10 Acoustically "large" and "small" rooms

The concept of critical frequency (also known as the "Schroeder frequency" or "large room frequency") allows us to define the difference between rooms which are "large" and "small" in acoustical terms. In an acoustically large room the critical frequency is below the lowest frequency of the sound that will be generated in the room whereas in an acoustically small room the critical frequency will occur within the frequency range of the sounds being produced in it. Examples of acoustically large rooms would be concert halls, cathedrals and large recording studios. Most of us listen to and produce music in acoustically small rooms, such as bedrooms, bathrooms and living rooms, and there is an increasing trend—due to the effect of computer recording and editing technology and because it's cheaper—to perform more and more music and sound production tasks in small rooms.

6.2.11 Calculating the critical frequency

How can the critical frequency be calculated? There are two main approaches. The first is to recognize that when the wavelength of the sound approaches the mean free path of the room then the likelihood of modal behavior increases, because a sound wave is "in touch" with all the walls in the room. This approach can be used to set an approximate lower frequency bound on the critical frequency below which it is likely to be difficult to prevent modal effects from dominating the acoustics without extreme measures being taken. This approach gives the following expression for calculating the critical frequency, which assumes that modal behavior dominates once the mean free path is equal to one and a half wavelengths:

$$f_{\text{critical}} = \left(\frac{3}{2}\right)\frac{c}{MFP} = \left(\frac{3}{2}\right)\frac{344 \text{ ms}^{-1}}{MFP}$$

This expression is useful for making a rapid assessment of the likelihood of achieving a particular critical frequency in a given room. However, the real critical frequency may well be higher because a room can have significant modal behavior at high frequencies if the absorption is low. Because of this the accepted definition of critical frequency is based on the mode bandwidth, although this can result in a chicken and egg situation at the initial design stages, hence the earlier equation. The rationale for this is as follows. The main consequence of modal behavior is the frequency and spatial variation caused by it. This means that if a given frequency excites only one mode, then this variation will be very strong. However, if a given frequency excites more than one mode, both the spatial and frequency variation will be reduced.

Figure 6.40 shows the effect of adding three adjacent modes together; it shows that once more than three adjacent modes are added together the variation is considerably reduced. The way to excite adjacent modes with a single frequency is to increase their bandwidth until the three bandwidths associated with the three modes overlap a given frequency point, as shown in Figure 6.41. The critical frequency is defined as when the modal overlap equals three, so at least three modes are excited by a given frequency, and is given by:

$$f_{\text{critical}} = 2102\sqrt{\left(\frac{T_{60}}{V}\right)}$$

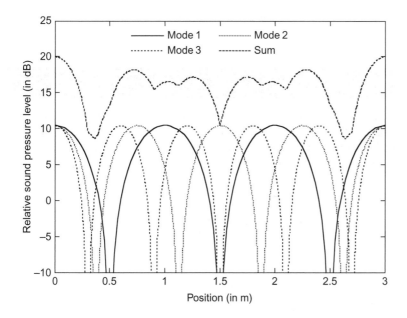

FIGURE 6.40

The composite effect of adjacent modes on the spatial variation in a room.

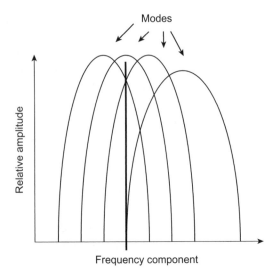

This equation shows that the critical frequency is inversely proportional to the square root of the room volume and is proportional to the square root of the reverberation time, which is also proportional to the cube root of the volume, if the absorption remains constant, as discussed earlier. The net result of this is that, as expected, the critical frequencies for larger rooms are generally lower than those of smaller ones. Thus big rooms are acoustically "large" as well.

As an example let us calculate the critical frequency of our typical living room.

EXAMPLE 6.12

What is the critical frequency of a room whose surface area is 75 m², and whose volume is 42 m³? What is the critical frequency of the same room if the average absorption coefficient is 0.2?

Using the first equation calculate the lowest bound on the critical frequency using the mean free path as:

$$f_{critical} = \left(\frac{3}{2}\right)\frac{c}{MFP} = \left(\frac{3}{2}\right)\frac{c}{\frac{4V}{S}} = 1.5 \times \frac{344\,\text{ms}^{-1}}{\frac{4 \times 42\,\text{ms}^3}{75\,\text{ms}^2}}$$

$$= 1.5 \times \frac{344\,\text{ms}^{-1}}{2.24\,\text{m}} = 230\text{Hz}$$

Using the second equation calculate the critical frequency using the reverberation time. Firstly, calculate the reverberation time as:

$$T_{60} = \frac{-0.161V}{S \ln(1-\alpha)} = \frac{-0.161 \times 42m^3}{75m^2 \times \ln(1-0.2)} = 0.43 \text{ s}$$

Then, using the second equation, calculate the critical frequency as:

$$f_{critical} \approx 2102\sqrt{\left(\frac{T_{60}}{V}\right)} = 2102\sqrt{\left(\frac{0.43 \text{ s}}{42 \text{ m}^3}\right)} = 213 \text{ Hz}$$

The second equation predicts a slightly lower critical frequency compared with the first one. However, the agreement is surprisingly good. Although the modal overlap has been calculated using a reverberation time, and hence a diffuse field assumption, this is probably just valid at this frequency, which represents the boundary between the two regions. The critical frequency results show that, for this room, frequencies below 213 Hz must be analyzed using modal decay time rather than reverberation time.

6.3 ABSORPTION MATERIALS

Absorption materials are clearly important in their effects on the acoustics, and this section briefly looks at the factors which affect the performance of these materials and their effects on an acoustic space.

There are two basic forms of absorption materials—porous absorbers and resonant absorbers—which behave differently because their mechanisms of absorption are different.

6.3.1 Porous absorbers

Porous absorbers, such as carpets, curtains and other soft materials, work due to frictional losses caused by the interaction of the velocity component of the sound wave with the surface of the absorbing material. In Chapter 1 we saw that the velocity component arose because the air molecules had to move between the compression and rarefaction states. A given pressure variation will require a greater pressure gradient, and hence higher peak velocities, as the wavelength gets smaller with rising frequency. Because the pressure gradient of a sound wave increases with frequency, the friction due

FIGURE 6.42

Typical absorption curves for porous absorbers.

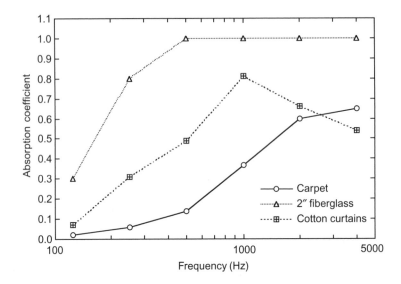

FIGURE 6.42

Typical absorption curves for porous absorbers.

to interaction with a surface will also increase with frequency and therefore the absorption of these types of material also rises with frequency. Clearly the larger the surface area available for interaction, the higher the friction and therefore the absorption. This means that porous materials, which consist of a large number of fibers per unit volume, such as high-density rock-wool or fiberglass, and plush carpets, will tend to have a high-level absorption. This also explains why curtains which are draped to a fraction of their cloth area absorb more strongly than ones which are flat.

Typical absorption curves for porous absorbers are shown in Figure 6.42. Because porous absorbers interact with the velocity component of the sound wave, they are affected by the space between them and the wall and their thickness. This is due to the fact that at the exposed surface of a hard surface, such as a wall, the velocity component is zero whereas at a quarter of a wavelength away from the wall the velocity component will be at a maximum, as shown in Chapter 1, and so a porous material will absorb more strongly at frequencies whose quarter wavelength is less than either the spacing of the material from the wall, or the thickness of the material if it is bonded directly to the surface.

This effect is shown in Figure 6.43. Although in principle there could be a variation in the absorption coefficient as the frequency increases above the quarter wavelength point, due to the inherent variation of the velocity component as a function of wavelength at a fixed distance from a surface, in practice this does not occur unless the material is quite thin.

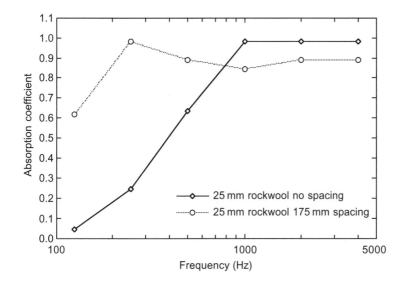

FIGURE 6.43

The effect of spacing a porous absorber away from a hard surface.

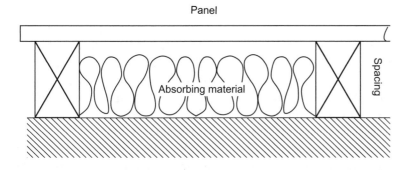

FIGURE 6.44

Typical construction of a panel absorber.

6.3.2 Resonant absorbers

Resonant absorbers, such as wood paneling, work because the incident sound energy causes vibrations in the absorber and these are converted to frictional losses within the absorbing structure itself. This makes them sensitive to the pressure component of the sound wave and so they work well when attached to walls. The typical construction of a panel absorber is shown in Figure 6.44. In the case of wood panels the absorption is due to the internal frictional losses in the wood. In the perforated absorber, discussed later, it is due to the enhancement of velocity that happens in the perforations at resonance. Because the absorbers are resonant their absorption increases at low frequencies, as shown in Figure 6.45.

FIGURE 6.45

*Typical absorption curves
for panel absorbers.*

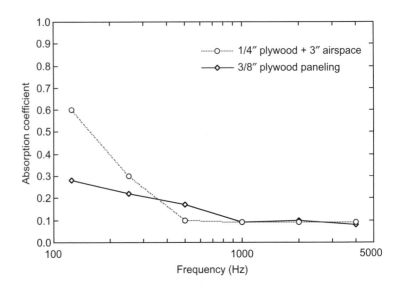

The resonant characteristics of these absorbers enable them to be tuned to low frequencies and so allow them to have absorption characteristics which complement those of porous absorbers. The peak absorption frequency of a resonant absorber is a function of the space behind the absorber and the effective mass of the front panel. To use an analogy with a spring and weight, the rear cavity acts like a spring whose stiffness is inversely proportional to the depth of the cavity and the effective mass per unit area of the front panel determines the size of the weight. As the spring gets less stiff and the effective mass becomes greater, the resonant frequency drops. Thus deeper rear cavities result in lower resonances for both types. For the panel absorbers the mass per unit area of the panel is directly related to the effective mass, so heavier front panels result in a lower resonant frequency. The resonance frequency of panel absorbers can be calculated using the following equation:

$$f_{resonance} = \frac{60}{\sqrt{Md}}$$

where M = panel's mass per unit area (in kg m^{-2})
and d = depth of the airspace (in m)

However, this equation must be applied with some caution because it assumes that the panel has no stiffness. This assumption is valid for thin panels but becomes less applicable as the panel becomes thicker and thus more stiff.

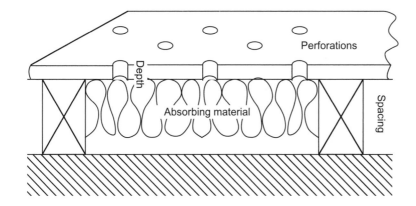

Typical construction of a Helmholtz resonant absorber.

6.3.3 Helmholtz absorbers

Another form of resonant absorber is based on the use of the resonance that occurs when air is trapped in a tube above an air space. This type of resonance is called a Helmholtz resonance and is the resonance that occurs in a beer bottle when you blow across it. The cavity acts like a spring and the air in the tube above the cavity acts like the mass. The construction of this type of absorber consists of a perforated panel above an airspace, as shown in Figure 6.46.

For the perforated panels the effective mass is a function of both the depth of the perforations and their effective area as a percentage of the total area. Their effective mass increases as the depth increases and the percentage hole area reduces. Typical absorption curves for this type of absorber are shown in Figure 6.47. This type of absorber is often used to add extra absorption at high-bass and low-midrange frequencies.

6.3.4 Wideband absorbers

It is possible to combine the effects of porous and resonant absorbers to form wideband absorbers. A typical construction is shown in Figure 6.48 and its performance is shown in Figure 6.49. As with all absorbers using rock-wool or fiberglass one must take precautions to prevent the egress of irritating fibers from the absorber into the space being treated.

An alternative means of achieving wideband absorption is to use a large depth of porous absorber, for example one meter, and this can provide effective absorption with a flat frequency response, but at the cost of considerable depth.

6.3.5 Summary

With these basic types of absorption structures it is possible to achieve a high degree of control over the absorption coefficient in a room as a function

FIGURE 6.47

*Typical absorption curves
for Helmholtz resonant
absorbers.*

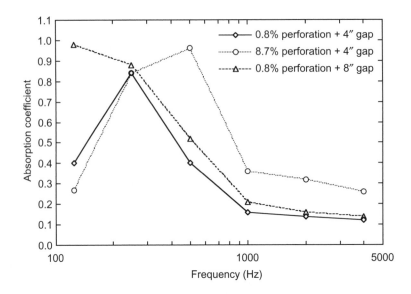

FIGURE 6.48

*Construction of a wideband
absorber.*

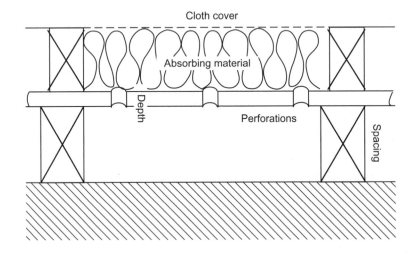

of frequency. In many cases much of the required absorption can be
achieved by using materials which fit naturally in the room. For example,
much baroque music was performed in the halls of mansions which had a
balanced acoustic due to the extensive use of wood paneling in their decoration. This paneling acted as an effective low-frequency absorber and, in
conjunction with the flags, drapes and tapestries that also decorated these
spaces, provided the necessary acoustic absorption.

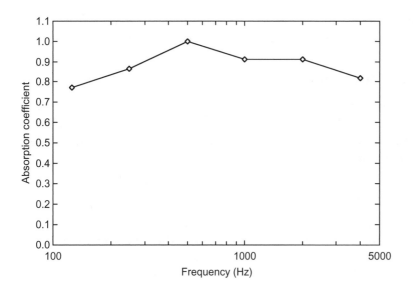

FIGURE 6.49

Typical absorption curves for wideband absorbers.

6.4 DIFFUSION MATERIALS

As well as absorption it is essential that the sound be diffused when it strikes a surface. Ideally we want the acoustic equivalent of a matt surface. Unfortunately most surfaces, including large areas of absorbing material, act like acoustic mirrors, with varying shades of darkness. In order to have a matt surface one needs a "bumpy wall" and many things can be used to provide this. Unfortunately the bumps need to be at least an eighth, and preferably a quarter, of a wavelength in size to be effective. This results in the requirement for very large objects at low frequencies (1.25–2.5 m at 34 Hz) and very small objects at higher frequencies (1.25–2.5 cm at 3.4 kHz). If the objects are too small, that is, less than one eighth of a wavelength, they will not diffuse properly; if they are too big, that is, greater than about a half a wavelength, they will behave as acoustic mirrors in their own right and so will not diffuse effectively.

Clearly effective diffusion is a difficult thing to achieve in an ad hoc manner. Curved and angled structures can help at mid and high frequencies, and at very high frequencies, greater than about 4 kHz, the natural rough textures of materials such as brick and rough cut stone are effective. Because of the need to achieve well-defined diffusion characteristics, diffusion structures based on patterns of wells whose depths are formally defined by an appropriate mathematical sequence have been proposed and used (Schroeder, 1975 and D'Antonio and Konnert, 1984). The design of these structures is quite involved and the reader is directed to the references if they want more information. However, a brief description of how they work is as follows.

6.4.1 How diffusers work

Consider a hard surface consisting of bumps of height d. Also consider an acoustic wavefront approaching it from a normal direction. The way this wavefront is reflected will depend on the height of the bumps relative to its wavelength. Let us consider three cases:

- In the case of $d \ll \lambda$ the surface will behave like a flat surface and specularly reflect the wavefront.

- In the case of $d = \frac{\lambda}{4}$ the wavefronts which are reflected from the front of the bumps are reflected $\frac{\lambda}{2}$ earlier than those from the surface. This means that in the normal direction the wavefronts cancel and so no sound pressure is propagated in this direction. However, there has been no energy loss in the system so the wavefront must be reflected in some direction. In fact as one moves away from the normal direction the relative path lengths between the bump and the surface become less and the amplitude of the wavefront increases. This is the basic principle behind diffusion using hard reflectors. That is, the diffusing surface modifies the phase of the wavefronts so that the reflected wave must propagate in directions other than the specular direction.

- In the case of $d = \frac{\lambda}{2}$ the wavefronts from the bumps and surface are delayed by λ and so arrive back in phase. Thus the bumps disappear and the surface behaves as if it were flat. That is, it behaves like a specular reflector.

So, one has a problem: a regular sequence of bumps will diffuse but only at frequencies at which it is an odd multiple of $\frac{\lambda}{4}$. Note also that these frequencies will depend on the angle of incidence of the incoming wavefront.

What is required is a pattern of bumps which alter the phases of the incident waves in such a way that two objectives are satisfied:

1. The sound is scattered in some "optimum" manner.

2. The scattering is optimum over a range of frequencies.

These objectives can be satisfied by several different sequences, but they share two common properties:

- The Fourier transform of the sequence is constant except for the d.c. component which may be the same or lower. This satisfies objective (1) because it can be shown that reflection surfaces with such a property scatter energy equally in all directions. The effect of a reduced d.c. component is to further reduce the amount of energy which is reflected in the specular direction.

■ The second desirable property of these sequences is that the Fourier transform is unaffected if the wavelength of the incident sound varies. This has the effect of changing the scale of the sequence, but again one can show that the resulting sequence still has the same properties as the original sequence.

Both the above properties arise because the sequences work by perturbing the wavefronts over a full cycle of the waveform. Such sequences are called phase reflection gratings because they perturb the phase of the wavefront.

To make this a little clearer, let us consider two sequences which are used for diffusers.

1. *Quadratic residue sequences* well depth $= n^2 \bmod p$ where p is a prime number. If $p = 5$ this gives a set of well depths of

    0, 1, 4, 4, 1, 0, 1, . . ., etc.

so the sequence repeats with a period of 5.

2. *Primitive root sequences* well depth $= a^n \bmod p$ where p is a prime and a is a suitable constant called a primitive root. For $a = 2$ and $p = 5$ we get the sequence

    1, 2, 4, 3, 1, 2, . . ., etc.

Here we have a sequence which has a period of 4 (5 − 1).

At the lowest design frequency for these examples a well of depth 5 would correspond to $\frac{\lambda}{2}$. At higher frequencies the sequences still have the same properties and thus scatter sound effectively. However, when the frequency gets high enough so that $\frac{\lambda}{2}$ becomes equal to the minimum difference in depths (1) then the surface again becomes equivalent to a flat surface.

The typical construction of these structures is shown in Figure 6.50 and their performance is shown in Figure 6.51.

6.4.2 Discussion

As we have seen, these sequences achieve their performance by spreading the phase of the reflected wavefront over at least one cycle of the incident wavefront. In order to do this, their maximum depth must be $\frac{\lambda}{2}$ at the lowest design frequency. This means that to achieve diffusion a reasonable depth is required. For example, to have effective diffusion down to 500 Hz a depth of 34 cm (13.5 inches) is required. To get down to 250 Hz one would

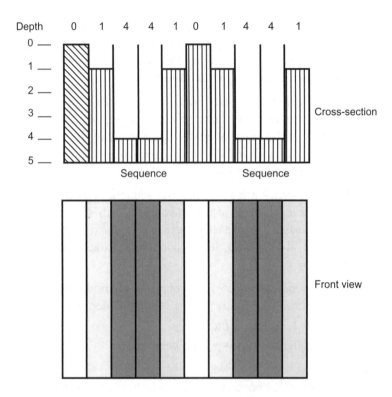

FIGURE 6.50

Typical construction of a quadratic residue diffuser.

need to double this depth. However, as we have seen, a simple bump of ¾ can provide diffusion, albeit somewhat frequency dependently. This is half the depth of the above sequences and represents the ultimate limit for a diffusing object.

It is possible to have sequences which achieve the phase scatter required for good diffusion using a depth closer to ¼ at the lowest frequency (4) and so allow better performance diffusers in restricted spaces. However, even ¼ at low frequencies is often too large to be useful. What one really requires is a diffuser which is effective without using any depth!

6.4.3 Amplitude reflection gratings

It is not just physically observable bumps on the wall that can cause diffusion of the sound. In fact any change in the reflecting characteristics of the surface will cause diffusion. The change from an absorbing region on a wall to a reflecting one is an example of a change that will cause the sound to scatter. Thus it is always better to distribute the absorption in small random amounts around a room rather than concentrate it in one particular area. As well as encouraging diffusion this strategy will avoid the possibility

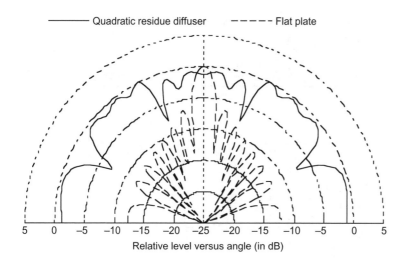

—————— Quadratic residue diffuser   — — — — Flat plate

Relative level versus angle (in dB)

FIGURE 6.51

Typical performance of a quadratic residue diffuser compared with a flat plate. (Note that for convenience the responses have normalized to 0 dB in the specular direction. In practice the flat plate's output would be greater in this direction.)

that some modes might shuttle between surfaces with minimum absorption. There are also mathematically-based procedures for the optimum placement of absorbing materials to encourage diffusion; for more details on this see Angus (1995). What is required is an amplitude weighting, that is, a pattern of absorbers, which gives a flat Fourier transform.

The most obvious sequences to consider are binary, that is, they contain only levels 0 and 1 where 1 represents reflection from a hard surface and 0 represents absorption from some form of absorbing material. Clearly not all acoustic absorbers are 100% absorbing but this can be simply allowed for by using (1-absorption) instead of zero in the sequence. The net effect of less than 100% absorption would be to increase the level of the specular component. Of the many possible binary sequences, m-sequences would seem to be a good starting point as they have desirable Fourier properties. There are many other bi-level sequences which have flat Fourier transforms but m-sequences are well documented.

Thus amplitude reflection gratings consist of a surface treatment which consists of strips of absorbing material, whose width is less than $\frac{\lambda}{2}$ at the highest frequency of use, laid out in a pattern in which strips of absorber represent zero and strips of reflecting wall represent 1 (see Figure 6.52). Note that because we are not depending on depth we do not have a low-frequency limit to the range of diffusion, only a high-frequency limit which is a function of the width of the strips.

A two-dimensional example of an amplitude reflection grating is shown in Figure 6.53. Amplitude gratings provide some diffusion although they cannot be as good at diffusing as phase reflection gratings. But, because of

> m-sequences are pseudo-random binary sequences that can be easily generated by software or hardware.

FIGURE 6.52

Simple implementation of a length 15 one-dimensional Binary Amplitude Diffuser.

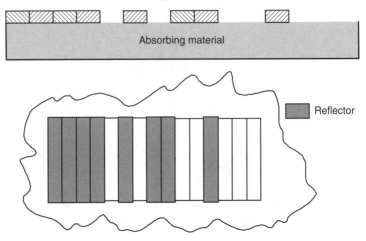

FIGURE 6.53

An implementation of a length 1023 two-dimensional Binary Amplitude Diffuser (white dots are holes over absorber).

their size, they are useful at low frequencies. It also is possible to develop curved diffusion structures, although there are no simple mathematical recipes for them. For further details see Cox (1996). Other structures are possible and the reader is referred to the references for more information.

6.5 SOUND ISOLATION

No discussion of the quality of sound in a room would be complete without a brief discussion of how to keep unwanted sound from entering a room, or how to keep the wanted sound in, so as not to disturb the pleasure of people inside or outside it.

   The first thing to note is that just because a material is a good absorber of sound doesn't mean that it is a good isolator of sound. In fact most absorbing materials are terrible at sound isolation. This is because, in the sound isolation case, we are interested in the amount of sound that travels through a structure rather than the amount that is absorbed by it, as shown in Figure 6.54. A poor value of sound isolation would be around 20 dB yet it corresponds to only one hundredth of the sound being transmitted. A good absorber with an absorption coefficient of 0.9 would let one tenth of the sound through, which corresponds to a sound isolation of only 10 dB! As we are more interested in sound isolations of 40 dB as a minimum, absorption is clearly not the answer.

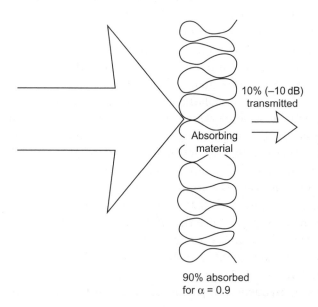

10% (−10 dB)
transmitted

Absorbing
material

90% absorbed
for α = 0.9

FIGURE 6.54

Sound transmission versus sound absorption in a material.

FIGURE 6.55

*Sound transmission as a
function of frequency for a
partition.*

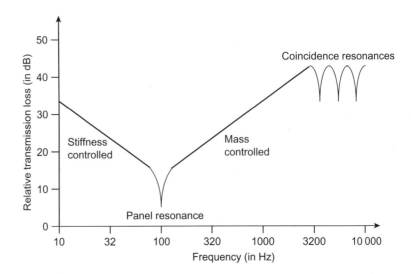

6.5.1 Ways of achieving sound isolation

There are only two ways to achieve sound isolation: using either stiffness or mass. Figure 6.55 shows the attenuation of a partition as a function of frequency and from it one can see that stiffness is effective at low frequencies due to the fact that the sound wave must push against the stiffness of the partition. This is known as the "stiffness-controlled isolation region." As the frequency rises, the partition needs to move less distance to re-radiate a given level of sound and so it gets less effective until at the resonant frequency of the partition its level of attenuation is at its lowest value. This is due to the fact that at resonance the partition can be moved easily by the incident sound wave and so re-radiates the sound effectively.

As the frequency rises above the partition's resonant frequency, the mass-controlled region of isolation is entered. In this region, it is the fact that the sound must accelerate a heavy mass that provides the isolation. Because more force is required to move the partition at higher frequencies, the attenuation rises as the frequency rises. At even higher frequencies there are resonances in which both the thickness of the partition and the way sound propagates within it interact with the incident sound to form coincident resonances that reduce the attenuation of the partition. Damping can be used to reduce the effect of these resonances.

Most practical partitions operate in the mass-controlled region of the isolation curve with coincident resonances limiting the performance at higher frequencies. Figure 6.56 shows the attenuation of a variety of single partitions as a function of frequency. In particular note that the plaster board wall has

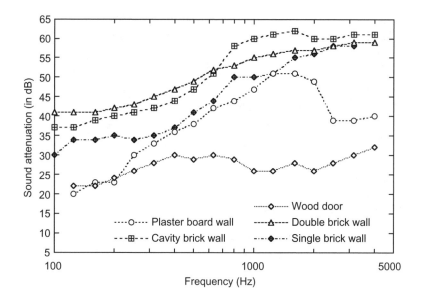

FIGURE 6.56

Sound transmission versus frequency for typical partitions.

a significant coincidence resonance. The performance of a single partition increases by 3 dB every time its mass is doubled but the coincidence resonances move lower in frequency as well. These coincidence resonances limit the ultimate performance of single partitions. In addition the cost, and size, of single partitions get unreasonable for large attenuations.

6.5.2 Independent partitions

The solution is to have two or more partitions which are independent of each other. If the two partitions are truly independent then the total attenuation, or effective sound isolation, is the product of the attenuations of individual partitions, that is, the dB attenuation is the sum of the dB attenuations of the individual partitions. In practice the partitions are not independent although the isolation is improved dramatically, but not as much as would be predicted by simply summing the dB attenuations. Coincidence resonances also reduce the effectiveness of a partition and it is important to ensure that the two partitions have different resonances. This is most easily assured by having them made with either a different thickness, or a different material.

As an amusing example Figure 6.57 shows the measured results, from Inman (1994), for single and double-glazing made with similar and different thicknesses of glass and spacing. Because of the effect of the coincidence resonances the double-glazed unit with 4 mm glass is actually worse than a

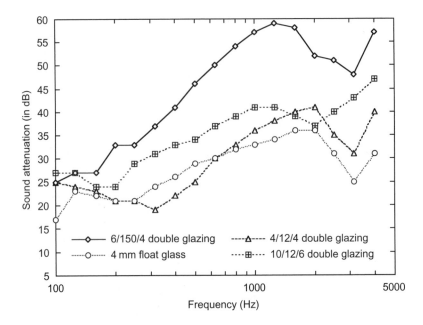

FIGURE 6.57

Sound transmission versus frequency for single and double-glazing.

single pane of 4 mm glass! As the other two curves show, if the glass is dissimilar the result is much improved, and is further improved if the spacing is increased so as to reduce the coupling between the individual partitions. Often, absorbing material is placed in the cavity between the two partitions to reduce the effect of coincidence resonances but it is important to ensure that the absorbing materials do not make contact with the two partitions or else flanking may occur.

6.5.3 Flanking paths

Flanking paths, which are the main limitation to sound isolating structures, arise when there are other paths that the sound can travel through in order to get round, that is, flank, the sound isolating structure, as shown in Figure 6.58. Typical paths for flanking are the building structure, heating pipes, and, most commonly, ventilation systems or air leaks. The effect of the building structure can be reduced by building a "floating room," as shown in Figure 6.59, which removes the effect of the building structure by floating the room on springs away from it.

In practice, ensuring that no part of the building is touching the floating room by any means (plumbing pipes and electrical wiring conduits are popular offenders in this respect) is extremely difficult. The effect of ventilation systems and air leaks is also a major source of flanking in many

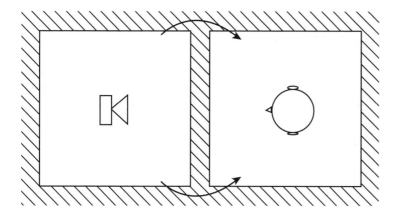

FIGURE 6.58

Flanking paths in a structure.

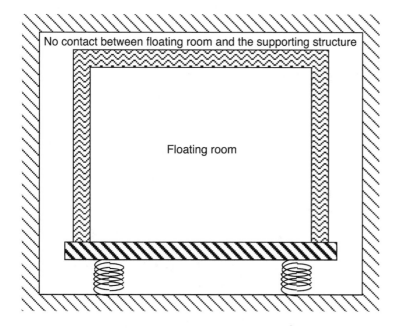

No contact between floating room and the supporting structure

Floating room

FIGURE 6.59

Floating room construction.

cases. In fact in the domestic situation the sound isolation is almost entirely dominated by air leaks and draught paths, and it is the removal of these that allow double-glazing salesmen to advertise a dramatic improvement in sound isolation, despite having two 4 mm panes of glass in the double-glazing.

So in order to have good sound isolation one needs good partitions and an airtight, draught-free, structure. Achieving this in practice while still allowing the occupants to breathe is a challenge.

6.6 THE EFFECT OF ROOM BOUNDARIES ON LOUDSPEAKER OUTPUT

If we wish to listen to sound from loudspeakers we must also consider the effect the room has on them. We have already seen that the reverberant field colors the sound of the loudspeakers depending on its level relative to the direct sound. However, the walls of the room also have an effect.

The effect of the presence of boundaries, such as walls, on the low-frequency output of loudspeakers is well known (Allison, 1980). Many methods of ameliorating these effects have been proposed, ranging from careful positioning to special speaker designs that try to remove the effect of the boundaries.

When a speaker is placed near a reflecting boundary an image source is formed due to that reflecting boundary, as shown in Figure 1.28 in Chapter 1. Thus when a speaker is placed near three reflecting boundaries three image sources are formed, as shown in Figure 6.60. The effect of these image sources, which represent reflections from the boundary, is to modify the local acoustic impedance seen by the loudspeaker. Because a dynamic speaker is mass controlled in its normal working region it forms

FIGURE 6.60

Images formed at the nearest three reflecting boundaries to a loudspeaker.

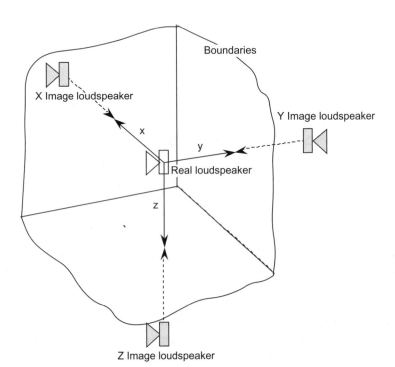

a high impedance source and therefore the radiated power is affected by the local impedance variations.

Waterhouse (1955) extended work by Rayleigh (1964) and confirmed experimentally that the power output of a source is affected the presence of boundaries, for example the walls in a room. The effect of this is plotted in Figures 6.61 and 6.62 for two different boundary conditions. In one case, all the dimensions are the same (Figure 6.61); in the other, they are different to minimize the effects (Figure 6.62). It does not matter from which surface a speaker has a particular spacing. However, it is important that the dimensions are in the correct ratio. Note that in these graphs the x-axis is the number

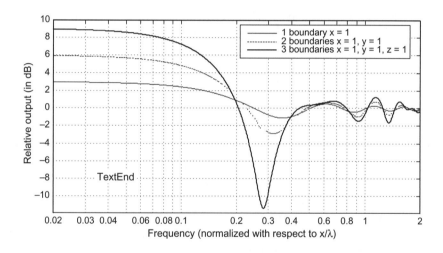

FIGURE 6.61

Loudspeaker boundary interaction "non-optimum" placement.

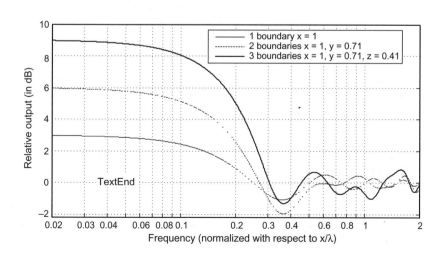

FIGURE 6.62

Loudspeaker boundary interaction "optimum" placement.

of wavelengths with respect to the distance "x" from the boundary. This is analogous to frequency, but scaled as with respect to distance. Thus if "x" equaled one meter a value of "1" on the x-axis would correspond to a wavelength of one meter (344 Hz); if "x" equaled half a meter a value of "1" on the x-axis would correspond to a wavelength of half a meter (688 Hz), and so on.

From these graphs, we can see two main things:

1. If the dimensions are the same then one has a low-frequency response that rises 9 dB above the mid-band response, and that has significant variation over the frequency range – in particular when the distance between the loudspeaker and the boundaries is just above a quarter of a wavelength

2. By placing the speaker carefully, with respect to the boundaries, one can eliminate some of the variation in the response. However, one still has a 9 dB rise in response at low frequencies. This is well known and the 9 dB rise at low frequencies is inherent in operating in an environment with three boundaries. In general, the low-frequency responses of loudspeakers designed to be placed close to boundaries is tailored to compensate for it.

In the case of flush mounting the speaker in the wall, the result is slightly different. Now the effect of the mounting surface extends to very high frequencies and therefore the number of surfaces that affect the low-frequency response reduces to two. This means the maximum low-frequency rise becomes 6 dB and the optimum ratios of placement change.

In the modal region the whole room affects the loudspeaker's output. In this region the output of the speaker is strongly affected by how the modal shapes in the room interact with the loudspeaker. As discussed with regard to bowed instruments in Chapter 4, the point of excitation strongly affects which modes are excited. This effect also happens with loudspeakers, except, unlike the bowed string, it is now a three-dimensional problem! It is sometimes possible to find compromise positions for the loudspeakers that excite the room modes as evenly as possible but, needless to say, this is very room-, and speaker-, dependent.

6.7 REDUCTION OF ENCLOSURE DIFFRACTION EFFECTS

The other aspect of loudspeakers is the effect of diffraction around the enclosure shape. If the speaker is wall mounted this is not important, but for free-standing speakers it is an issue.

In 1950, Olson (1969) demonstrated, experimentally, that the enclosure shape and size significantly affect the frequency response of a loudspeaker, due to diffraction effects. These diffraction effects can occur at any frequency where the wavelength is of similar size, or smaller than the box. This means that they can occur in any small enclosure, such as a top mounted tweeter. They can also occur in the main cabinets if there are drivers whose frequency ranges produce wavelengths of appropriate dimensions relative to the box sizes.

In the context of loudspeaker design it is the bounded to unbounded case that is important as this is what happens when a sound wave from a loudspeaker reaches the edge of the front face of an enclosure. In this situation, the wave is making a transition from being bounded by the front of the enclosure to being unbounded, as shown in Figure 6.63. The consequential reflections cause disturbances to the frequency response of the loudspeaker because they interfere with the main output.

Some of Olson's results are shown in Figure 6.64. His research showed that the shape of an enclosure should be, ideally, curved and small with respect to the wavelengths in use. He also found that the shape with the smoothest response was spherical; see Figure 6.64(b). However, he also showed that non-spherical shapes, such as the truncated pyramid and rectangular parallelepiped combination shown in Figure 6.64(c), could also give good performance.

Although an ideal shape is spherical, as shown in Figure 6.65(a), other curved shapes offer some practical advantages. In particular, a truncated ellipsoidal shape, as shown in Figure 6.65(b), offers a good diffraction performance and some flexibility in the volume for a given frontal area. This is important for several reasons.

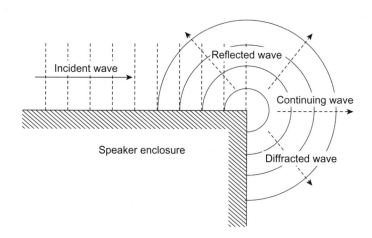

FIGURE 6.63

Reflection due to bounded–unbounded transition at a loudspeaker cabinet edge.

Incident wave

Reflected wave

Continuing wave

Speaker enclosure

Diffracted wave

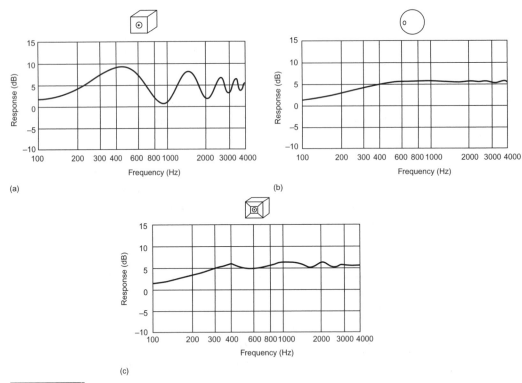

FIGURE 6.64 *Diffraction effects due to different enclosure (from Olson, 1969)*

Firstly, in order to maintain an approximately spherical shape, a spherical enclosure must have a diameter that is significantly larger than the diameter of the drive unit. In Figure 6.65a the sphere's diameter is twice that of the drive unit. If the diameter of the sphere equaled that of the drive unit the enclosure shape would be a half sphere, which would have a poorer performance than a sphere.

Secondly, the volume of the enclosure interacts with the drive unit to set a lower frequency limit for the loudspeaker. Clearly, a sphere cannot have a diameter that is less than the diameter of the drive unit. Truncated ellipsoidal shapes, as shown in Figure 6.65(b), have an additional degree of freedom that, within reason, allows one to set the volume independently of the diameter of the drive unit.

Finally, a sphere suffers from coincident resonances that can affect the frequency and time response of the loudspeaker. A truncated ellipsoidal

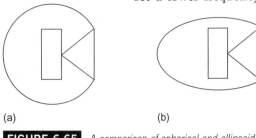

FIGURE 6.65 *A comparison of spherical and ellipsoidal loudspeaker enclosures*

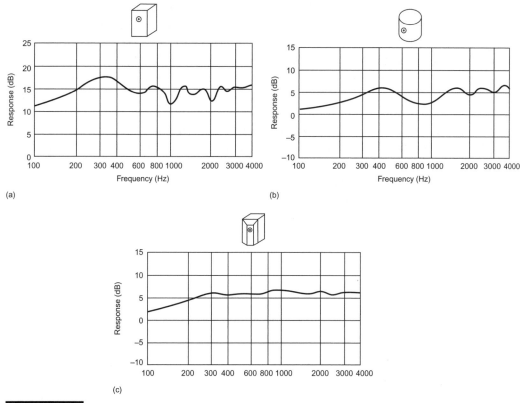

FIGURE 6.66 *Diffraction effects due to enclosures with at least two flat sides (from Olson, 1969).*

shape, as shown in Figure 6.65(b), has non-coincident resonances that can result in a smoother response.

Olson's results indicate that truncated ellipsoidal shapes, as shown in Figure 6.65, can achieve nearly as good results as a sphere.

Diffraction at bass frequencies is generally less of a problem, because the wavelengths are large compared with the cabinet size. However, if the main cabinet contains midrange units then it must also take account of diffraction effects. As this enclosure must also contain the bass units, and must sit on a surface, it is not possible to have a spherical enclosure. Instead, the curvature can only be arranged in the vertical axis.

Olson (1969) also examined this condition and some of his results are shown in Figure 6.66. From Figure 6.66(a) we can see that the rectangular parallelepiped enclosure has significant diffraction effects. This is despite the fact that the driver is mounted off-center, a strategy that reduces the effect of diffraction by reducing the number of coincidence effects. The cylindrical enclosure, Figure 6.66(b), does not seem to offer much improvement. This is due to both the presence of two edges, and to the coincidence effects caused by the central location of the driver between those two edges. An off-center driver would have improved the results.

The rectangular truncated pyramid and rectangular parallelepiped combination enclosure with an off-center driver, Figure 6.66(c), offers a considerable improvement. If the front surfaces are smooth curves instead of facets, the response will be even better, and will have only been limited by the abrupt transitions from the rear of the cabinet. The main consequence of these results is that wherever possible the enclosure should be curved near the drive-units and that the drive-units should not be mounted in the center of the enclosure. In practice, they will have to be mounted centrally in the horizontal direction, but they do not need to be so in the vertical direction. Under these circumstances, the most appropriate shape is circular. However, for the reasons outlined previously, an ellipsoidal shape is often more appropriate.

In this chapter we have examined how the space in which the sound is reproduced affects both its reproduction and perception. We have seen how the room affects the output of loudspeakers, and other sound sources. We have also analyzed various situations and examined various techniques for achieving a good acoustic environment for hearing music.

REFERENCES

Allison, R.F., 1980. The influence of room boundaries on loudspeaker power output. In: Cooke, R.E. (Ed.), Loudspeakers – An Anthology, vol. 1. The Audio Engineering Society, New York, pp. 353–359.

Angus, J. A. S. (1995). Sound diffusers using reactive absorption gratings. Audio Eng. Soc. Conv. 98, 25–28 Feb, Paris, preprint #3953.

Cox, T.J., 1996. Designing curved diffusers for performance spaces. J. Audio Eng. Soc. 44 (5), 354.

Cox, T.J., D'Antonio, P., Avis, M.R., June 2004. Room sizing and optimization at low frequencies. J. Audio Eng. Soc. 52 (6), 640–651.

D'Antonio, P., Konnert, J.H., April 1984. The reflection phase grating diffusor: design theory and application. J. Audio Eng. Soc. 32 (4), 228–238.

Inman, C., Sept.–Oct. 1994. A practical guide to the selection of glazing for acoustic performance in buildings. Acoust. Bull. 19 (5), 19–24.

Olson, H.F. (1969). Direct radiator loudspeaker enclosures. Journal of the Audio Engineering Society, January. Vol. 17, Number 1, 22–29. Note, originally presented at the Second Annual Convertion of the Engineering Society, 27 October, 1950.

Rayleigh, Lord., 1964. Work Done by Detached Sources. In: Scientific Papers, vols. 5–6. Dover, New York, pp. 135–141.

Schroeder, M.R., Jan. 1975. Diffuse sound reflection by maximum length sequences. J. Acoust. Soc. Am. 57 (1), 149–150.

Walker, R., 1996. Optimum dimension ratios for small rooms. Audio Eng. Soc. Conv. 100, preprint #4191.

Waterhouse, R.V., Mar. 1955. Output of a sound source in a reverberation chamber and other reflecting environments. J. Acoust. Soc. Am. 27 (2).

FURTHER READING

Angus, J.A.S., McManmon, C.I., Dec. 1998. Orthogonal sequence modulated phase reflection gratings for wide-band diffusion. J. Audio Eng. Soc. 46 (12).

D'Antonio, P., Cox, T., Nov. 1998. Two decades of diffuser design and development, Part 1: applications and design. J. Audio Eng. Soc. 46 (11).

Newell, P., 2003. Recording Studio Design. Focal Press, Oxford.

CHAPTER 7

Applications: Acoustics and Psychoacoustics Combined

So far in this book we have considered acoustics and psychoacoustics as separate topics. However, real applications often require the combination of the two because although the psychoacoustics tells us how we might perceive the sound, we need the acoustic description of sound to help create physical, or electronic, solutions to the problem. The purpose of this chapter is to give the reader a flavor of the many applications that make use of acoustics and psychoacoustics in combination. Of necessity these vignettes are brief and do not cover all the possible applications. However, we have tried to cover areas that we feel are important, and of interest. The level of detail also varies but, in all cases, we have tried to provide enough detail for the reader to be able to read, and understand, the more advanced texts and references that we provide, and any that the reader may discover themselves, for further reading. The rest of this chapter will cover listening room design, audiometry, psychoacoustic testing, filtering and equalization, public address systems, noise reducing headphones, acoustical social control devices, and last, but by no means least, audio coding systems.

7.1 CRITICAL LISTENING ROOM DESIGN

Although designing rooms for music performance is important, we often listen to recorded sound in small spaces. We listen to music, and watch television and movies, in both stereo and surround, in rooms that are much smaller that the recording environments. If one wishes to evaluate the sound in these environments then it is necessary to make them suitable for this purpose. In Chapter 6 we have seen how to analyze existing rooms and predict their performance. We have also examined methods for improving their acoustic characteristics. However, is there anything else we can do to make rooms better for the purpose of critically listening to music? There are a variety of approaches to achieving this and this section examines: optimal speaker placement, IEC rooms, room energy evolution, LEDE rooms, non-environment rooms, and diffuse reflection rooms.

7.1.1 Loudspeaker arrangements for critical listening

Before we examine specific room designs, let us first examine the optimum speaker layouts for both stereo and 5.1 surround systems. The reason for doing this is that most modern room designs for critical listening need to know where the speakers will be in order to be designed. It is also pretty pointless having a wonderful room if the speakers are not in an optimum arrangement.

Figure 7.1 shows the optimum layout for stereo speakers. They should form an equilateral triangle with the center of the listening position. If one has a greater angle than this the center phantom image becomes unstable—the so-called "hole in the middle" effect. Clearly, having an angle of less than 60° results in a narrower stereo image.

5.1 surround systems are used in film and video presentations. Here the objective is to provide both clear dialog and stereo music and sound effects, as well as a sense of ambience. The typical speaker layout is shown in Figure 7.2. Here, in addition to the conventional stereo speakers there are some additional ones to provide the additional requirements. These are as follows:

- *Center dialog speaker*: The dialog is replayed via a central speaker because this has been found to give better speech intelligibility over a stereo presentation. Interestingly the fact that the speech is not in stereo is not noticeable because the visual cue dominates so that we hear the sound coming from the person speaking on the screen even if their sound is coming from a different direction.

- *Surround speakers*: The ambient sounds, and sound effects, are diffused via rear mounted speakers. However they are, in the

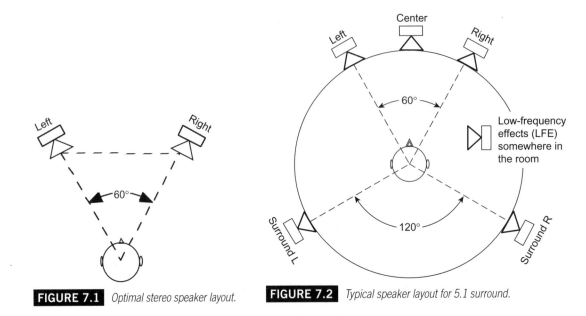

FIGURE 7.1 *Optimal stereo speaker layout.*

FIGURE 7.2 *Typical speaker layout for 5.1 surround.*

main, not supposed to provide directional effects and so are often deliberately designed, and fed signals, to minimize their correlation with each other and the front speakers. The effect of this is to fool the hearing system into perceiving the sound as all around with no specific direction.

- *Low-frequency effects*: This is required because many of the sound effects used in film and video, such as explosions and punches, have substantial low-frequency and subsonic content. Thus, a specialized speaker is needed to reproduce these sounds properly. Note: this speaker was never intended to reproduce music signals, notwithstanding their presence in many surround music systems.

More recently systems using six or more channels have also been proposed and implemented; for more information see Rumsey (2001).

As we shall see later the physical arrangement of loudspeakers can significantly affect the listening room design.

7.1.2 IEC listening rooms

The first type of critical listening room is the IEC listening room (IEC, 2003). This is essentially a conventional room that meets certain minimum requirements: a reverberation time that is flat, and between 0.3 and 0.6 seconds above 200 Hz, a low noise level, an even mode distribution and a recommended floor area. In essence this is a standardized living room that provides a consistent reference environment for a variety of listening tasks. It is the type of room that is often used for psychoacoustic testing as it provides results that correlate well with that which is experienced in conventional domestic environments. This type of room can be readily designed using the techniques discussed in Chapter 6.

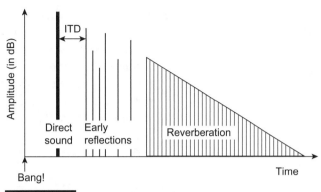

FIGURE 7.3 *An idealized energy–time curve.*

However, for critically listening to music mixes, etc. something more is required and these types of room will now be discussed. All of them don't only control reverberation, but also the time evolution and level of early reflections. They also all take advantage of the fact that the speakers are in specific locations to do this and very often have an asymmetric acoustic that is different for the listener and the loudspeakers. Although there are many different implementations, they fall into three basic types: *reflection controlled* rooms, *non-environment* rooms, and *diffuse reflection* rooms. As they all control the early reflections within a room we shall look at them first.

7.1.3 Energy–time considerations

A major advance in acoustical design for listening to music has arisen from the realization that, as well as reverberation time, the time evolution of the first part of the sound energy build-up in the room matters, that is, the detailed structure and level of the early reflections, as discussed in Chapter 6. As it is mostly the energy in the sound that is important as regards perception, the detailed evolution of the sound energy as a function of time in a room matters. Also there are now acoustic measurement systems that can measure the energy–time curve of a room directly, thus allowing a designer to see what is happening within the room at different frequencies, rather than relying on a pair of "golden ears." An idealized energy–time curve for a typical room is shown in Figure 7.3. It has three major features:

- A gap between the direct sound and first reflections. This happens naturally in most spaces and gives a cue as to the size of the space. The gap should not be too long—less than 30 ms—or the early reflections will be perceived as echoes. Some delay, however, is desirable as it gives some space for the direct sound and so improves the clarity of the sound, but a shorter gap does add "intimacy" to the space.

■ The presence of high-level diffuse early reflections, which come to the listener predominantly from the side, that is, lateral early reflections. This adds spaciousness and is easier to achieve over the whole audience in a shoebox hall rather than a fan-shaped one. The first early reflections should ideally arrive at the listener within 20 ms of the direct sound. The frequency response of these early reflections should ideally be flat and this, in conjunction with the need for a high level of lateral reflections, implies that the side walls of a hall should be diffuse reflecting surfaces with minimal absorption.

■ A smoothly decaying diffuse reverberant field which has no obvious defects, and no modal behavior, and whose time of decay is appropriate to the style of music being performed. This is hard to achieve in practice so a compromise is necessary in most cases. For performing acoustic music a gentle bass rise in the reverberant field is desirable to add "warmth" to the sound but in studios this is less desirable.

7.1.4 Reflection-controlled rooms

For the home listener, or sound engineer in the control room of a studio, the ideal would be an acoustic that allows them to "listen through" the system to the original acoustical environment that the sound was recorded in. Unfortunately the room in which the recorded sound is being listened to is usually much smaller than the original space and this has the effect shown in Figure 7.4. Here the first reflection the listener hears is due to the wall in the listening room and not the acoustic space of the sound that has been recorded. Because of the precedence effect this reflection dominates, and the replayed sound is perceived as coming from a space the size of the listening room, which is clearly undesirable. What is required is a means of making the sound from the loudspeakers appear as if it is coming from a larger space by suppressing the early reflections from the nearby walls, as shown in Figure 7.5. Examples of this approach are: "live end dead end" (LEDE) (Davies and Davies, 1980), "Reflection free zone" (RFZ) (D'Antonio and Konnert, 1984), and controlled reflection rooms (Walker, 1993, 1998).

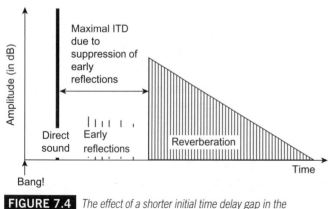

FIGURE 7.4 *The effect of a shorter initial time delay gap in the listening room.*

One way of achieving this is to use absorption, as shown in Figure 7.6. The effect can also be achieved by using angled or shaped walls, as shown in Figures 7.7 and 7.8. This is known as the "controlled reflection technique" because it relies on the suppression of early reflections in a particular area of the room to achieve a larger initial time delay gap. This effect can only be achieved over a limited volume of the room unless the room is made anechoic, which is undesirable. The idea is simple: by absorbing, or reflecting away, the first reflections from all walls except the furthest one away from the speakers the initial time delay gap is maximized. If this gap is larger than the initial time delay gap in the original recording space then the listener will hear the original space, and not the listening room.

However, this must be achieved while satisfying the need for even diffuse reverberation, and so the rear wall in such situations must have some

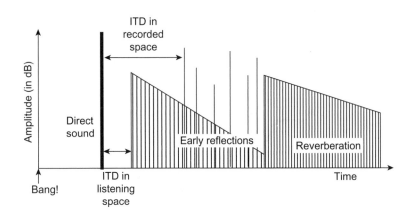

FIGURE 7.5

Maximizing the initial time delay by suppressing early reflections.

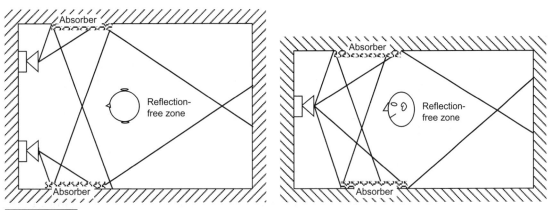

FIGURE 7.6   *Achieving a reflection-free zone using absorption.*

Controlled reflection room (in the style of Bob Walker) for free-standing loudspeakers (from Newell 2008).

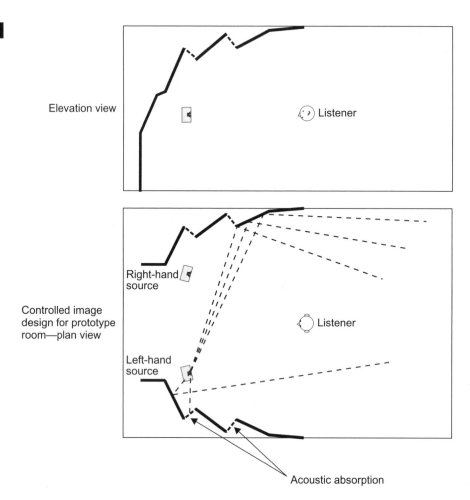

Elevation view

Listener

Controlled image design for prototype room—plan view

Right-hand source

Left-hand source

Listener

Acoustic absorption

explicit form of diffusion structure on it to assure this. The initial time delay gap in the listening should be as large as possible, but is clearly limited by the time it takes sound to get to the rear wall and back to the listener. Ideally this gap should be 20 ms but it should not be much greater or it will be perceived as an echo. In most practical rooms this requirement is automatically satisfied and initial time delay gaps in the range of 8 ms to 20 ms are achieved.

Note that if the reflections are redirected rather than being absorbed, then there will be "hot areas" in the room where the level of early reflections is higher than normal. In general it is often architecturally easier to use absorption rather than redirection, although this can sometimes result in a room with a shorter reverberation time.

FIGURE 7.8

An example controlled reflection room, Sony Music M1, New York, NY. (Photo by Paul Ellis of The M Network Ltd; Acoustician: Harris, Grant Associates)

7.1.5 The absorption level required for reflection-free zones

In order to achieve a reflection-free zone it is necessary to suppress early reflections, but by how much? Figure 7.9 shows a graph of the average level that an early reflection has to be at in order to disturb the direction of a stereo image. From this we can see that the level of the reflections must be less than about 15 dB to be subjectively inaudible. Allowing for some reduction due to the inverse square law, this implies that there must be about 10 dB, or $\alpha = 0.9$ of absorption on the surfaces contributing to the first reflections. In a domestic setting it is possible to get close to the desired target using carpets and curtains, and bookcases can form effective diffusers, although persuading the other occupants of the house that carpets, or curtains, on the ceiling is chic can be difficult. In a studio more extreme treatments can be used. However, it is important to realize that the overall acoustic must still be good and comfortable, that it is not anechoic, and that, due to the wavelength range of audible sound, this technique is only applicable at mid to high frequencies where small patches of treatment are significant with respect to the wavelength.

7.1.6 The absorption position for reflection-free zones

Figure 7.10 shows one method of working out where absorption should be placed in a room to control early reflections. By imagining the relevant walls to be mirrors it is possible to create "image rooms" that show the

FIGURE 7.9

*The degree of reflection
suppression required to
assure a reflection-free
zone (data from Toole,
1990).*

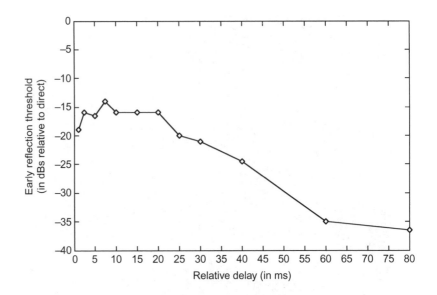

FIGURE 7.9

*The degree of reflection
suppression required to
assure a reflection-free
zone (data from Toole,
1990).*

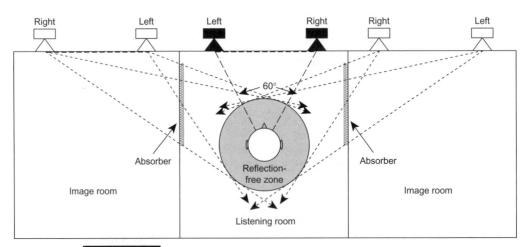

FIGURE 7.10 *The image method for controlled reflection room absorption placement.*

direction of the early reflections. By defining a reflection-free space around the listening position, and by drawing "rays" from the image speaker sources, one can see which portions of the wall need to be made absorbent, as shown in Figure 7.11. This is very straightforward for rectangular rooms, but a little more complicated for rooms with angled walls. Nevertheless, this technique, can still be used. It is applicable for both stereo and surround systems, the only real difference being the number of sources.

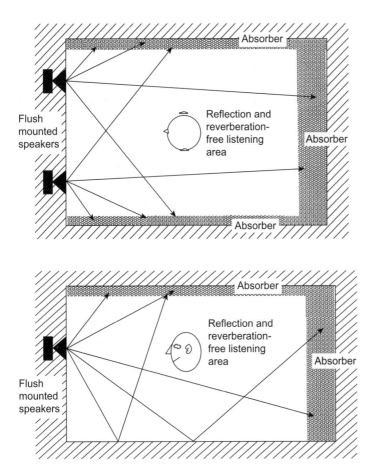

FIGURE 7.11

Non-environment room principles.

In Figure 7.11 the rear wall is not treated because normally some form of diffusing material would be placed there. However, absorbing material could be so placed, in the places determined by another image room created by the rear wall, if these reflections were to be suppressed. One advantage of this technique is that it also shows places where absorption is unnecessary. This is useful because it shows you where to place doors and windows that are difficult to make absorptive. To minimize the amount of absorption needed one should make the listening area as small as possible because larger reflection free volumes require larger absorption patches. The method is equally applicable in the vertical as well as the horizontal direction.

7.1.7 Non-environment rooms

Another approach to controlling early reflections, which is used in many successful control rooms, is the "non-environment" room. These rooms control both the early reflections and the reverberation. However, although they are quite dead acoustically, they are not anechoic. Because for users in the room there are some reflections from the hard surfaces, there are some early reflections that make the room non-anechoic. However, sound that is emitted from the speakers is absorbed and is never able to contribute to the reverberant field. How this is achieved is shown in Figure 7.11.

These rooms have speakers, which are flush mounted in a reflecting wall, and a reflecting floor. The rear wall is highly absorbent, as are the side walls and ceiling. The combined effect of these treatments is that sound from the loudspeakers is absorbed instead of being reflected so that only the direct sound is heard by the listener, except for a floor reflection. However, the presence of two reflecting surfaces does support some early reflections for sources away from the speakers. This means that the acoustic environment for people in the room, although dead, is not oppressively anechoic. Proponents of this style of room say that the lack of anything but the direct sound makes it much easier to hear low-level detail in the reproduced audio and provides excellent stereo imaging. This is almost certainly due to the removal of any conflicting cues in the sound, as the floor reflection has very little effect on the stereo image.

These rooms require wide-band absorbers as shown in Figure 7.12. These absorbers can take up a considerable amount of space. As one can see in Figure 7.12, the absorbers can occupy more than 50% of the volume. However, it is possible to use wide-band membrane absorbers, as discussed in Chapter 6, with a structure similar to that shown in Figure 6.48 with a limp membrane in place of the perforated sheet. Using this type of absorber it is possible to achieve sufficient wide-band absorption with a depth of 30 cm, which allows this technique to be applied in much smaller rooms whose area is approximately 15 m². Figure 7.13 shows a typical non-environment room implementation: "The Lab", at the Liverpool Music House

Because non-environment rooms have no reverberant field, there is no reverberant room support for the loudspeaker level, as discussed in Section 6.1.7. Only the direct sound is available to provide sound level. In a normal domestic environment, as discussed in Chapter 6, the reverberant field is providing most of the sound power and is often about 10 dB greater than the direct sound. Thus in a non-environment room one must use either 10 times the power amplifier level, or specialist loudspeaker systems with a greater efficiency, to reproduce the necessary sound levels (Newell, 2008).

7.1.8 The diffuse reflection room

A novel approach to controlling early reflections is not to try to suppress or redirect them, but instead diffuse them. This results in a reduced reflection level but does not absorb them.

Plan of "non-environment" control room

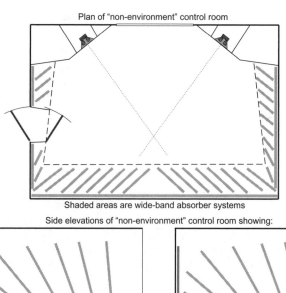

Shaded areas are wide-band absorber systems

FIGURE 7.12

A non-environment control room. Shaded areas are wide-band absorbers (from Newell, 2008)

Side elevations of "non-environment" control room showing:

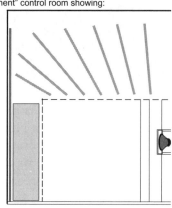

(a) Horizontal rear absorbers   (b) Vertical rear absorbers

In general most surfaces absorb some of the sound energy and so the reflection is weakened by the reflection. Therefore the level of direct reflections will be less than that which would be predicted by the inverse square law, due to surface absorption. The amount of energy, or power, removed by a given area of absorbing material will depend on the energy, or power, per unit area striking it. As the sound intensity is a measure of the power per unit area this means that the intensity of the sound reflected is reduced in proportion to the absorption coefficient. Therefore the intensity of the early reflection is given by:

$$I_{\text{direct sound}} = \frac{QW_{\text{Source}}(1 - \alpha)}{4\pi r^2} \qquad (7.1)$$

From the above equation (7.1), which is Equation 1.18 with the addition of the effect of surface absorption, it is clear that the intensity reduction of a specular early reflection is inversely proportional to the distance squared.

Diffuse surfaces on the other hand scatter sound in other directions than the specular. In the case of an ideal diffuser the scattered energy polar pattern would be in the form of a hemisphere. A simple approach to calculating the effect of this can be to model the scattered energy as a source whose initial intensity is given by the incident energy. Thus, for an ideal scatterer, the intensity of the reflection is give by the product of the equation describing the intensity from the source and the one describing the sound intensity radiated by the diffuser. For the geometry shown in Figure 7.14 this is given by:

$$I_{\text{diffuse reflection}} = \left(\frac{W_{\text{Source}}}{4\pi r_s^2} \right) \times \left(\frac{2}{4\pi r_d^2} \right) \tag{7.2}$$

The factor 2 in the second term represents the fact that diffuser only radiates into half a hemisphere and therefore has a "Q" of 2. From Equation 7.2 one can see that the intensity of a diffuse reflection is inversely proportional to the distance to the power of four. This means that the intensity of an individual diffuse reflection will be much smaller than that of a specular reflection from the same position.

So diffusion can result in a reduction of the amplitude of the early reflection from a given point. However, there will also be more reflections, due to the diffusion, arriving at the listening position from other points on the wall, as shown in Figure 7.15. Surely this negates any advantage of the technique? A closer inspection of Figure 7.15 reveals that although there are

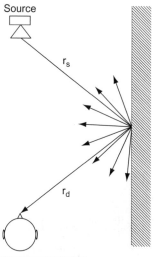

FIGURE 7.14 *The geometry for calculating the intensity of an early reflection from a diffuse surface.*

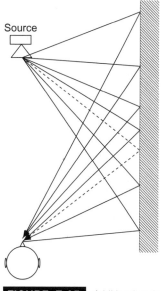

FIGURE 7.15 *Additional early reflection paths due to a diffuse surface.*

many reflection paths to the listening point they are all of different lengths, and hence time delay. The extra paths are also all of a greater length than the specular path, shown dashed in Figure 7.15. Furthermore the phase reflection diffusion structure will add an additional temporal spread to the reflections. As a consequence the initial time delay gap will be filled with a dense set of low-level early reflections instead of a sparse set of higher level ones, as shown in Figure 7.16. Of particular note is that, even with no added absorption, the diffuse reflection levels are low enough in amplitude to have no effect on the stereo image, as shown earlier in Figure 7.9.

The effect of this is a large reduction of the comb filtering effects that high-level early reflections cause. This is due to both the reduction in amplitude due to the diffusion and the smoothing of the comb filtering caused by the multiplicity of time delays present in the sound arriving from the diffuser. As these comb filtering effects are thought to be responsible for perturbations of the stereo image (Rodgers, 1981), one should expect improved performance even if the level of the early reflections is slightly higher than the ideal.

The fact that the reflections are diffuse also results in an absence of focusing effects away from the optimum listening position and this should result in a more gradual degradation of the listening environment away from the optimum listening position. Figure 7.17 shows the intensity of the

The intensity–time plots at the listener position of a diffuser walled room compared with the direct sound.

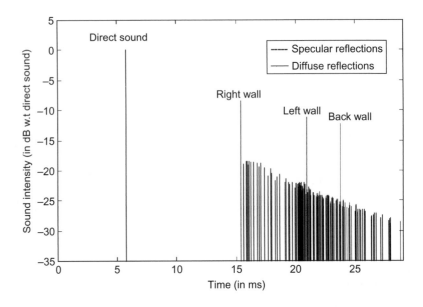

The intensity of the largest diffuse side wall reflection relative to the direct sound as a function of room position; contours are in dB relative to the direct sound.

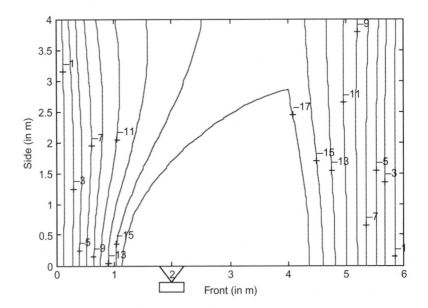

largest diffuse side wall reflection relative to the largest specular side wall reflection as a function of room position for the speaker position shown. From this figure we can see that over a large part of the room the reflections are less than 15 dB below the direct sound.

FIGURE 7.18

A diffuse reflection room implementation: "Studio C," at Blackbird Studio, Nashville. (Photo by Max Crace courtesy of George Massenburg and Blackbird Studio.)

Figure 7.18 shows one of the few examples of such a room. The experience of this room is that one is unaware of sound reflection from the walls: it sounds almost anechoic, yet it has reverberation. Stereo and multi-channel material played in this room has images that are stable over a wide listening area, as predicted by theory. The room is also good for recording in as the high level of diffuse reflections and the acoustic mixing it engenders, as shown in Figure 7.15, helps to integrate the sound emitted by acoustic instruments.

Summary

In this section we have examined various techniques for achieving a good acoustic environment for hearing both stereo and multi-channel music. However, the design of a practical critical listening room requires many detailed considerations regarding room treatment, sound isolation, air conditioning, etc. that are covered in more detail in Newell (2008).

7.2 PURE-TONE AND SPEECH AUDIOMETRY

In this section, a number of acoustic and psychoacoustic principles are applied to the clinical measurement of hearing ability. Hearing ability is described in Chapter 2 and summarized in Figure 2.10 in terms of the frequency and amplitude range typically found. But how can these be measured in practice, particularly in the clinic where such information can provide medical professionals with critical data for the treatment of hearing problems?

The ability to detect sound and the ability to discriminate between sounds are the two aspects of hearing that can be detrimentally affected by age, disease, trauma or noise-induced hearing loss. The clinical tests that are available for the diagnosis of these are:

- Sound detection: *pure-tone audiometry.*
- Sound discrimination: *speech audiometry.*

Pure-tone audiometry is used to test a subject's hearing threshold at specific frequencies approximately covering the speech hearing range (see Figure 2.10). These frequencies are spaced in octaves as follows: 125 Hz, 250 Hz, 500 Hz, 1 kHz, 2 kHz, 4 kHz and 8 kHz. The range of sound levels that are tested usually start 10 dB below the average threshold of hearing and they can rise to 120 dB above it; recall that the average threshold of hearing varies with frequency (see Figure 2.10).

A clinical audiometer is set up to make diagnosis straightforward, and quick and easy to explain to patients. Because the threshold of hearing is a non-uniform curve and therefore not an easy reference to use on an everyday basis in practice, a straight line equating to the average threshold of hearing is used instead to display the results of a hearing test on an *audiogram*. A dBHL (hearing level) scale is defined for hearing testing, which is the number of dBs above the average threshold of hearing.

Figure 7.19 shows a blank audiogram which plots frequency on the x axis (the octave values between 125 Hz and 8 kHz inclusive as shown above) against dBHL between −10 dBHL and +120 dBHL on the y axis. Note that the dBHL scale increases *downwards* to indicate greater hearing *loss* (a higher amplitude or greater dBHL value needed for the sound to be detected). The 0 dBHL (threshold of hearing) line is thicker than the other lines to give a visual focus on the average threshold of hearing as a reference against which measurements can be compared.

A pure-tone audiometer has three main controls: (1) frequency; (2) output sound level; and (3) a spring-loaded output key switch to present the sound to the subject. When the frequency is set, the level is automatically altered to take account of the average threshold of hearing, which enables the output

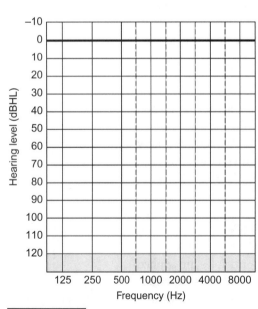

FIGURE 7.19 *A blank audiogram.*

sound level control to be calibrated in dBHL directly. The output sound level control usually works in 5 dB steps and is calibrated in dBHL. It is vitally important that the operator is aware that an audiometer can produce very high sound levels which could do permanent damage to a normal hearing system (see Section 2.5). When testing a subject's hearing, a modest level around +30 dBHL should be used to start with, which can be increased if the subject cannot hear it.

The spring-loaded output key is used to present the sound, thereby giving the operator control of when the sound is being presented and removing any pattern of presentation that might allow the subject to predict when to expect the next sound. Such unpredictability adds to the overall power of the test, but, in the context of hearing measurement, it is particularly important when hearing is being tested in the context of, at one extreme, a legal claim for damages being made for hearing loss due to noise-induced hearing loss or, at the other, a health screening for normal hearing as part of a job interview.

When a sound is heard the subject is asked to press a button, which illuminates a lamp, or light emitting diode (LED), on the front panel of the audiometer. The subject should be visible to the tester, but the subject should not be able to see the controls. When carrying out an audiometric test, local sound levels should be below the levels defined in BS EN ISO 8253-1, which are shown for the test frequencies in Figure 7.20. Generally, the local level should be below 35 dBA.

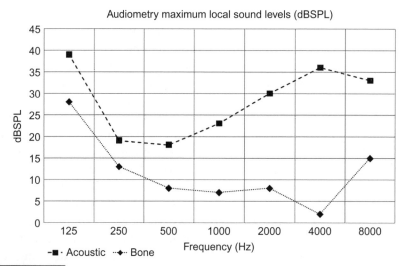

FIGURE 7.20 *Local maximum sound levels at audiometric test frequencies for acoustic and bone conduction measurements (adapted from BS EN ISO 8253-1).*

During audiometry, test signals are presented in one of two ways:

- air conduction
- bone conduction.

For air conduction audiometry, sound is presented acoustically to the outer ear and thereby tests the complete hearing system. Three types of air conduction transducers are available:

1. circum-aural headphones
2. supra-aural headphones
3. ear canal insert earphones.

Circum-aural headphones surround and cover the pinna (see Figure 2.1) completely thereby providing a degree of sound isolation. Supra-aural headphones rest on the pinna and are the more traditional type in use, but they are not particularly comfortable since they press quite heavily on the pinna in order to keep the distance between the transducer itself and the tympanic membrane constant. Both circum- or supra-aural headphones can be uncomfortable and somewhat awkward and they can in certain circumstances deform the ear canal. As an alternative, ear canal insert earphones that have a disposable foam tip can be used which will not distort the ear canal. They have the added advantage that less sound leaks to the other ear, which reduces the need to consider presenting a masking signal to it. There are, however, situations, such as infected or obstructed ear canals, when the use of ear canal insert earphones is not appropriate.

For bone conduction audiometry, sound is presented mechanically using a bone vibrator which is placed just behind but not touching the pinna on the bone protrusion known as the "mastoid prominence." It is held in place with an elastic headband. The sound presented when using bone conduction bypasses the outer and middle ears since it vibrates the temporal bone in which the cochlea lies directly. Thus it can be used to assess inner function and the presence or otherwise of what is known as "sensorineural hearing loss" with no hindrance from any outer or middle ear disorder. Bone conduction is carried out in the same way as air conduction audiometry except that only frequencies from 500 Hz to 4 kHz are used due to the limitations of the bone conduction transducers themselves. When a bone conduction measurement is being made for a specific ear it is essential that the other ear is masked using noise. Specific audiometric guidelines exist for the use of masking.

The usual audiometric procedure for air or bone conduction measurements (recalling the one difference for bone conduction that the frequencies used are from 500 Hz to 4 kHz only) is to test frequencies (in the following order: 1 kHz, 2 kHz, 4 kHz, 8 kHz, 500 Hz, 250 Hz, 125 Hz, 1 kHz). (Note that the starting frequency is 1 kHz which is a mid frequency in the hearing range and is therefore likely to be heard by all subjects to give them confidence at the start and end of a test.) If the retest measurement at 1 kHz has changed by more than 5 dB, other frequencies should be retested and the most sensitive value (lowest dBHL value) recorded. When testing one ear, consideration will be given as to whether masking should be presented to the other ear to ensure that only the test ear is involved in the trial. This is especially important when testing the poorer ear.

Tests are started at a level that can be readily heard (usually around +30 dBHL), which is presented for 1–3 s using the output key switch, and then involve watching for the subject to light the lamp or LED. If this does not happen, the level is increased in 5 dB steps (5 dB being a minimum practical value to enable tests to be carried out in a reasonable time)—presenting the sound and awaiting a response each time. Once a starting level has been established, the sound level is changed using the "10 down, 5 up" method as follows:

1. Reduce the level in 10 dB steps until the sound is not heard.
2. Increase the level in 5 dB steps until the sound is heard.
3. Repeat 1 and 2 until the subject responds at the same level at least 50% of the time, defined as two out of two, two out of three or two out of four responses.
4. Record the threshold as the lowest level heard.

There are a number of degrees of hearing loss, which are defined in Table 7.1. These descriptions are used to provide a general conclusion about a subject's hearing and they should be interpreted as such. They consist of a single value which is the average dBHL value across frequencies 250 Hz to 4 kHz. The values are used to provide a general guideline as to the state of hearing and it must be remembered that there could be one or more frequencies for which the hearing loss is worse than the average.

Consider, for example, the audiogram for damaged hearing given in Figure 2.19. Here, the average dBHL value for frequencies 250 Hz to 4 kHz

Table 7.1 Definitions of different degrees of hearing loss	
Description	**dBHL**
No hearing handicap	<20
Mild hearing loss	20–40
Moderate hearing loss	41–70
Severe hearing loss	71–95
Profound hearing loss	>95

would be $\{((10 + 5+5 + 15 + 60)/5) = 19\,\text{dBHL}\}$ which indicates "no hearing handicap" (see Table 7.1), which is clearly not the case.

The upper part of Figure 7.21 shows audiograms for a young adult with normal healthy hearing within both ears based on air and bone conduction tests in the left and right ears (a key to the symbols used on audiograms is given in the figure). Notice that the bone and air conduction results lie in the same region (in this case $<20\,\text{dBHL}$) and a summary statement of "no hearing handicap" (see Table 7.1) would be entirely appropriate in this case. Pure-tone audiometry is the technique that enables the normal deterioration of hearing with age, or *presbycusis* (see Section 2.3 and Figure 2.11) to be monitored.

The lower part of Figure 7.21 shows example audiograms for two hearing loss conditions. The audiogram in the lower left position shows a conductive hearing loss in the left ear because the bone conduction plot is normal, but the air conduction plot shows a significant hearing loss that would be termed a "moderate hearing loss" (see Table 7.1). This indicates a problem between the outside world and the inner ear, and a hearing aid, tailored to the audiogram, could be used to correct for the air conduction loss.

The audiogram in the lower right position shows the effect on hearing of congenital rubella syndrome which can occur in a developing fetus of a pregnant woman who contracts rubella (German measles) from about 4 weeks before conception to 20 weeks into pregnancy. One possible effect on the infant is profound hearing loss ($>95\,\text{dBHL}$—see Table 7.1), which is sensorineural (note that both the air and bone conduction results lie in the same region indicating an inner ear hearing loss). Sadly, there is no known cure; in this example, a hearing aid would not offer much help because there is no usable residual hearing above around $500\,\text{Hz}$.

Pure-tone audiometry tests a subject's ability to *detect* different frequencies, and the dBHL values indicate the extent to which the subject's hearing is reduced at different frequencies. It thus indicates those frequency regions in which a subject is perhaps less sensitive than normally hearing listeners. This could, for example, be interpreted in practice in terms of timbral differences between specific musical sounds that might not be heard, or vowel or other speech sounds that might be difficult to perceive. However, pure-tone audiometry does not provide a complete test of a subject's hearing ability to *discriminate* between different sounds. Discrimination of sounds does start with the ability to detect the sounds, but it also requires appropriate sound processing to be available. For example, if the critical bands (see Section 2.2) are widened, they are less able to separate the components of complex sounds—the most important to us being speech. In order to test

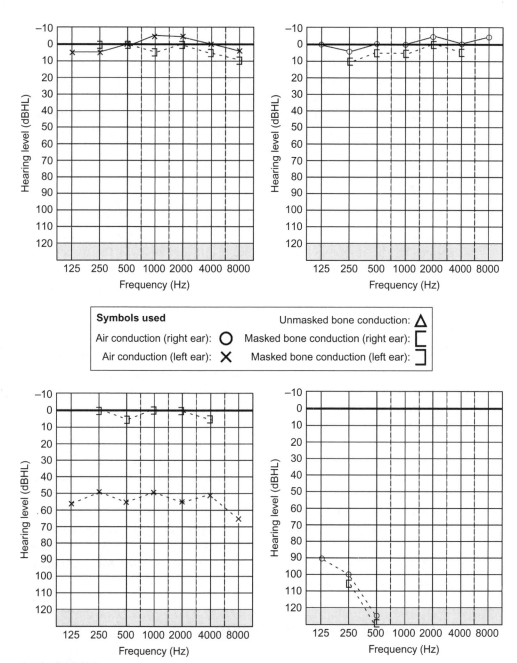

FIGURE 7.21 *UPPER: Example audiograms for the left and right ears (left- and right-hand plots respectively) of a young adult with normal healthy hearing along with a key to the symbols commonly used on audiograms. LOWER: Example audiograms for (left) a left ear conductive hearing loss, and (right) a right ear hearing loss due to congenital rubella syndrome.*

hearing discrimination, *speech* audiometry is employed which makes use of spoken material.

Speech audiometry is carried out for each ear separately and tests speech discrimination performance against the pure-tone audiograms for each ear and normative data. When testing one ear, consideration will be given as to whether masking should be presented to the other ear to ensure that only the test ear is involved in the trial. This is particularly important when testing the poorer ear.

Speech audiometry involves the use of an audiometer and speech material that is usually recorded on audio compact disc (CD). Individual single syllable words such as *bus*, *fun*, *shop* are played to the subject, who is asked to repeat them, providing part words if that is all they have heard. Each spoken response is scored phonetically in terms of the number of correct sounds in the response (for example, if *bun* or *boss* was the response for *bus*, the subject would score two out of three). Words are presented in sets of 10, and if a total phonetic score of 10% or better is achieved for a list, the level is reduced by 10 dB and a new set of 10 words is played, repeating the process until the score falls below 10%. The Speech Reception Threshold (SRT) is the lowest level at which a 10% phonetic score can be achieved. The Speech Discrimination Score (SDS) is the percentage of single syllable words that can be identified at a comfortable loudness level.

The results from speech audiometry indicate something about the ability to discriminate between sounds whereas pure-tone audiometry indicates ability to detect the presence of particular frequency components. Clearly detection ability is basic to being able to make use of frequency components in a particular sound, but how a listener might make use of those components depends on their discrimination ability. Discrimination will change if, for example, a listener's critical bands are widened, which can result in an inability to separate individual components. This could have a direct effect on pitch, timbre and loudness perception. In addition, the ability to hear separately different instruments or voices in an ensemble might be impaired—something that could be very debilitating for a conductor, accompanist or recording engineer.

7.3 PSYCHOACOUSTIC TESTING

Knowledge of psychoacoustics is based on listening tests in order to find out how humans perceive sounds in terms, for example, of pitch, loudness and timbre. Direct measurements are not possible in this context since direct connections cannot be made for ethical as well as practical reasons, and, in many cases, there is a cognitive dimension (higher-level processing) that is unique to each and every listener. Our knowledge of psychoacoustics therefore is based on listening tests, and this section presents an overview of procedures that are typically used in practice. Apart from offering this as a background to the origins of the psychoacoustic information presented in this book, it also enables readers to think through aspects of the creation of their own listening tests to progress psychoacoustic knowledge in the future.

When carrying out a psychoacoustic test, it is important to note that the responses will be from the opinions of listeners; that is, they will be subjective, whereas an objective test involves a direct physical measurement such as dB SPL, Hz, or spectral components. There is no right answer to a subjective test since it is the opinion of a particular listener and each listener will have an opinion that is unique; the process of psychoacoustic testing is to collect these listener opinions in a non-judgmental manner. Subjective testing is unlike objective testing where direct measurements can be made of physical quantities such as sound pressure level, sound intensity level or fundamental frequency; in a subjective test a listener is asked to offer an opinion in answer to questions such as "Which sound is louder?", "Does the pitch rise or fall?", "Are these two sounds the same or different?", "Which chord is more in-tune?", or "Which version do you prefer?".

Psychoacoustic testing involves careful experimental design to ensure that the results obtained can be truly attributed to whatever aspect of the signal is being used as the controlled variable. This process is called controlled experimentation. A starting point for experimental design may well be a hunch or something we believe to be the case from our own listening experience, or from anecdotal evidence. A controlled experiment allows such listening experiences to be carefully explored in terms of which aspects of a sound affect them and how. Psychologists call a behavioral response, such as a listening experience, the dependent variable, and those aspects that might affect it are called the independent variables. Properly controlled psychoacoustic testing involves controlling all the independent variables so that any effects observed can be attributed to changes in the specific independent variable under test.

7.3.1 Psychoacoustic experimental design issues

One experimental example might be to explore what aspects of sound affect the perception of pitch. The main dependent variable would be f_0, but other aspects of sound can affect the perceived pitch such as loudness, timbre and duration (see Section 3.2). Experimentally, it would also be very appropriate to consider other issues that might affect the results—some of which may not initially seem obvious—such as the fact that hearing abilities of the subjects can vary with age (see Section 2.3) and general health, or that perhaps subjects' hearing should be tested (see Section 7.2).

The way in which sounds are presented to subjects can also make a difference since the use of loudspeakers would mean that the acoustics of the room will alter the signals arriving at each ear (see Chapter 6) whereas the use of headphones would not. There may be background acoustic noise in the listening room that could affect the results and this may even be localized, perhaps to a ventilation outlet. Subjects can become tired (listener fatigue), distracted, or may perform better at different times of the day. The order in which stimuli are presented can have an effect—perhaps of alerting the listener to specific features of the signal, which prepares them better for a following stimulus. These are all potential independent variables and would need proper controlling.

Part of the process of planning a controlled experiment is thinking through such aspects (the ones given here are just examples and are not presented as a definitive list) before carrying out a full test. It is common to try a pilot test with a small number of listeners to check the test procedure and for the presence of any additional independent variables. Some independent variables can be controlled by ensuring they remain constant (for example, the ventilation might be turned off, and measures could be taken to reduce background noise). Others can be controlled through the test procedure (for example, any learning effect could be explored by playing the stimuli in a different order to different subjects or asking each listener to take the test twice with the stimulus order being reversed the second time).

7.3.2 Psychoacoustic rating scales

For many psychoacoustic experiments the request to be given to the listeners is straightforward. In the pitch example above one might ask listeners to indicate which of two stimuli has the highest pitch or whether the pitch of a single stimulus was changing or not. In experiments where the objective is to establish the nature of change in a sound, such as whether one synthetic sound is more natural than another, it is not so easy. A simple "yes" or "no" would not be very informative since it would not indicate the nature of the difference. A number of rating scales have been produced that are commonly used in such cases, where the listener is invited to choose the point on the scale that best describes what they have heard. Some examples are given below.

When speech signals are rated subjectively by listeners, perhaps for the evaluation of the signal provided by a mobile phone or the output from a speech synthesis system, it is usually the quality of the signal that is of interest. The number of listeners is important since each will have a personal opinion and it is generally suggested that at least 16 are used to ensure that statistical analysis of the results is sufficiently powerful. However, the greater the number of listeners the more reliable the results are. It is also most appropriate to use listeners who are potential users of whatever system might result from the work and listeners who are definitely not experts in the area. A number of rating scales exist for the evaluation of the quality of a speech signal and the following are examples:

- absolute category-rating (ACR) test;
- degradation category-rating (DCR) test;
- comparison category-rating (CCR) test.

The absolute category-rating (ACR) test requires listeners to respond with a rating from the five-point ACR rating scale shown in Table 7.2. The results from all the listeners are averaged to provide a *mean opinion score* (MOS) for the signals under test. Depending on the purpose of the test, it might be of more interest to present the results of the listening test in terms of the percentage of listeners that rated the presented sounds in one of the categories such as *good* or *excellent* or *poor* or *bad*.

A comparison is requested of listeners in a degradation category-rating (DCR) test, and this usually involves a comparison between a signal before and after some form of processing has been carried out. The assumption here is that the processing is going to degrade the original signal in some way, for example after some sort of coding scheme such as MP3 (see Section 7.8) has been applied, where one would never expect the signal to be improved. Listeners use the DCR rating scale (see Table 7.2) to evaluate the extent to which the processing has degraded the signal when comparing

Table 7.2 Rating scale and descriptions for the absolute category-rating (ACR) test, which produces a mean opinion score (MOS), and degradation category-rating (DCR) test, which produces a degraded mean opinion score (DMOS)

Rating	ACR description – MOS	DCR description – DMOS
5	Excellent	Degradation not perceived
4	Good	Degradation perceived but not annoying
3	Fair	Degradation slightly annoying
2	Poor	Degradation annoying
1	Bad	Degradation very annoying

the processed version with the unprocessed original. The results are analyzed in the same way as for the ACR test and these are sometimes referred to as the "degradation mean opinion score" (DMOS).

In situations where the processed signal could be evaluated as being either better or worse, a comparison category-rating (CCR) test can be used. Its rating scale is shown in Table 7.3 and it can be seen that it is a symmetric two-sided scale. Listeners are asked to rate the two signals presented in terms of the quality of the second signal relative to the first. The CCR test might be used if one is interested in the effect of a signal processing methodology being applied to an audio signal, such as noise reduction, in terms of whether it has improved the original signal or not.

7.3.3 Speech intelligibility: Articulation loss

Psychoacoustic experiments may be used to define thresholds of perception and rating scales for small degradations, such as the quality of the sound. However, at the other end of the quality scale is the case where the degradation, due to noise distortion, reverberation, etc., is so severe that it affects the intelligibility of speech. This is also measured and defined by the results of psychoacoustic experiments, but in these circumstances, instead of annoyance, the *dependent variable* is the proportion of the words that are actually heard correctly.

Two parameters are found to be important by those who work on speech: the "intelligibility" and the "quality," or "naturalness," of the speech. Both reflect human perception of the speech itself, and while they are most directly measured subjectively with panels of human listeners, research is being carried out to make equivalent objective measurements of these, because of the problems of setting up listening experiments and

Table 7.3	Rating scale and descriptions for the comparison category-rating (CCR) test
Rating	**Description**
3	Much better
2	Better
1	Slightly better
0	About the same
−1	Slightly worse
−2	Worse
−3	Much worse

the inherent inter- and intra-listener variability. The relationship between intelligibility and naturalness is not fully understood. Speech that is unintelligible would usually be judged as being unnatural. However, muffled, fast or mumbling speech is natural, but less intelligible, and speech that is highly intelligible may or may not be unnatural.

Subjective measures of intelligibility are often based on the use of lists of words that rhyme, differing only in their initial consonant. In a diagnostic rhyme test (DRT), listeners fill in the leading consonants on listening to the speech, and often the possible consonants will be indicated. In a modified rhyme test (MRT) each test consists of a pair-wise comparison of acoustically close initial consonants such as *feel–veal, bowl–dole, fought–thought, pot–tot*. The DRT identifies quite clearly in what area a speech system is failing, giving the designers guidance on where they might make modifications. DRT tests are widely accepted for testing intelligibility, mainly because they are rigorous, accurate and repeatable. Another type of testing is "Logatom" testing.

In Logatom testing nonsense words such as "shesh" and "bik" are placed into a carrier phrase such as "Can con buy <Logatom, e.g., "shesh" here> also." to ensure that they are all pronounced with the same inflection. The listener then has to identify the nonsense word and write it down. Using nonsense words has the advantage of removing the higher language processing that we use to resolve words with degraded quality and so provides a less biased measure. The errors listeners make show how the system being tested damages the speech, such as particular letter confusions, and provide a measure of intelligibility. Typically lists of 50 or 100 Logatoms are used as a compromise between accuracy and fatigue, as discussed earlier. Although in theory any consonant-vowel-consonant may be used, it has been the authors' experience that rude or swear words must be excluded, because the talker usually cannot pronounce them with the same inflection as normal words.

All of these tests result in a measure of the number of correctly identified words. This, as a percentage of the total, can be used as a measure of intelligibility, or articulation loss, respectively. As consonants are more important in Western languages than vowels, this measure is often focused on just the consonants to form a measure called %ALcons (**A**rticulation **L**oss; **cons**onants) which is the percentage of consonants that are heard incorrectly. If this is greater than 15% then the intelligibility is considered to be poor. Although articulation loss is specific to speech—an important part of our auditory world—much music also relies on good articulation for its effect.

Subjective testing is a complex subject and could fill a complete book just by itself! For more details see Bech and Zacharov (2006).

7.4 FILTERING AND EQUALIZATION

One of the simplest forms of electronic signal processing is to filter the signal in order to remove unwanted components. For example, often low-frequency noises, such as ventilation and traffic rumble, need to be removed from the signal picked up by the microphone. A high-pass filter would accomplish this and mixing desks often provide some form of high-pass filtering for this reason. High frequencies also often need to be removed to either ameliorate the effects of noise and distortion or remove the high-frequency components that would cause alias distortion in digital systems. This is achieved via the use of a low-pass filter. A third type of filter is the notch filter, which is often used to remove tonal interference from signals. Figure 7.22 shows the effect of these different types of filter on the spectrum of a typical music signal.

In these cases the ideal would be to filter the signal in a way that minimized any unwanted subjective effect on the desired signal. Ideally, in these cases the timbre of the sound being processed should not change after filtering, but in practice there will be some effects. What are these effects and how can they be minimized in the light of acoustic and psychoacoustic knowledge?

The first way of minimizing the effect is to recognize that many musical instruments do not cover the whole of the audible frequency range. Few instruments have a fundamental frequency that extends to the lowest frequency in the audible range and many of them do not produce harmonics or components that extend to the upper frequencies of the audible range. Therefore, in theory one can filter these instruments such that only the frequencies present are passed with no audible effect. In practice this is not easily achieved for two reasons:

- *Filter shape*: Real filters do not suddenly stop passing signal components at a given frequency. Instead there is a transition from passing the signal components to attenuating them, as shown in Figure 7.23. The cut-off frequency of a filter is usually expressed as the point at which it is attenuating the signal by 3 dB relative to the pass-band; see Figure 7.23. Thus if a filter's cut-off is set to a given frequency there will be a region within the pass-band that affects the amplitude of the frequency components of the signal. This region can extend as far as an octave away from the cut-off point. Therefore, in practice, the filter's cut-off frequency must be set beyond the pass-band that one would expect from a simple consideration of the frequency range of the instruments, in order to avoid any tonal

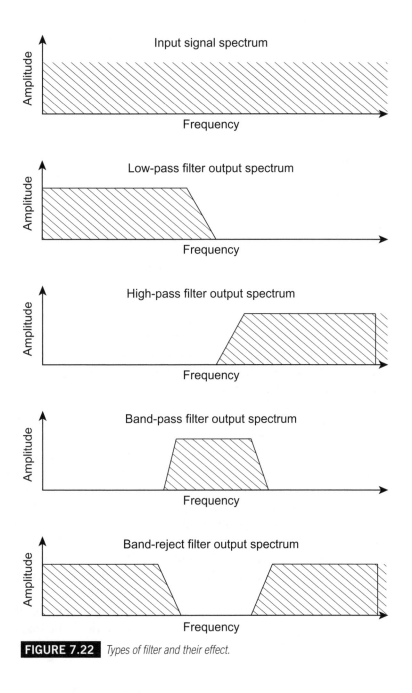

FIGURE 7.22   *Types of filter and their effect.*

FIGURE 7.23

The specifications of a real filter.

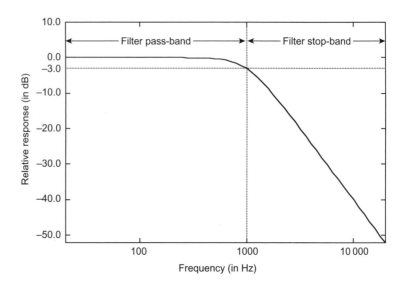

change due to change in frequency content caused by the filter's transition region. As the order of the filter increases, both the slope of the attenuation as a function of frequency and the sharpness of the cut-off increase; this reduces the transition region effects, but unfortunately increases the time domain effects.

■ *Time domain effects:* Filters also have a response in the time domain. Any form of filtering which reduces the bandwidth of the signal will also spread it over a longer period of time. In most practical filter circuits these time domain effects are most pronounced near the cut-off frequency and become worse as the cut-off becomes sharper. Again, as in the case of filter shape, these effects can extend well into the pass-band of the filter. Note that even the notch filter has a time response, which gets longer as the notch bandwidth reduces. Interestingly, particular methods of digital filtering are particularly bad in this respect because they result in time domain artifacts that precede the main signal in their output. These artifacts are easily unmasked and so become subjectively disturbing. Again, the effect is to require that the filter cut-off be set beyond the value that one would expect from a simple consideration of the frequency range of the instruments.

Because of these effects the design of filters that achieve the required filtering effect without subjectively altering the timbre of the signal is difficult.

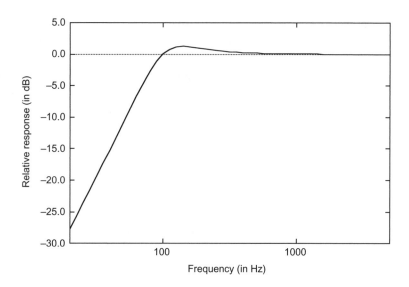

FIGURE 7.24

Partially compensating for filtered components with a small boost at the band-edge.

The second way of minimizing the subjective effects is to recognize that the ear uses the spectral shape as a cue to timbre. Therefore the effect of removing some frequency components by filtering may be partially compensated by enhancing the amplitudes of the frequency components nearby, as discussed in Chapter 5. Note that this is a limited effect and cannot be carried too far. Figure 7.24 shows how a filter shape might be modified to provide some compensation. Here a small amount of boost, between 1 dB and 2 dB, has been added to the region just before cut-off in order to enhance the amplitude of the frequencies near to those that have been removed.

7.4.1 Equalization and tone controls

A related, and important, area of signal processing to filtering is equalization. Unlike filtering, equalization is not interested in removing frequency components but in selectively boosting, cutting or reducing them to achieve a desired effect. The process of equalization can be modeled as a process of adding or subtracting a filtered version of the signal from the signal, as shown in Figure 7.25. Adding the filtered version gives a boost to the frequencies selected by the filter whereas subtracting the filtered output reduces the frequency component amplitudes in the filter's frequency range. The filter can be a simple high- or low-pass filter, which results in a treble or bass tone control, or it can be a band-pass filter to give a bell-shaped response curve. The cut-off frequencies of the filters may be either fixed or variable depending on the implementation. In addition the bandwidths of

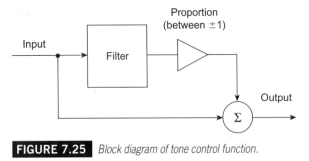

FIGURE 7.25 *Block diagram of tone control function.*

the band-pass filters and, less commonly, the slopes of the high- and low-pass filters can be varied.

An equalizer in which all the filter's parameters can be varied is called a parametric equalizer. However, in practice many implementations, especially those in mixing desks, only use a subset of the possible controls for both economy and simplicity of use. Typically in these cases, only the cut-off frequencies of the band-pass, and in some cases the low- and high-pass, filters are variable. There is an alternative version of the equalizer structure that uses a bank of closely spaced fixed frequency band-pass filters to cover the audio frequency range. This approach results in a device known as the "graphic equalizer" with typical bandwidths of the individual filters ranging from one-third of an octave to 1 octave. For parametric equalizers the bandwidths can become quite small.

Because a filter is required in an equalizer the latter also has the same time domain effects that filters have, as discussed earlier. This is particularly noticeable when narrow-bandwidth equalization is used, as the associated filter can "ring," as shown in Figures 1.62 and 1.63, for a considerable length of time in both boost and cut modes.

Equalizers are used in three main contexts (discussed below) which each have different acoustic and psychoacoustic rationales.

7.4.2 Correcting frequency response faults due to the recording process

This was one of the original functions of an equalizer in the early days of recording, which to some extent is no longer required because of the improvement in both electroacoustic and electronic technology. However, in many cases there are effects that need correction due to the acoustic environment and the placement of microphones. There are three common acoustic contexts that often require equalization:

- *Close miking with a directional microphone*: The acoustic bass response of a directional microphone increases, as it is moved close to an acoustic source, due to the proximity effect. This has the effect of making the recorded sound bass heavy; some vocalists often deliberately use this effect to improve their vocal sound. This effect can be compensated for by applying some bass-cut to the microphone signal and this often has the additional benefit of further reducing low-frequency environmental noises. Note that some microphones have this equalization built in but that in general a variable equalizer is required to fully compensate for the effect.

- *Compensating for the directional characteristics of a microphone*: Most practical microphones do not have an even response at all angles as a function of frequency. In general they become more directional as the frequency increases. As most microphones are designed to give a specified on-axis frequency response, in order to capture the direct sound accurately, this results in a response to the reverberant sound which falls with frequency. For recording contexts in which the direct sound dominates, for example close miking, this effect is not important. However, in recordings in which the reverberant field dominates, for example classical music recording, the effect is significant. Applying some high-frequency boost to the microphone signal can compensate for this.

- *Compensating for the frequency characteristics of the reverberant field*: In many performance spaces the reverberant field does not have a flat frequency response, as discussed in Section 6.1.7, and therefore subjectively colors the perceived sound if distant miking is used. Typically the bass response of the reverberant field rises more than is ideal, resulting in a bass heavy recording. Again the use of some bass-cut can help to reduce this effect. However, if the reverberation is longer at other frequencies, for example in the midrange, then the reduction should be applied in a way that complements the increase in sound level this causes. As in these cases the bandwidth of the level rise may vary, this must also be compensated for—usually by adjusting the bandwidth, or "Q," of the equalizer.

All the above uses of equalization compensate for limitations imposed by the acoustics of the recording context. To make intelligent use of it in these contexts requires some idea of the likely effects of the acoustics of the space at a particular microphone location, especially in terms of the direct to reverberant sound balance.

7.4.3 Timbre modification of sound sources

A major role for equalizers is the modification of the timbre of both acoustically and electronically generated sounds for artistic purposes. In this context the ability to boost or cut selected frequency ranges is used to modify the sounds spectrum to achieve a desired effect on its timbre. For example boosting selected high-frequency components can add "sparkle" to an instrument's sound whereas adding a boost at low frequencies can add "weight" or "punch." Equalizers achieve these effects through spectral modification only: they do not modify the envelope or dynamics of a music signal. Any alteration of the timbre is purely due to the modification, by the equalizer, of the long-term spectrum of the music signal. There is also a limit to how far these modifications can be carried before the result sounds odd, although in some cases this may be the desired effect.

When using equalizers to modify the timbre of a musical sound it is important to be careful to avoid "psychoacoustic fatigue"—this arises because the ear and brain adapt to sounds. This has the effect of dulling the effect of a given timbre modification over a period of time. Therefore one puts in yet more boost, which one adapts to, and so on. The only remedy for this condition is to take a break from listening to that particular sound for a while and then listen to it again later. Note that this effect can happen at normal listening levels and so is different to the temporary threshold shifts that happen at excessive sound levels.

7.4.4 Altering the balance of sounds in mixes

The other major role is to alter the balance of sounds in mixes—in particular the placing of sound "up-front" or "back" in the mix. This is because the ability of the equalizer to modify particular frequency ranges can be used to make a particular sound become more or less masked by the sounds around it. This is similar to the way the singer's formant is used to allow a singer to be heard above the orchestra as mentioned in Chapter 4. For example suppose one has a vocal line that is being buried by all the other instrumentation going on. The spectrum of such a situation is shown in Figure 7.26 and from this it is clear that the frequency components of the instruments are masking those of the vocals. By selectively reducing the frequency components of the instruments at around 1.5 kHz, while simultaneously boosting the components in the vocal line over the same frequency range, the frequency components of the vocal line can become unmasked, as shown in Figure 7.27. This has the subjective effect of bringing the vocal line out from the other instruments.

Similarly, performing the process in reverse would further reduce the audibility of the vocal line in the mix. To achieve this effect successfully requires the presence of frequency components of the desired sound within the frequency range of the equalizer's boost and cut region. Thus different instruments require different boost and cut frequencies for this effect. Again it is important to apply the equalization gently in order to avoid substantial changes in the timbre of the sound sources.

Equalizers therefore have a broad application in the processing of sound. However, despite their utility, they must be used with caution—firstly to avoid extremes of sound character, unless that is desired, and secondly to

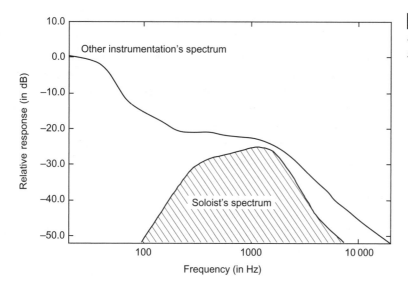

FIGURE 7.26

Spectrum of a masked soloist in the mix.

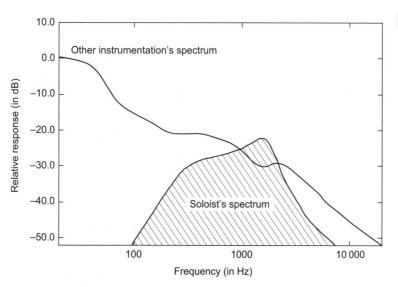

FIGURE 7.27

Spectrum after the use of equalization to unmask the soloist.

avoid unwanted interactions between different equalizer frequency ranges. As a simple example consider the effect of adding treble, bass and midrange boost to a given signal. Because of the inevitable interaction between the equalizer frequency responses, the net effect is to have the same spectrum as the initial one after equalization. All that has happened is that the gain is higher. Note that this can happen if a particular frequency range is boosted and then, because the result is a little excessive, other frequency ranges are adjusted to compensate.

7.5 PUBLIC ADDRESS SYSTEMS

Sound reinforcement of speech is often taken for granted. However, as anyone who has tried to understand an announcement in a reverberant and noisy railway station knows, obtaining clear and intelligible speech reinforcement in a real acoustic environment is often difficult. The purpose of this section is to review the nature of the speech reinforcement problem from its fundamentals in order to clarify the true nature of the problem. We will examine the problem from the perspective of the sound source, the listener, and the acoustics. At the end you should have a clear appreciation of the difficulties inherent in reinforcing one of our most important, and sensitive, methods of communication.

There are several aspects of an acoustic space that affect the intelligibility of speech within it.

7.5.1 Reverberation

As discussed in Chapter 6 (see Section 6.1.12), bigger spaces tend to have longer reverberation times and well-furnished spaces tend to have shorter reverberation times. Reverberation time can vary from about 0.2 of a second for a small well-furnished living room to about 10 seconds for a large glass and stone cathedral.

There are two main aspects of the sound to consider:

- *The direct sound*: This is the sound that carries information and articulation. For speech it is important that the listener receive a large amount of direct sound if they are to comprehend the words easily. Unfortunately, as discussed in Chapter 1, the direct sound gets weaker as it spreads out from the source. Every time you double your distance from a sound source the level of the *direct sound* goes down by a factor of four, that is, an inverse square law. Thus the further away you are from a sound source, the weaker the direct sound component.

- *The reverberant sound*: The second main aspect of the sound is the reverberant part. This behaves differently to the direct sound, as discussed in Chapter 6. The reverberant sound is the same in all parts of the space.

The effect of these two aspects is shown in Figure 7.28. As one moves away from a source of sound in a space, the level of direct sound reduces but the reverberant sound stays constant. This means that ratio of direct sound to reverberant sound becomes less and so the reverberant sound becomes more dominant. The critical distance, where the reverberant sound dominates,

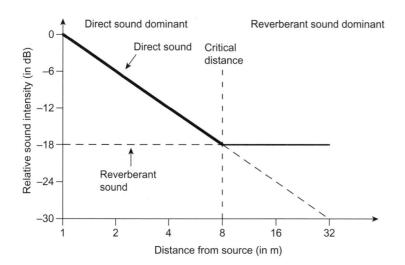

FIGURE 7.28

Regions of dominance for direct and reverberant sound.

is dependent on both the absorption of the space and the directivity of the source. As the absorption and directivity increase so does the critical distance, but only proportionally to the square root of these factors. As discussed in Chapter 6 the critical distance is:

$$D_{crit} = 0.141\sqrt{RQ}$$

where $R = \dfrac{S\alpha}{(1 - \alpha)}$ the room constant (in m^2)　　　　(7.3)

and Q = the directivity of the source (compared to a sphere)

7.5.2 The effect of reverberation on intelligibility

The effect of reverberation, and early reflections, is to mask the stops and bursts associated with consonants. They can also blur the rapid formant transitions that are also important cues to different consonant types. Clearly the degradation will depend on both the reverberation time and the relative level of the reverberation to the direct sound. One would expect longer reverberation times to be more damaging than short ones.

Because of the importance of consonants to intelligibility, it is therefore important to maintain a high level of direct to reverberant sound; ideally one should operate a system at less than the critical distance. There is an empirical equation that links the number of consonants lost to the characteristics of the room (Peutz, 1971). As consonants occupy frequencies above 1 kHz, and have very little energy above 4 kHz, this equation is based on the average reverberation time in the 1 kHz and 2 kHz octave bands.

Up to D, $3.5Dc$ (at which point the Direct/Reverberation ratio $= -11\,\text{dB}$)

$$\%ALcons = \frac{200D^2 T_{60}^2 N}{VQ} + a \qquad (7.4a)$$

where D = the distance from the nearest loudspeaker (in m)

T_{60} = the reverberation time of the room (in s)

N = the number of equal power sources in the room

V = the volume of the room (in m^3)

Q = the directivity of the nearest loudspeaker

a = is a listener factor; because we aways make some errors it can range from 1.5% to 12.5% where 1.5% is an excellent listener

For $D > 3.5Dc$ (where the Direct/Reverberation ratio is always worse than $-11\,\text{dB}$)

$$\%ALcons = 9T_{60} + a \qquad (7.4b)$$

Note that when D is greater than $3.5Dc$ the intelligibility is constant.

The %*ALcons* is related to intellibility as follows. If:

%*ALcons* is less than 10% then the intelligibility will be very good;

%*ALcons* is between 10% and 15% then the intelligibility will be acceptable;

%*ALcons* is greater than 15% then the intelligibility will be poor.

In order to achieve this one might think that placing more loudspeakers in the space would be better, because this would place the loudspeakers closer to the listeners.

Notice that the %*Alcons* increases as the number of sources increases. This is counterintuitive because you would think that more loudspeakers would mean they are closer to the listener and therefore should be clearer.

7.5.3 The effect of more than one loudspeaker on intelligibility

Unfortunately increasing the number of speakers decreases the intelligibility, because only the loudspeaker that is closest to you provides the direct sound. All the other loudspeakers contribute to the reverberant field, and not to the direct sound! The net effect of this is to reduce the critical

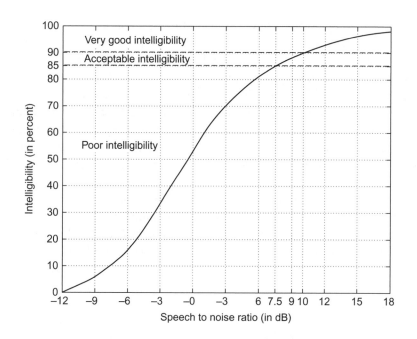

FIGURE 7.29

Intelligibility of Logatoms and monosyllabic words versus speech to noise ratio (data from ISO TR 4870 1991).

distance and make the problem worse. If one assumes that all the loudspeakers radiate the same power then the critical distance becomes:

$$D_{crit} = 0.141\sqrt{\frac{RQ}{N}}$$ \hfill (7.5)

where N = the number of equal power sources in the room

So, in this case, more is not better! Ideally one should have the minimum number of speakers, preferably one, needed to cover the space. When this is not possible, it is possible to regain the critical distance by increasing the "Q" of each loudspeaker in proportion to their number. This has its own problems, which will be discussed later.

The need to minimize the number of sources in the space has led to a design called the central cluster in which all the speakers required to cover an area are concentrated at one coherent point in the space. In general such an arrangement will provide the best direct to reverberant ratio for a space. Unfortunately it is not always possible, especially for large spaces.

If the sources do not have equal power, then N is the ratio of the total source power to the power of the source producing the direct sound. That is:

$$N = \frac{W_{reverb}}{W_{direct}}$$

$$= \frac{\sum\limits_{all\ sources} W_{each\ source}}{W_{direct}}$$

Where N = the ratio of all power sources in the room, to the direct power

7.5.4 The effect of noise on intelligibility

The effect of noise, like reverberation, is to mask the stops and bursts associated with consonants. This is because the consonants are typically 20 dB quieter than the vowels. They can also blur the rapid formant transitions, which are also important cues to different consonant types. Because of the importance of consonants to intelligibility it is therefore important to maintain a high signal to noise ratio.

Figure 7.29 shows how the intelligibility of speech varies according to the signal to noise ratio. From this figure we can see that a speech to background noise ratio of greater than +7.5 dB is required for adequate intelligibility. Ideally, a signal to noise ratio of greater than 10 dB is required for very good intelligibility. This assumes that there is minimal degradation due to reverberation.

Different types of noise have different effects on speech. For example, background noise that is hiss-like can be spectrally very similar to the initial consonants in *sip* or *ship*; periodic sounds such as the low-frequency drone of machines or vehicle tire noise can mask sounds with predominantly low-frequency energy such as the vowels in *food* or *fun*; sounds such as motor noise that exhibit a continuous whine can mask a formant frequency region and reduce vowel intelligibility; short bursts of noise can either mask or insert plosive sounds such as the initial consonants in *pin*, *tin*, or *kin*; and broad-band noise can contribute to the masking of all sounds, particularly those which depend on higher-frequency acoustic cues (see Howard, 1991) such as the initial consonants in *fun*, *shun*, *sun*, and *thump*.

High levels of noise can mask important formant information. This is especially true of high levels of low-frequency noise that, as shown earlier in Chapter 5, can mask the important lower formants. As high levels of low-frequency and broad-band noise are often associated with transport noise, this can be a serious problem in many situations. More subtly, it is possible for speech that is produced at high levels to mask itself. That is, if the speech is too loud then, notwithstanding the improvement in signal to noise ratio, the intelligibility is reduced, because the low-frequency components of the speech mask the high frequency components, due to the upward spread of masking.

There may be situations where acoustic treatment may be essential before sound reinforcement is attempted. Interfering noises which have similar rates of variation as speech are particularly difficult to deal with as they fool our higher order processing centers into attending to them as if they are speech. Because of this their effect is often more severe than a simple measurement of level would indicate.

There may also be high levels of noise that cannot be controlled. In these circumstances it can sometimes be possible to increase intelligibility by boosting the speech spectrum in the frequency regions where the interfering sound is weakest, as discussed in Section 7.4, thus causing the desired speech to become unmasked in those regions and so enhancing the speech intelligibility.

7.5.5 Requirements for good speech intelligibility

In general, for good intelligibility we require the following:

- The direct sound should be greater than, or equal to, the reverberant sound. This implies that the listener should be no further away than the critical distance.

- The speech to interference ratio should, ideally, be greater than 10 dB and no worse than 7.5 dB.

- The previous two requirements have the implication that the level of the direct sound should be above a certain level, that is, at least 10 dB above the background noise and equal to the reverberant sound level. For both efficiency and the comfort of the audience, this implies that the direct sound should be constant at this level throughout the coverage area.

Usually the only way of achieving this is to make use of the directivity of the loudspeakers used. This is because any other technique, such as reducing T_{60}, tends to require major architectural changes and therefore considerable cost. However, sometimes this may be the only way of achieving a usable system. Sometimes communication can be assisted by using speakers with good elocution, especially female ones because their voices' higher pitch tends to be less masked by the reverberation and noise typically present. Another possibility is to "chant" the message, which gives an exaggerated pitch contour that assists intelligibility. As a last resort one can use the *international radiotelephony-spelling alphabet* (Oscar, Bravo, Charlie . . .) to facilitate communication. Paradoxically electronic volume compression does not improve intelligibility; in many cases it makes it worse, because it can distort the syllabic amplitude variations that help us understand words.

These simple rules, outlined previously, must be considered in the light of the actual context of the system. Their apparent simplicity belies the care, analysis and design that must be used in order for practical systems to achieve their objectives.

7.5.6 Achieving speaker directivity

If the major way of achieving a good quality public address system is to use directional loudspeakers, it is worth considering how this might be achieved. Ideally, speakers for public address systems should have a directivity that is constant with frequency; that is, the angles into which they radiate their sound energy remain the same over the whole audio spectrum.

There are two main ways of achieving this:

■ *The Array Loudspeaker*: One way is to use a large number of speakers together as an "array loudspeaker." With appropriate signal processing of the audio into these systems a good constant directivity performance can be achieved. With the advent of technologies that make this processing much easier such speakers are becoming more popular because of the flexibility they allow.

■ *The Constant Directivity Horn*: This is the other main technique. It has a constant directivity above a specific frequency, or frequencies, and is simple and very efficient. It can convert about 25% of its electrical energy input into acoustic energy and is able to sustain outputs of around 10 W (130 dB) of acoustic power for long periods of time. Its main limitation is its low-frequency performance, which typically limits its frequency range to frequencies that are greater than 500 Hz.

However, irrespective of the technology used, there is a fundamental lower limit on the frequencies at which the speakers are directive that is determined by their size.

Recall that in Chapter 1 sound diffraction and scattering was discussed, in Sections 1.5.9 and 1.5.10, and we saw that the size of an object depended on its size in wavelengths. That is, sound is diffracted around objects that are small with respect to wavelength and is reflected from objects that are large with respect to wavelength. The same thing applies to loudspeakers.

Although a standard loudspeaker looks like it should radiate sound in the direction that its drivers are pointing, in practice it doesn't. This is because at many frequencies it is small with respect to the wavelength; for example a 200 mm (8") loudspeaker will only start becoming directive at about 1 kHz! Note that the size of the box has very little influence; it is the size of the part that radiates the sound that matters. So we can consider any small loudspeaker to be similar to a torch or flashlight bulb without a reflector, irrespective of what it looks like!

The equation that relates the minimum size of the radiating size of a speaker for a given directivity is:

$$F_{\theta_{min}} = \frac{219}{L\sin\theta}$$

where $F_{\theta_{min}}$ = the minimum frequency that $\pm\theta$ can equal -6 dB

$\qquad L$ = the length of the radiating source (in m)

$\qquad \theta$ = the $-6\,dB$ angle for the speaker. This is on either side of the center line so the coverage would be 2θ

e.g. $\theta = 45°$ means $90°$ of coverage

(7.6)

As an example, if one wanted a constant directivity horn that had a coverage of $90° \times 40°$ from 500 Hz then the mouth would have to be at least 0.62 m for the 90° direction and 1.28 m for the 40° direction. This is not small! You can have a smaller horn but you must recognize that it will not have this coverage angle at 500 Hz. The sine of the -6 dB θ point is inversely proportional to the frequency below $F_{\theta_{min}}$; thus at an octave below $F_{\theta_{min}}$ θ will be approximately double that desired, and so on, until θ becomes greater than 90°, which implies the speaker is omnidirectional and has no directivity. In practice many commercially available directional speakers have to make some form of compromise in the frequency range of the desired directivity.

However, beware of specmanship. One of the authors has seen a loudspeaker advertised as having a constant directivity of $90° \times 60°$ using a horn whose mouth dimensions were 130×130 mm. This gives an $F_{\theta_{min}}$ of 2.4 kHz and 3.4 kHz respectively. As most human speech energy is between 100 Hz and 5 kHz this speaker's directivity is going to have very little influence on the intelligibility of the speech!

7.5.7 A design example: How to get it right

Now that we know, let's look at how we would go about designing a PA system that has a good coverage. In order to do this we need to work out what level the sound will be some distance away from the loudspeaker. Fortunately this can be done very easily, as manufacturers provide a parameter called the speaker's sensitivity. This measures the sound pressure level (SPL) emitted by a loudspeaker at one meter for 1 W of electrical power input. Using this measure the sound pressure level (SPL) at a given distance from the loudspeaker is given by:

$$SPL = 10\log(P_{in}) + Sensitivity - 20\log(r)$$

where P_{in} = the power sent to the speaker (in W), (7.7)

$Sensitivity$ = is the speaker's sensitivity in (in dB W^{-1} @1 m),

r = is the distance from the speaker (in m)

Consider a room that is 30 m long by 12 m wide by 9 m high and with a T_{60} = of 1.5 s at 1 kHz and 2 kHz. The audience's ears start at 2.2 m away from the front wall and extend all the way to the back.

Making use of Equation 7.7 let us look at an example public address problem for several speaker arrangements.

EXAMPLE 7.1

A small speaker mounted at ear height.

Figure 7.30 shows a simple full-range speaker mounted at ear height and orientated to cover the entire audience area. The loudspeaker sensitivity is 94 dB per watt at 1 m. As the entire audience is "on-axis" there will be no variation in level due to speaker directivity; therefore, at the positions shown, the levels will be as shown in Table 7.4.

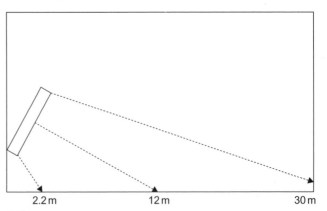

| 2.2 m | 12 m | 30 m |

FIGURE 7.30 *A simple speaker mounted at ear height.*

There is a huge SPL variation from front to back—22.7 dB! People in the front row are being deafened (one of the authors has seen people close to the speakers wince in a York church installation), while people in the rear row are straining to hear anything.

Table 7.4	Calculated SPL for a small full-range loudspeaker		
Distance from front	2.2 m	12 m	30 m
Distance from speaker	2.2 m	12 m	30 m
Output for 1 watt	94 dB	94 dB	94 dB
Path attenuation	−6.8 dB	−21.6 dB	−29.5 dB
SPL at the listener	87.2 dB	72.4 dB	64.5 dB

Unfortunately this is not an unusual situation. Furthermore, the critical distance for this speaker is 2.63 m! So the %Alcons for this space will be 15.5%. Note that adding more speakers does not make the problem any better. Instead it makes it worse, because the extra loudspeakers further reduce the critical distance. In fact in a typical multi-speaker installation it is possible to show that an unaided human voice can be more intelligible than the PA system, provided the speaker can project their voice sufficiently!

The problem is that the path length variation is too large (2.2–30 m) and this results in a very high SPL variation. The arrangement would also have a very poor direct to reverberant ratio over most of the audience and would be prone to feedback. Let us see if using a more directive speaker mounted higher up can do any better.

EXAMPLE 7.2

A single constant directivity horn speaker 7.5 m above ear height.

Mounting the speaker higher up, as shown in Figure 7.31, reduces the path variations between the front and back row. By making the speaker directive we can ensure that more of the sound goes to the audience. However, we have to handle the low frequencies with a separate loudspeaker because of the frequency limitations of the constant directivity horn.

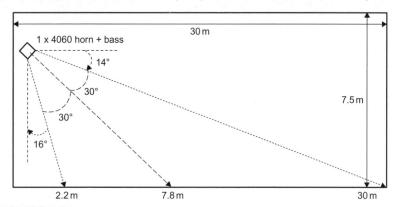

FIGURE 7.31 *A single 40° × 60° constant directivity horn speaker mounted 7.5 m above ear height.*

Table 7.5 shows the results of doing this. Because of the lower variation in path length the path loss variation is lower: only 12 dB. However, the directivity of the loudspeaker both helps and hinders the total SPL at the listener. For the front row it overcompensates for the shorter path length and provides a slightly lower SPL there; unfortunately for the back row it adds to the path loss resulting in a lower SPL. The peak SPL variation is therefore 15.1 dB, which is better than the non-directive loudspeaker's case, but still unacceptable.

Table 7.5	Calculated SPL for a single 40° × 60° constant directivity horn speaker		
Distance from front	2.2 m	7.8 m	30 m
Distance from speaker	7.8 m	10.8 m	30.9 m
Output for 1 watt	113 dB	113 dB	113 dB
Path attenuation	−17.8 dB	−20.7 dB	−29.8 dB
Directivity effect	−6 dB	0 dB	−6 dB
SPL at the listener	89.2 dB	92.3 dB	77.2 dB

Although better, the directivity of one speaker is not precise enough to obtain an even distribution. What we need is more control over the directivity. Ideally we want an "on-axis" performance further down the audience. However, if we use a more directive horn then the front will suffer. So let's see if by combining two horns to make a "central cluster" we can do better.

EXAMPLE 7.3

Two constant directivity horn speakers 7.5 m above ear height.

We need to cover 60 degrees of angle using two constant directivity horns. This can be achieved by using a 60° × 40° combined with 40° × 20° horn, as shown in Figure 7.32; the horns are aligned so that the −6 dB points of the two horns are at the same angle. Because the signals are adding coherently these −6 dB points can become equivalent to an on-axis level. Therefore the combine horn has three "on-axis" points on the audience: one at 5.4 m, due to the 60° × 40° horn; one at 16.8 m due to the 40° × 20° horn; and finally one at 11.1 m due to the two horns combined. Note that we still have only one source of sound in the room despite using two speakers. This is because they both radiate sound energy from a single point.

Table 7.6 shows the results of doing this. Note that the 40° × 20° horn has a higher sensitivity, because it is concentrating the sound into a smaller solid angle. Note that the level is now much more uniform—to about halfway down the audience the maximum variation

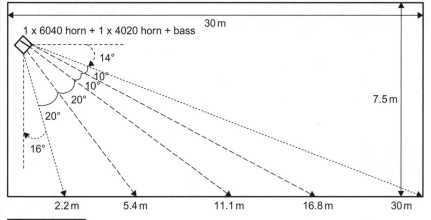

FIGURE 7.32 *A 60° × 40° horn combined with 40° × 20° horn at 7.5m above ear height.*

Table 7.6	Calculated SPL for a 60° × 40° horn combined with 40° × 20° horn				
Distance from front	2.2m	5.4m	11.1m		16.8m	30m
Distance from speaker	7.8m	9.2m	13.3m	13.3m	18.4m	30.9m
Output for 1 watt	113dB	113dB	113dB	115dB	115dB	115dB
Path attenuation	−17.8dB	−19.3dB	−22.5dB	−22.5dB	−25.3dB	−29.8dB
Directivity effect	−6dB	0dB	−6dB	−6dB	0dB	−6dB
SPL for each speaker	89.2dB	93.7dB	84.5dB	86.5dB	89.7dB	79.2dB
SPL at the listener	89.2dB	93.7dB	91.6dB		89.7dB	79.2dB

is 4.5dB, or ±2.25dB. However, the rear row at 30m is still 14.5dB lower than the level at 5.4m, and this is still too much variation. One might be able to do better by using three horn loudspeakers, but this is getting more complicated and expensive. The main problem seems to be that the levels in the audience covered by the 40° × 20° horn are lower than those in the area covered by the other horn. So, as a more economical solution, let's try raising the power fed to the 40° × 20° horn to 2.5W, which is a +4dB increase in level.

7.5.8 More than one loudspeaker and delays

Sometimes it is necessary to use more than one loudspeaker, for example to "fill in" a shaded under-balcony area, or cover a very wide area. In these situations the speakers should be as directive as possible so as to cover only the area required. This will minimize the amount of extra energy

EXAMPLE 7.4

Two constant directivity horn speakers 7.5 m above ear height with more power to the horn covering the back.

Table 7.7 shows the results of doing this. Now the sound is more even within ±0.15 dB from 5 m to 19 m in the audience. It drops at the front and the back, and the maximum variation is 10.8 dB. In practice the front row is likely to receive additional sound from the

Table 7.7	Calculated SPL for 1 W to the 60° × 40° horn combined with 2.5 W to the 40° × 20° horn					
Distance from front	2.2 m	5.4 m	11.1 m		16.8 m	30 m
Distance from speaker	7.8 m	9.2 m	13.3 m	13.3 m	18.4 m	30.9 m
Output for 1 watt	113 dB	113 dB	113 dB	115 dB	115 dB	115 dB
Effect of 2.5 W to 40° × 20° horn	0 dB	0 dB	0 dB	4 dB	4 dB	4 dB
Path attenuation	−17.8 dB	−19.3 dB	−22.5 dB	−22.5 dB	−25.3 dB	−29.8 dB
Directivity effect	−6 dB	0 dB	−6 dB	−6 dB	0 dB	−6 dB
SPL for each speaker	89.2 dB	93.7 dB	84.5 dB	90.5 dB	93.7 dB	83.2 dB
SPL at the listener	89.2 dB	93.7 dB	94.0 dB		93.7 dB	83.2 dB

stage, thus boosting its level. In general it is not a good idea to try to make the rear row the same level because this means the rear wall must be on-axis to a loudspeaker, which means half that speaker's energy is splashing off the back wall and contributing to either the reverberant field or causing interference effects at the back of the venue. The calculated %Alcons for this design is 11%, which is on the boundary between acceptable and very good intelligibility.

We should also consider the lateral coverage. In the final arrangement the horn covering the front of the audience covers a wider angular width. This matches the trapezoidal audience shape seen by a high central cluster.

that is fed into the reverberant field. Electronic delay of the signal will often be required in order to match the acoustic delay. When this is used it should be set at about 15 ms greater than the acoustic delay. This has two advantages:

- Firstly, it allows the near speaker to be louder than the further one whilst giving the illusion that the sound is still coming from the further source via the Haas effect, as discussed in Chapter 2.

- Secondly, any combing effects due to the combination of the near and far signal are of a close enough frequency spacing (67 Hz) to be averaged by the ear's critical bands and so not affect the intelligibility.

7.5.9 Objective methods for measuring speech quality

%Alcons is a subjective/empirical formula that can be used as a means of estimating the speech intelligibility from architectural data. However, for specifying systems it is better to have an objective measure that can be assessed by some form of acoustic instrumentation. Objective methods for measuring speech quality make use of either an auditory model or a measure that is based on a measure that is sensitive to speech spectral variations.

The articulation index (AI), now called the speech intelligibility index (SII) (Pavlovic, 1987), objectively measures "articulation" in individual critical bands, which is defined as that fraction of the original speech energy perceivable (i.e., between the threshold of hearing and the threshold of pain, and above the background noise). The AI can be measured by averaging the signal-to-noise ratio across all the bands. The validity of AI depends on the noise being non-signal dependent, which may not be the case with some processing. Also used are various measurements based on comparing the smoothed spectrum of the processed version with that of the original.

Another objective measure is the speech transmission index (STI) (IEC, 2003) (Houtgast and Steeneken, 1985), which uses a modulation transfer function approach to measure the effect of a given situation on speech intelligibility. It works for most forms of speech degradation and shows good correlation with subjective tests (Steeneken and Houtgast, 1994). It also has the advantage of being relatively easy to calculate, and so can be used in simulations to predict likely improvements in performance.

The criteria for STI and intelligibility are:

- $0.0 < STI < 0.4$ intelligibility is *poor*
- $0.4 < STI < 0.6$ intelligibility is *fair*
- $0.6 < STI < 0.8$ intelligibility is *good*
- $0.8 < STI < 1.0$ intelligibility is *excellent*.

It can also be measured, and the simpler STIpa and RaSTI methods that it replaces, are available as simple handheld instruments. For more details about public address system design see Ahnert and Steffen (1999).

7.6 NOISE-REDUCING HEADPHONES

One important feature relevant to obtaining good quality audio listening is the relative levels of the wanted sound and unwanted sound such as local background acoustic noise. In environments such as aircraft cabins where the ambient acoustic noise level is high, it is not easy to obtain good quality audio. This is particularly the case with the headphones provided to economy class travelers, which are generally of a low quality that distorts the sound when the volume is turned up sufficiently high to hear the music or film soundtrack. There is also the danger that the overall sound level being presented (wanted signal plus the unwanted background noise) could even cause noise-induced hearing loss (see Section 2.5) depending on the volume of the wanted signal that is set.

One way of reducing the overall sound level being presented in such situations is to reduce the level of the unwanted acoustic background noise being experienced. This has the added advantage of perhaps improving the sound quality of the wanted sound because it can then be presented at a lower level, thereby possibly avoiding any distortion issues due to a high signal presentation level.

There are two common methods that are used for acoustic background noise reduction:

1. active noise cancellation
2. passive noise cancellation.

Active noise cancellation is based on the fact that if a waveform is added to an equal and opposite (antiphase) version of itself, cancellation results (see Figure 1.13). Active noise cancelation is designed into headphones which have a microphone on the outside of each earpiece to pick up the local acoustic background noise on each side of the head. This is essentially the acoustic noise that is reaching each ear. This microphone signal is phase reversed and added at the appropriate level to cancel the background noise. The wanted signal is also added in and the result is that the background noise is significantly reduced in level, and the overall volume of the wanted signal can be reduced to a more comfortable listening level.

Passive noise cancellation can be achieved with in-ear earphones that seal well in the ear canal. The principle here is to block the ear canal to reduce the level of acoustic background noise that enters the ear canal using the same technique used for in-ear ear defenders. Usually there are a number (two or three) of soft rubber flanges which form a seal with the border of the ear canal to attenuate the level of acoustic background noise that can enter the ear. The wanted sound is played via the earphones which are mounted in the body of the earphones, and the level of the wanted sound can be reduced as with active noise cancelation. Once again this allows a more comfortable overall acoustic level to be achieved and lessens the likelihood of sound distortion.

Both active and passive noise cancelation systems can work very effectively and many people (including both authors!) tend to prefer the passive type because only the wanted sound is being presented to the ear and the sound itself is not being modified in any way.

In the active case, the wanted signal is having noise added to it – not something one really wants to be doing if one can avoid it. If the phase shift is not absolutely correct, complete noise cancelation will not occur, with the result that the wanted signal is then being further contaminated. Furthermore, because of the phase shift requirement, higher frequencies are not attenuated as effectively—something the passive types do very well.

However, both kinds can sound very good in practice and whilst the sound is being modified in the active version, this needn't modify or distort the wanted sound. The passive types are somewhat intrusive as their presence in the ear canal is felt physically since the seal has to be complete to enable them to function well. One advantage of a good seal with the ear canal is that transmission of the low-frequencies components is significantly improved.

7.7 "MOSQUITO" UNITS AND "TEEN BUZZ" RING TONES

Sound is being used to deter young people from congregating in particular areas via devices termed "mosquito units" or "teen deterrents." These devices play sounds at relatively high levels that can only be heard by young people, thereby making it acoustically unpleasant for them to remain in a particular area. They can be placed outside shops, restaurants and other public places to discourage young people from hanging around there. The devices exploit the natural change in hearing response that occurs with age known as "presbycusis" (see Section 2.3) which causes a significant reduction in the ear's ability to hear high frequencies (see Figure 2.11). The mosquito was invented by Howard Stapleton in 2005 and it was first marketed in 2006.

The nominal range of human hearing is usually quoted as being from 20 Hz to 20 kHz, but during a person's twenties the upper frequency region reduces greatly in the range above about 16 kHz. The mosquito unit exploits this by emitting a sound between 16 and 19 kHz, which can only be heard by those less than 20–25 years of age. The usual range over which the mosquito operates as a deterrent is around 15–25 m. The UK National Physical Laboratory (NPL) conducted a test of the Mosquito (Ref E05110518, December 2005) in which they reported that the device presented:

- a mean f_0 of 16.8 kHz;
- a maximum f_0 of 18.6 kHz;

- an A-weighted sound pressure level of 76 dBA at 3 m;
- no hearing hazard under the UK Control of Noise at Work Regulations (April, 2006).

These units have proved themselves to be successful deterrents, saving considerable police time and effort in moving on gatherings of young people, and they are now used quite widely in public spaces. The unit will work in the presence of other sounds such as music, which typically contains no high-level energy at these high frequencies. Young people report these high-level high-frequency sounds to be very annoying, unpleasant and irritating.

However, young people now benefit from this basic idea because it has now been used to provide mobile ring tones, sometimes known as "teen buzz," that cannot be heard by older people. The original teen buzz was created by recording the output from a mosquito device, but nowadays there are plenty of synthesized downloadable teen buzz ringtones available on-line. In general, these ring tones are not audible to adults over 25 years of age, but this does depend on the overall amplitude of the sound and the rate of presbycusis change for given individuals.

An example mosquito tone is provided on the accompanying CD to demonstrate what it sounds like; the track has four pure tones rising in octaves to the mosquito average frequency of 16.8 kHz as follows: 2.1 kHz, 4.2 kHz, 8.4 kHz and 16.8 kHz. Bear in mind when listening to this track that most youngsters will find the 16.8 kHz tone unpleasant; be cautious with the listening level.

7.8 AUDIO CODING SYSTEMS

Many of the advances in the distribution of audio material via film, DVD, television and the Internet, and even on DVD-Audio or Super Audio CD (SA-CD) have been made possible because of developments in audio coding systems. Audio coding systems are methods that reduce the overall data rate of the audio signal so that it may be transmitted via a limited data rate channel, such as the Internet, or stored in a data limited storage medium, like a DVD. In all cases the ability to reduce the data rate is essential for the system. There are two types of audio coding system:

- *Lossless audio coding systems*: In these systems the data rate, or data quantity, is reduced but in such a way that *no information is lost*. That is, after coding and subsequent decoding, the signal that comes out is identical to the signal that went in. This is like audio computer data file "zip" compression. Examples of such systems are: *Shorten*, an early lossless format; *MPEG4 ALS, FLAC, Apple*

Lossless, examples of newer ones; and *Direct Stream Transfer* (DST) and *Meridian Lossless Packing* which are proprietary methods that are used in the SA-CD and DVD audio formats respectively. Lossless methods typically achieve a compression ratio of around 2 to 1. That is, the data size after compression is typically half the size of the input. The actual amount of compression achieved is dependent on the nature of the input audio signal itself (i.e., what kind of music it is or if it is speech) and can vary from about 1.5 to 1 to 3 to 1.

■ *Lossy audio coding systems*: These systems make use of similar techniques to that of lossless coding but, in addition, reduce the data further by quantizing the signal to minimize the number of bits per sample. In order to achieve this with the minimum of perceived distortion, some form of psychoacoustic model is applied to control the amount of quantization that a particular part of the signal suffers. Because this type of coding removes, or loses, information from the audio signal it is known as a "lossy coding system." Unlike the lossless systems these coders do not preserve the input signal; the output signal is *not* identical to the original input. The output signal is thus distorted but hopefully in a way that does not disturb and is inaudible to the listener. Examples of such systems are *MPEG 1*, *MPEG 2*, *mp3*, and *MPEG 4*, which are used on the Internet, in broadcasting and for DVDs. In the film world *DTS*, *Sony-SDDS* and *Dolby AC3* are used to fit multiple channels onto standard film stock. The advantage of lossy coding is that it can compress the signal much more than a lossless system. For example, to achieve the 128 k bit data rate stereo audio that is used for mp3 coding of material on many music download sites, the audio signal has to be compressed by a factor of 11.025! This is considerably more than can be achieved by lossless compression.

So how do these systems work? What aspects of the audio signal allow one to losslessly compress the signal, and how can one effectively further reduce the data rate by doing psychoacoustic quantization?

7.8.1 The archetypical audio coder

Figure 7.33 shows the three archetypical stages in an audio coder. All modern audio coders perform these operations. The decoder essentially operates in reverse. The three stages are:

1. *A signal redundancy removal stage* which removes any inter-sample correlations in the *signal*. In order to do this, the coder may have to send additional *side information* to the receive end. This stage

FIGURE 7.33

Block diagram of an archetypical audio coder.

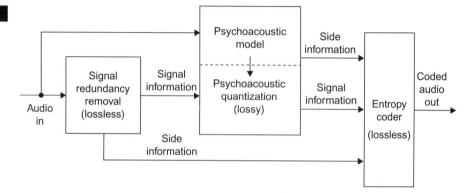

does not remove any information from the signal, it merely makes it more efficient and can therefore be considered to be *lossless*.

2. *A psychoacoustic quantization stage* which allocates bits to the various components in the audio signal in a manner that has the minimum subjective distortion. Again, in order to do this it has to send additional *side information* to the decoder, such as the number of bits allocated to each signal component. This is the only stage that removes information from the signal and is therefore the only *lossy* stage in the whole process. It is this stage that makes the difference between a *lossy* and a *lossless* audio coding system. Note that although the decoder stage works in reverse to provide real audio levels at the levels quantized by the coder, it cannot restore the information that has been thrown away by the encoding quantizer.

3. *An entropy coding stage* which tries to use the most efficient bit arrangement to transmit both the *signal information* and the *side information* to the decode end. This stage also does not remove any information from the signal and can also be considered to be *lossless*.

The purpose of these three stages is to maximize the amount of audio information transmitted to the receive end. So in order to understand how these stages work we need to understand what we mean by information and how it is related to the audio signal. Then we can unpack the function of the three stages in more detail.

7.8.2 What exactly is information?

What characteristics of a signal or data stream indicate information? For example, you are currently reading text in this book; what is it about the text that is carrying information? The answer is that, as you read it, you are seeing new combinations of words that are telling you something you didn't know before. Another way of looking at it is there is an element of novelty or surprise in the text. On the other hand, if, during a web chat, you got a message that said "hhhhhhhhhhhhhhhhhhhhhhh . . ." it would be carrying no information, other than the possibility that the other person has fallen asleep on the keyboard! So the more surprising a thing is the more information it carries. That is, the less probable something is the more information it carries. So:

$$\text{Information} \propto \text{Surprise} \propto \frac{1}{\text{Probability of Occurring}} \qquad (7.8)$$

How does this relate to an audio signal? Well, consider a sine wave; it sounds pretty boring to listen to because it is very predictable. On the other hand, a piece of music that is jumping around, or an instrument whose texture is continuously changing, is much more interesting to listen to because it's more unpredictable or surprising. If we looked at the spectrum of a sine wave we would find that all the energy is concentrated at one frequency whereas for the more interesting music signal it's spread over lots of frequencies. In fact the audio signal that carries the most information is either random noise, or a single spike that happens at a time you don't or can't predict. Interestingly, in both these cases the spectrum of the signal contains an equal amount of energy at all frequencies. Again, just like text, the more surprising a signal is the more information it carries.

So to maximize the information carried by our coded audio signal we first need to maximize the surprise value of the audio signal. This is done by the signal redundancy removal stage.

7.8.3 The signal redundancy removal stage

In Chapter 4 we saw that all musical instruments, including the human voice, could be modeled as a *sound source* followed by *sound modifiers*, which apply a filtering function. In the limit the source may be regular spaced impulses, for pitched instruments, a single pulse for percussion instruments, random noise for fricatives, or a combination of them all. A filter that combines the effect of the source with the acoustic effect of the instrument and output to shape the final sound then follows this. The effect of this filtering is to add correlation to the audio signal. Correlation implies that information from previous samples is carried over to the current sample, as shown in Figure 7.34. The basic principle behind redundancy

FIGURE 7.34

Information in a sample at a given time due to previous samples.

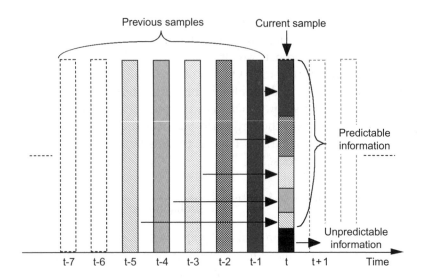

removal is that at a given time instant the audio signal will consist of two elements (as shown in Figure 7.34).

■ *Information about the previous signals that have passed through the filter*. In principle the contribution of this to the overall signal can be calculated from knowledge of the signal that has already passed through the filter. That is, this contribution is predictable; this is shown by the different hatchings in the sample at time "t" in Figure 7.34 that correspond to the similarly hatched earlier samples.

■ *Information that is purely due to the excitation of the filter by the source*. The contribution of this to the overall signal cannot be calculated from knowledge of the signal that has already passed through the filter and hence this contribution is not predictable, and is shown in black on Figure 7.34.

To maximize the information content of the signal we should aim to remove the predictable parts, because they can be recalculated at the receive end, and only encode the unpredictable part because this represents the new information.

This is what the correlation removal stage does. There are two different ways of doing this:

■ *Time domain prediction*: This is the method used by most lossless encoding schemes, which calculates an inverse filter to one that has filtered the unpredictable parts of the signal. It is possible, using the method of *linear prediction*, to calculate the necessary inverse filter from the input data. This filter is then used to remove the correlated components from the input signal, prior to coding the signal. At the decoder the original signal is recovered by feeding the decoded signal into the complementary filter, which then puts the correlations back into the signal. To do this the coder must send additional *side information* that specifies the necessary filter coefficients so that the decoder restores the correlation correctly. In addition, because the correlations within the signal vary with time, due to different notes and instruments, it is necessary to recalculate the required filter coefficients and resend them periodically. Typically this happens at about 50 to 100 times a second. This type of system is known as a "forward adaptive predictor" because it explicitly sends the necessary reconstruction information *forward* to the decoder. There is an alternative known as a "backward adaptive predictor" that sends the necessary side information implicitly in the data stream, but it is seldom used.

■ *Frequency domain processing*: This is the method used by most of the lossy audio coding schemes, which splits the audio spectrum into many small bands. In principle the smaller the better, but there is a limit to how far one can do this because smaller bands have a longer time response, as discussed in Chapter 1. It practice, due to the limitations imposed by temporal masking (see Chapter 5) the time extent of the signal is limited to 8–25 ms, which implies a minimum bandwidth of 125–40 Hz. This technique removes correlation from the signal because although each band will have a different amplitude level, which implies correlation. These can be normalized by applying an appropriate scale factor to each band, which effectively removes correlation, and the narrower the bands are the more effective this removal can be. It also allows the coder to flag bands that contain no energy at all and therefore no information needs to be transmitted. These scale factors and unused band information need to be transmitted to the decoder and therefore represent additional *side information* to be encoded and transmitted to the decoder. The signal within the bands is also less correlated because, as the band gets narrower in frequency range, the spectrum within the band is more likely to be uniform, and thus approaches the desired white noise spectrum with zero correlation between samples. A further advantage of frequency domain processing is that it converts the audio signal into a form that makes it easy for *psychoacoustic quantization* to be efficiently applied.

The use of either a time or frequency domain method is possible, and their relative strengths and weaknesses are primarily determined by the application. For example, the frequency domain approach fits well with psychoacoustic quantization algorithms – hence its choice for lossy coding systems.

However, all signal redundancy removal schemes have to tread an uneasy balance between increasing their effectiveness, which requires more *side information*, and having sufficient resources to effectively encode the actual audio signal information.

7.8.4 The entropy coding stage

Although the entropy stage is the final stage in the coder, it is appropriate to consider it now because in conjunction with the redundancy removal stage it forms the structure of a lossless encoder system. Entropy encoding works by maximizing the information carried by the bit patterns that represent the audio signal and the side information. In order to understand how it works we need to understand a little bit about how we measure information and how *entropy* relates to information.

7.8.5 How do we measure information?

In order to measure the information content of a signal we need to know how likely it is to occur. However, we also want to be able to relate the information content to something real, such as the number of bits necessary to transmit that information. So how could we measure information content in such a way that it is related to the number of bits needed to transmit it?

Consider a 3-bit binary digit. It has eight possible bit patterns, or symbols, as shown in Table 7.8. Furthermore let's assume that each possible symbol has the same one in eight probability of happening occurring ($P_{symbol} = 0.125$). If we use the following equation:

$$I_{symbol} = \log_2\left(\frac{1}{P_{symbol}}\right) \qquad (7.9)$$

where I_{symbol} = the self-information of the symbol (in bits)

P_{symbol} = the probability of that symbol occurring

$\log_2$ = the logarithm base 2, because we are dealing in bits

One can calculate $\log_2(x)$ simply as follows:

$$\log_2(x) = \frac{\log_{10}(x)}{\log_{10}(2)}$$

Where: $\log_{10}(x)$ is the logarithm base 10, the standard log key on most calculators

For any of the eight symbols shown in Table 7.8, the self-information is 3 bits because they all have the same probability. However, if the probability of a particular symbol was one, i.e., it was like our repeating "h" discussed earlier, the self-information would be zero. On the other hand, if the probability of one of the symbols was lower than one-eighth, for example, one-hundredth or (0.01), then that symbol's self-information would be

Table 7.8	Symbols associated with a 3-bit binary code
Binary code	**Symbol**
000	0
001	1
010	2
011	3
100	4
101	5
110	6
111	7

$\log_2(1/0.01)$, which would equal 6.64 bits, which would mean that symbol was carrying more information than the other symbols.

Because the total probability of all the symbols must add up to one, if one symbol has a very low probability then the probability of the other symbols must be slightly greater to compensate. In the case of one symbol having a probability of one-hundredth, the other symbols taken all together will have to have a probability of $P_{symbol} = (99/(7 \times 100)) = 0.141$. This gives a self-information for all the other symbols of 2.83 bits per symbol. So in this case one symbol is worth more than 3 bits of information, but the other symbols are worth less than 3 bits of information. So one of the symbols is using the bits more efficiently than the other seven symbols.

To unravel this we must look at more than the self-information carried by each symbol. Instead we must look at the total information carried by all the symbols used in the data stream.

7.8.6 How do we measure the total information?

In the previous section we had the case of one symbol in the whole set of possibilities being very small and as a consequence the other seven symbols had to have a higher probability, because the probability of all symbols being used has to be one. As an extreme case consider our person asleep on the keyboard who is sending the same symbol all the time. In this case the probability of the symbol being sent is one, and all other symbols have zero probability of being sent. The total information being sent by this source is zero bits. How can this be? Surely, if the probability is zero, then the self-information of these symbols is infinite? In theory this might be true but, as these symbols are never sent, the total information is zero.

So in order to find out the total information of the source we need to incorporate not only the self-information of each symbol but also their probability of being sent. This gives the proportion of information that symbol actually carries as part of the whole data set. To do this we simply multiply the self-information of the symbol by the probability of that symbol actually occurring. This gives the amount of information that symbol carries in proportion to the other symbols in the data source.

$$E_{symbol} = P_{symbol} \times \log_2 \left(\frac{1}{P_{symbol}} \right) \tag{7.10}$$

where E_{symbol} = the entropy of the symbol (in bits)

This is also known as the "entropy" because the equation is analogous to one used to calculate entropy in physics. In general we are interested in the total information content, or entropy of the data source, because this gives us the minimum number of bits required to encode it. This is given by:

$$E_{source} = \sum_{All\ symbols} P_{symbol} \times \log_2 \left(\frac{1}{P_{symbol}} \right) \tag{7.11}$$

where E_{source} = the entropy of the data source (in bits)

Equation 7.11 shows that the way to calculate the entropy of a data source is to add up all the individual symbol entropies of that source.

Table 7.9 shows the calculated entropy for four different data sources—the first three correspond to the examples that have been discussed already and the fourth represents the output of a 5-level audio signal. Notice that, except for the case of all symbols having equal probability, all the other

Table 7.9 Source entropies for a 3-bit binary code with different symbol probabilities

Binary code	Symbol	Uniform probability	One P_{symbol} equals 1	One P_{symbol} equals 0.01	P_{symbol} for 5 levels
000	0	0.125	0	0.01	0.60
001	1	0.125	0	0.141	0.15
010	2	0.125	0	0.141	0.05
011	3	0.125	1	0.141	0.00
100	4	0.125	0	0.141	0.00
101	5	0.125	0	0.141	0.15
110	6	0.125	0	0.141	0.05
111	7	0.125	0	0.141	0.00
Source entropy (in bits)		3.00	0.00	2.79	1.70

sources have a total source information, or entropy, of less than 3 bits. This is always true: the most efficient information source is one that uses all its possible symbols with equal probabilities. If we encoded the sources shown in Table 7.9 using a 3-bit binary code, then we would be wasting bits. In principle we could code the audio signal using only 1.7 bits, but how?

This is the idea behind entropy coding; what we need is a transmitted code that relates more closely to the entropy of each symbol.

7.8.7 Entropy coding

In order to implement entropy coding we have to more closely match the number of bits we use for each symbol to the amount of information it carries. Furthermore, each codeword associated with the symbol must have an integer number of bits. The net result of this is that instead of using a fixed number of bits for each symbol we need to use a number of bits that is related to the information carried by that symbol. For example, we know that in a real audio signal the signal spends much more time at low amplitudes than high amplitudes, so low amplitudes carry less information and should be encoded with fewer bits than high signal levels. This way fewer bits would be used, on average, to transmit the information. But how do we generate codes that have this desirable property? One simple way of doing it is by using a technique called Huffman coding, which is best illustrated by an example.

Consider the 5-level audio signal shown in column 6 of Table 7.9. At the moment we are using 3 bits of information to transmit an information source that only has 1.7 bits of information. How can we assign code words to the symbols such that the average data rate is closer to 1.7 bits? Firstly, we can recognize that three of the symbols are not used and hence can be ignored.

A *Huffman code* is generated by starting with the least probable symbol. The list of symbols is first sorted into a list of decreasing probability, as shown in Table 7.10. Then, working up from the lowest probability symbol, a code is constructed by combining the probabilities together to form different levels that correspond to bits in the code. These bits are then used to select either a code word or the next level down, except for the longest/lowest level, corresponding to the two lowest probability symbols in which the bit is used to select between a one and a zero.

By allocating a bit for each level, a variable length code is built up where a leading zero represents the beginning of a new code word, up to the maximum length of the codewords. This property makes the code *comma free*, which means it needs no additional bits to separate the variable code words

---

One can calculate the average bits per symbol, after coding, by simply using the actual bits it uses and multiplying that by its probability of occurring. This gives that symbol's average bit rate. Then, by adding up the average bit rate for all the symbols, you get the total average bit rate:

$$BitRate_{actual} = \sum_{\text{All symbols}} P_{symbol} \times Bits_{actual}$$

In our example the rate calculation becomes:

$$BitRate_{actual} = 0.6 \times 1 \\ + 0.15 \times 2 + 0.15 \times 3 \\ + 0.05 \times 4 + 0.05 \times 4$$

Which gives:
$$BitRate_{actual} = \\ 1.75 \text{ bits symbol}^{-1}$$

Table 7.10	Forming a Huffman code for a 5-level audio signal					
Symbol	Source probability	Step 1 (Level 4)	Step 2 (Level 3)	Step 3 (Level 2)	Step 4 (Level 1)	Huffman code
0	0.60	0.60	0.60	0.60	1.00	0
1	0.15	0.15	0.15	0.40	. . . 0	10
5 (−1)	0.15	0.15	0.25	. . . 0	. . . 10	110
2	0.05	0.1	. . . 0	. . . 10	. . . 110	1110
6 (−2)	0.05	. . . 1	. . . 1	. . . 11	. . . 111	1111
Source entropy (in bits)	1.70	Code formation; least significant bit first			Average bits/symbol	1.75

from each other, thus maximizing the efficiency. A code word of length one is assigned to the most probable symbol and longer length code words are optimally assigned to the lower probability symbols.

Huffman coding is optimal in that it gets the code rate within one bit of the source entropy. However, because bits only come in integer multiples the efficiency for small symbol sets is quite low. In MPEG_1 layer 3 coders, some symbols are Huffman coded as pairs to allow a greater coding efficiency. However, there is a practical limit to the size of a Huffman code because the computation and the tables blow up exponentially in size and so become unfeasible. A more subtle limitation is that, for large symbol sets, it becomes very difficult to gather enough symbol statistics data with sufficient accuracy to generate a Huffman code.

In order to have longer symbol lengths, which allow greater coding efficiencies, other approaches are used. *Golomb Rice Codes* use a predetermined statistical distribution to remove the need for tables, thus allowing longer code words; they are used in some lossless encoding schemes. Another approach is *Arithmetic Coding*, which does not need predetermined distributions but, by using the statistics of the symbols occurrence, can encode very long symbols and so approach the source entropy much more closely. For details about these and other entropy coding methods see Salomon (2007).

7.8.8 The psychoacoustic quantization stage

This is the stage that makes the difference between a lossless and a lossy coder. Again the actual signal processing can be carried out in either the time domain or the frequency domain. In both cases the process of *adaptive quantization* and *noise shaping* is used.

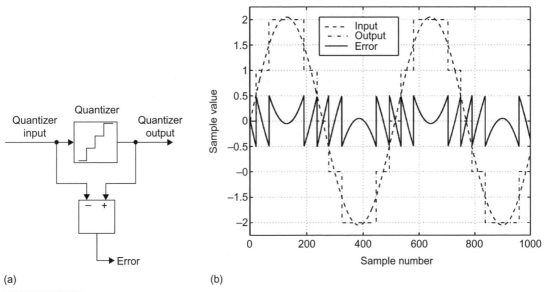

(a)   (b)

FIGURE 7.35 *The input, output and error of a quantizer.*

7.8.9 Quantization and adaptive quantization

Quantization is the process of taking an audio, or video, signal and converting it to a discrete set of levels. The input to this process may be from a continuous audio signal, like the one you get from a microphone, or may be from an already quantized signal, for example, the signal from a compact disc. An important parameter is the number of levels in the quantizer. This is due to the fact that the act of quantization is lossy because it throws away information. If the input signal is not exactly the same as the desired output then there will be an error between the input and the output, as shown in Figure 7.35.

Although the quantizer will pick the output level that causes minimum error, there will on average always be some error. This error adds noise and distortion to the signal and is often referred to as quantization noise. Ideally this noise should be random and often *dither* is added to ensure this. The effect of the error is to reduce the signal to noise ratio of the audio signal. If a binary word of N_{bits} bits is used to encode the audio signal, then the maximum signal to noise ratio is given by:

$$\left(\frac{S}{N}\right)_{Max} = 6 \times N_{bits} \text{ dB} \qquad (7.12)$$

For a 16-bit word (the CD standard), this gives a maximum signal to noise ratio of 96 dB. If you compare this signal to noise ratio with the idealized masked thresholds in Figure 5.9 in Chapter 5, you will see that the quantization noise will be masked over most of the audio band for loud signals. Although quantizers very often have a number of levels, that are powers of 2, such as 16, 256 or 65 536, because this makes best use of a binary word, there is no reason that other numbers of levels cannot be used, especially if entropy coding is going to be employed. In this case the maximum signal to noise ratio for an N level quantizer will be given by:

$$\left(\frac{S}{N}\right)_{\text{Max}} = 20\log(N) \text{ dB} \tag{7.13}$$

In either case, the quantization error is uniform over the frequency range in the ideal case.

These maximum signal to noise calculations are assuming that both the signal and noise have a signal probability distribution that is uniform; that is, all signal levels within the range of the quantizer are equally possible. If one assumes a sine wave input at maximum level then the maximum signal to noise ratio is improved, by $+1.76\,\text{dB}$, because a sine wave spends more time at higher levels. However, in general we do not listen to either sine waves or uniform random noise. We listen to signals that spend more time at low signal levels and this means that often the signal to noise ratio is worse than predicted by these equations.

Real signals often spend most of their time at low signal levels, but one has to design the quantizer to handle the maximum signal level, even if it isn't used very often. This means that more bits are used than is strictly necessary most of the time. One way of reducing the number of bits needed to quantize an audio signal is to make the quantizer adapt to the level of the signal because loud signals mask weaker signals, as discussed in Chapter 5. This type of quantizer is known as an "adaptive quantizer," and can save some bits. There are two main types of adaptive quantizer:

1. *Backward adaptive quantizers* (Jayant, 1973): These make use of the adapted output bits to drive the adaptation and so require no additional bits to be sent to the receive end. However, they are sensitive to errors and are not guaranteed to be overload-free. Although there are ways of mitigating these problems they are not often used in lossy compression systems.

2. *Forward adaptive quantizers*: Look at a block of the input signal and then set a scale factor that makes maximum use of the quantizer. This type of adaptive quantizer is guaranteed to be free from overload and is more robust to errors. However, it needs some *side*

information—the scale factor—to be sent to the receive end, which clearly robs bits from the quantized signal samples. Therefore there is an uncomfortable balance that must be struck between the amount of side information and the number of levels in the adaptive quantizer. In particular, if the block is made longer then less scale values have to be sent to the receiver. But if the block is too long then the quantization noise may become unmasked, due to *non-simultaneous masking*. So the block size is set typically within the range 8–25 ms to avoid this happening. There is also an issue about the scale precision. The minimum scale precision is 6 dB which allows for a very simple implementation, but there is a possibility that only just over half the quantizer's range is used because if the input has a sample that is just over half the quantizer's range then the scale would set to that case, as the next lower scale value would result in an overload. Increasing precision of the scale factor would allow a greater proportion of the quantizer to be used on average, but would require the scale value to have a longer word length to handle the increased precision. This increases the amount of side information that must be sent to the receiver. MPEG uses finer scale factors of 1.5 dB to 2 dB.

Adaptive quantizers attempt to make the input signal statistics match the ideal uniform distribution for maximum signal to noise. However, even with fine-scale steps this is rarely achieved and there is still a tendency for the small signal levels to be more probable. Furthermore, the higher levels may have a better signal to noise ratio than is strictly necessary for masking.

One solution to this is to use *non-uniform quantization*, sometimes called non-linear quantization, in which the levels are not equally spaced. By having the less probable higher levels further apart, the increased quantization error that results is more likely to be masked by the signal, and, because it's less probable, the average signal to noise level does not increase. In fact, for a given number of levels the average signal to noise reduces because the more likely levels are closer together and thus generate less quantization noise. MPEG-1 layer 3 uses a *non-uniform quantizer* to quantize the filtered signal samples.

However, even if an adaptive quantizer is used, then, because of the uniform spectrum of the quantization noise, the overall noise becomes unmasked if the signal to noise ratio falls below about 60 dB. This corresponds to 10 bits of quantizer precision. In order to do better we need to arrange for the noise to be shaped so that it is less perceptible.

7.8.10 Psychoacoustic noise shaping

As discussed in Chapter 2, our threshold of hearing is not constant with frequency, as shown in Figure 2.14. We are much more sensitive to sounds around 4 kHz than at the extremes of the frequency range. Therefore a uniform quantization spectrum may be more inaudible at low and high frequencies and yet still be heard if it is above the threshold of hearing in the most sensitive part of our hearing range. Ideally we need to have more noise where we are less sensitive to it and less noise where we are more sensitive. This is possible via a technique known as "noise shaping."

Figure 7.36 shows the block diagram of a noise-shaping quantizer. In it the quantization error is extracted, fed back via a noise-shaping filter and subtracted from the input. The effect of this can be analyzed as follows, assuming that the output of a quantizer can be considered to be the sum of the input signal and the quantization error.

$$\text{Quantizer}_{\text{output}} = \text{Quantizer}_{\text{input}} - N(z)\text{Error} + \text{Error} \qquad (7.14)$$

Which gives:

$$\text{Quantizer}_{\text{output}} = \text{Quantizer}_{\text{input}} + (1 - N(z))\text{Error} \qquad (7.15)$$

Equation 7.15 shows that the quantizer's error is shaped by the filter function $(1 - N(z))$. It is possible to design this filter to reduce the noise within the most sensitive bit of our hearing range. This technique is used to improve the quality of sounds on a CD. An example is Sony Super Bit Mapping (Akune *et al.*, 1992), which uses psychoacoustic noise shaping to give an effective signal to noise ratio of 120 dB (20 bits) in our most sensitive frequency range.

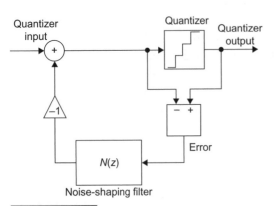

FIGURE 7.36 *Block diagram of a noise-shaping quantizer.*

7.8.11 Psychoacoustic quantization

Although noise shaping provides a means of psychoacoustically shaping the quantization noise, it is difficult to achieve very low bit rates using it. In particular one needs to be able to avoid transmitting information in frequency regions that either contain no signal, or are masked by other signal components. To do this easily one must work in the frequency domain.

Figure 7.37 shows the block diagram of a lossy audio coder. Ideally the time to frequency mapping splits the signal up into bands that are equal to, or smaller than, the width of our critical

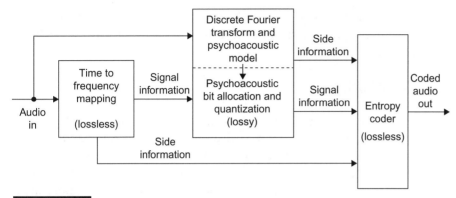

FIGURE 7.37 *Block diagram of a lossy audio coder.*

bands. Unfortunately for some audio coders this is not true at the lower frequencies.

The psychoacoustic quantization block is now replaced by a bit alloca-tion and quantization block, which is driven by a psychoacoustic model that allocates the number of quantization levels for each frequency band, including zero for no bits allocated. Psychoacoustic models can be quite complicated and are continuously evolving. This is because most lossy audio coding systems define how the receiver interprets the bit stream to form the audio output, but leave how those bits are allocated the bit alloca-tion to the encoder. This is clever because it allows the encoder to improve, as technology and knowledge get better, without having to alter the decoder standard. This is an important consideration for any audio delivery format.

However, regardless of the psychoacoustic bit allocation algorithm, they all effectively convert the linear frequency scale of the discrete Fourier transform into a perceptually based frequency scale similar to the ERB scale shown in Figure 5.10. Most use the Bark scale, which is similar but is based on the earlier work of Zwicker. Both scales are quasi-logarithmic and convert simultaneous masking thresholds into approximately straight lines.

Using some form of simultaneous masking model masking curves, which are different for tonal and non-tonal sounds, a *signal to mask-ing ratio* (SMR) is calculated and the bits allocated such that components that need a high SMR are given more bits than those that have a lower SMR. For components that have a negative SMR no bits are allocated because these components are masked and therefore need not be transmit-ted. The amount of bits that can be allocated depends on the desired bit rate. Also, the process interacts with itself so this process usually is inside some form of optimization loop that minimizes the total perceptual error.

For more information the book by Marina Bosi (Bosi and Goldberg, 2003), one of the developers of current lossy coding systems, provides a wealth of information.

7.9 SUMMARY

This chapter has looked at a variety of applications that combine both acoustic and psychoacoustic knowledge to achieve a specific audio objective. The need to combine both these aspects of this knowledge to achieve useful results, and the diversity of applications are what make this subject so exciting! As we write, acoustics is being applied to many more areas that affect our everyday life directly, for example the noise caused by wind farms, acoustic screening between individual working spaces in open office areas, and the design of urban public spaces that not only look nice but also sound good. Our hearing is one of our most precious senses, and it good to see that, as we become more technologically sophisticated, we are coming to realize that, for our human existence, sound really matters!

REFERENCES

Ahnert, W., Steffen, F., 1999. Sound Reinforcement Engineering: Fundamentals and Practice. Spon, London.

Bech, S., Zacharov, N., 2006. Perceptual Audio Evaluation – Theory, Method and Application. Wiley.

Bosi, M., Goldberg, R.E., 2003. Introduction to Digital Audio Coding and Standards, second edn. Springer, NY and London.

D'Antonio, P., Konnert, J.H., 1984. The RFZ/RPG approach to control room monitoring. Audio Engineering Society 76th Convention, October, New York, USA, preprint #2157.

Davies, D., Davies, C., 1980. The LEDE concept for the control of acoustic and psychoacoustic parameters in recording control rooms. J. Audio Eng. Soc. 28 (3), 585–595 (November).

Houtgast, T., Steeneken, H.J.M., 1985. The MTF concept in room acoustics and its use for estimating speech intelligibility in auditoria. Journal of the Acoustical Society of America. 77, 1069–1077.

Howard, D.M., 1991. Speech: measurements. In: Payne, P.A. (Ed.), Concise Encyclopaedia of Biological and Biomedical Measurement Systems. Pergamon Press, Oxford, pp. 370–376.

IEC 60268-13:1998, BS 6840-13:1998 (2003). Sound system equipment. Listening tests on loudspeakers IEC standard n. 60268-16. Sound System Equipments – Objective rating of speech intelligibility by speech transmission index, July 2003.

IEC standard n. 60268-16. Sound System Equipments – Objective rating of speech intelligibility by speech transmission index, July 2003.

ISO/TR 4870:1991 Acoustics—The construction and calibration of speech intelligibility tests.

Jayant, N.S., 1973. Adaptive quantization with a one-word memory. Bell Systems Technical Journal. 52, 1119–1144 (September).

Newell, P., 2008. Recording Studio Design, second edn. Focal Press, Oxford.

Pavlovic, C.V., 1987. Derivation of primary parameters and procedures for use in speech intelligibility predictions. Journal of the Acoustical Society of America. 82, 413–422.

Peutz, 1971. Articulation loss of consonants as a criterion for speech transmission in a room. J. Audio Eng. Soc. 19 (11), 915–919; December.

Rodgers, C.A.P., 1981. Pinna transformations and sound reproduction. J. Audio Eng. Soc. 29 (4), 226–234 April.

Rumsey, F., 2001. Spatial Audio (Music Technology Series). Focal, Oxford.

Salomon, D., 2007. Data Compression: The Complete Reference, fourth edn. Springer, New York & London.

Toole, F. E., 1990. Loudspeakers and Rooms for Stereophonic Sound Reproduction. In: The Proceedings of the Audio Engineering Society 8th International Conference, The Sound of Audio, Washington, DC, 3–6 May, pp. 71–91.

Walker, R., 1993. A new approach to the design of control room acoustics for stereophony. Audio Engineering Society Convention, preprint #3543, 94.

Walker, R., 1998. A controlled-reflection listening room for multichannel sound. Audio Engineering Society Convention, preprint #4645, 104.

FURTHER READING

Akune, M., Heddle, R., Akagiri, K., 1992. Super bit mapping: psychoacoustically optimized digital recording. Audio Engineering Society Convention 93, preprint # 3371.

Angus, J.A.S., 1997. Controlling early reflections using diffusion, Audio Engineering Society 102nd Convention, 22–25 March, Munich, Germany, preprint #4405.

Angus, J.A.S., 2001. The effects of specular versus diffuse reflections on the frequency response at the listener. J. Audio Eng. Soc. 49 (3), 125–133 (March).

ANSI S3.5-1997, American National Standard Methods for Calculation of the Speech Intelligibility Index. American National Standards Institute, New York.

Holman, T., 1999. 5.1 Surround Sound. Focal Press, Boston.

Newell, P., 1995. Studio Monitoring Design. Focal Press, Oxford.

Newell, P., 2000. Project Studios: A More Professional Approach. Focal Press, Oxford.

Schroeder, M.R., 1975. Diffuse sound reflection by maximum-length sequences. J. Acoust. Soc. Am. 57 (January), 149–151.

Schroeder, M.R., 1984. Progress in architectural acoustics and artificial reverberation: concert hall acoustics and number theory. J. Audio Eng. Soc. 32 (4), 194–203 April.

Steeneken, H.J.M. and Houtgast, T., 1994. Subjective and objective speech intelligibility measures. Proceedings of the Institute of Acoustics, 16(4), 95–112.

Toole, F., 2008. Sound Reproduction: the Acoustics and Psychoacoustics of Loudspeakers and Rooms. Focal Press, Oxford.

Walker, R., 1996. Optimum dimension ratios for small rooms. Audio Engineering Society Convention, preprint #4191, 100.

Appendix 1: The Fourier Transform

CHAPTER CONTENTS

A1.1 FOURIER'S THEOREM

Jean Baptiste Joseph Fourier (1768–1830) worked on a mathematical model of heat transfer in solid bodies (amongst many other topics). His thesis *On the Propagation of Heat in Solid Bodies* was published in 1807 and contained a novel idea for expanding a continuous function as a trigonometric series. Although Fourier developed this as a part of his model for heat transfer, it has a much wider application. Fourier's theorem is now usually stated as:

> *Any periodic function can be represented as an infinite sum of harmonic sinusoids multiplied by appropriate coefficients.*

Mathematically, this is expressed as:

$$f(t) = a_0 + \sum_{n=1}^{\infty} a_n \cos(n\omega_0 t) + b_n \sin(n\omega_0 t)$$

where $f(t)$ = the periodic time function to be represented
n = the harmonic number
ω_0 = the angular frequency $(2\pi f_0)$ of the periodic function
a_0 = the d.c. content of the periodic function
a_n = the level of the nth cosine harmonic of the periodic function
b_n = the level of the nth sine harmonic of the periodic function

(A1.1)

The periodic function $f(t)$ has a period T_0 such that:

$$\omega_0 = \frac{2\pi}{T_0}$$

(A1.2)

The sum of the sines and cosines on the *right-hand side* (RHS) of Equation A1.1 is called a Fourier series. The sinusoids are *harmonic*. This means that all their frequencies are integer multiples of the lowest, or *fundamental*, frequency ω_0.

Equation A1.1 states that, provided we know the correct values for the amplitudes of the sinusoids a_n and b_n, we can add them up and make any periodic signal $f(t)$ we like. Different signals will need different values for a_n and b_n. The act of summing the harmonic sinusoids together to make a signal is called Fourier synthesis. However, we may need an infinite number of harmonics to represent the signal properly. Examples of Fourier synthesis, with their corresponding spectra, were shown in Chapter 1 in Figures 1.50, 1.51 and 1.52 respectively.

These figures also show the effect of not having enough coefficients. We would need an infinite amount to have something that really looked like a square wave.

For the square wave shown in Figure 1.50, its Fourier series coefficients are:

$$a_0 = 0$$
$$a_n = 0$$
$$b_n = \begin{cases} 0, & \text{if } n \text{ is even} \\ \dfrac{4}{n\pi}, & \text{if } n \text{ is odd} \end{cases}$$

(A1.3)

A1.1.1 Frequency spectrum

Plotting the magnitude of the coefficients of the Fourier series gives us its *frequency spectrum*. This is shown in Figure 1.51.

The frequency spectrum tells us how much energy the signal has at any particular harmonic. We can convert Figure 1.51 to a real frequency in Hz simply by multiplying each value of relative frequency, F, by the fundamental frequency, f_0.

We can see that the square wave has a lot of energy at the fundamental frequency $(F = 1)$. But the amplitude of the other harmonics decreases quite slowly with frequency. In fact, the harmonic amplitude decreases as $1/F$. This slow decay rate is closely associated with the way the square wave looks and how it sounds. A spectrum that falls off as $1/F$ is always associated with a signal with a discontinuity, and, if its periodic, it will sound "buzzy."

Odd and even functions

There's something else that's interesting about the Fourier series of the square wave. Half of its coefficients are zero: all the values of $a_n = 0$. This is because the square wave is an *odd function*. An odd function is one in which $f(-t) = -f(t)$; at any given negative value of t the function is the negative of what it is at the corresponding positive value of t. Look at the square wave again and you'll see that this is true. Any cosine terms in the Fourier series would ruin this property because the cosine is not an odd function. On the other hand, an *even function* of time, satisfying $f(t) = f(-t)$, can have only cosine terms in its Fourier series.

However, there are many functions that are neither odd nor even. They will have both a sine and a cosine part to their Fourier series. So in order to plot the magnitude frequency spectrum in this case we must combine the two coefficients together as follows:

$$|C_n| = \frac{\sqrt{a_n^2 + b_n^2}}{2} \qquad (A1.4)$$

A1.2 FOURIER ANALYSIS

The Fourier coefficients discussed earlier are not found by trial and error. Fourier also developed equations for extracting them. For any periodic function, $f(t)$ the harmonic coefficients a_0, a_n and b_n can be found from:

$$a_0 = \frac{1}{T_0} \int_{\frac{-T_0}{2}}^{\frac{T_0}{2}} f(t)\, dt \qquad (A1.5a)$$

$$a_n = \frac{2}{T_0} \int\limits_{-\frac{T_0}{2}}^{\frac{T_0}{2}} f(t) \cos(n\omega_0 t) \, dt \qquad \text{(A1.5b)}$$

$$b_n = \frac{2}{T_0} \int\limits_{-\frac{T_0}{2}}^{\frac{T_0}{2}} f(t) \sin(n\omega_0 t) \, dt \qquad \text{(A1.5c)}$$

A1.3 THE COMPLEX FOURIER SERIES

Like many things in acoustics, dealing with the Fourier series in trigono-metric form is messy and inconvenient. For example calculating the Fourier series coefficients using the equations in A1.5 requires that we do three integrals! However, by using a complex number theory we can combine the sine and cosine together to form a complex exponential.

$$e^{j\theta} = \cos\theta + j\sin\theta \qquad \text{(A1.6)}$$

Where θ is in radians.

This allows us to express the Fourier series as a complex exponential as follows:

$$f(t) = \sum_{n=-\infty}^{\infty} C_n e^{jn\omega_0 t} \qquad \text{(A1.7)}$$

Where the now, in general complex coefficients C_n are calculated by:

$$C_n = \frac{1}{T_0} \int\limits_{-\frac{T_0}{2}}^{\frac{T_0}{2}} f(t) \, e^{-jn\omega_0 t} \, dt \qquad \text{(A1.8)}$$

Although in general the coefficients C_n are complex. If the waveform has odd symmetry then resulting coefficients C_n will be purely *imaginary* (the sine bit), whereas for even symmetry they will be purely *real* (the cosine bit).

Equations A1.7 and A1.8 can be used instead of Equations A1.1 and A1.5. They also have a further advantage in that when we do a Fourier analysis using Equation A1.4 on a periodic signal, which is neither odd nor even, we end up with two sets of coefficients: the a_n's and the b_n's. But when we listen to a periodic sound we only hear one frequency spectrum. Therefore to work out this spectrum we have to combine the a_n's and the

b_n's to get the total contribution at each frequency, as shown in Equation A1.4.

However, the complex exponential form of the Fourier series automatically gives us a single complex value for the coefficients and by finding their absolute values, or modulus, we get the magnitude of the spectrum, which is usually more perceptually relevant to the listener's appreciation of timbre. On the other hand the *phase* of the signal $f(t)$ at each harmonic frequency is given by the *argument* of each of the complex coefficients C_n.

A1.4 FREQUENCY ANALYSIS OF NON-PERIODIC SIGNALS: THE FOURIER TRANSFORM

The Fourier series is useful for analyzing periodic signals. However, many signals are non-periodic – that is, only occur once. This requires a different approach.

We can consider a single instance of a waveform as being like a periodic signal except that the period is infinity.

If we do this then it possible to state

$$F(\omega) = \int_{-\infty}^{\infty} f(t)\, e^{-j\omega t}\, dt \qquad (A1.9)$$

Where $F(\omega)$ is called the Fourier transform of the function $f(t)$.
Note that:

- $F(\omega)$ replaces the discrete Fourier series coefficients C_n with a continuous function of angular frequency ω.

- The limits of the integral are now theoretically infinite; however, in practice the integral only has to be calculated over the time range where the signal $f(t)$ is non-zero.

- As a matter of convention, time domain signals are represented using lower case letters whereas the frequency domain signals use capitals.

Equation (A1.9) transforms the time domain representation of the signal $f(t)$ into the frequency domain representation of the signal $F(\omega)$.

There is a complementary transform that reverses the process and converts the function from the frequency domain back into the time domain, which is:

$$f(t) = \frac{1}{2\pi} \int_{-\infty}^{\infty} F(\omega) e^{j\omega t}\, d\omega \qquad (A1.10)$$

Equations A1.9 and A1.10 are known as the Fourier transform pair and can be used on any type of signal, whether periodic or non-periodic. Together they form a powerful basis for analyzing and processing signals, in particular because it is easier to consider filtering in the frequency domain rather than the time domain.

One way of thinking about this is to remember that the complex exponential representation of a sine wave, e^{jwt}, is a spiral as a function of time, as shown in Figure A1.1, and integration is merely adding up all the numbers within the integration range. Equation A1.5a shows that to get the dc content, all one has to do is add up all the signal values. So what e^{-jwt} is doing, as a spiral in the other direction, is to untwist the waveform at the frequency specified. This converts it to dc where it can be simply extracted by adding up all the dc values in the integration range. The inverse transform re-twists the coefficient to the original frequency, amplitude and phase, and then all the re-twisted sine waves are added up to form the waveform.

A1.5 THE CONVOLUTION THEOREM

A powerful theorem behind the Fourier transform is the convolution theorem. Convolution is what filters do when they filter signals. However, we normally think about the action of the filter as multiplication of the spectrum of the signal by the filtering function. For example a low-pass filter

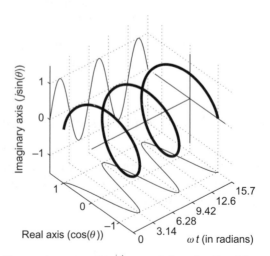

FIGURE A1.1 The complex exponential e^{jwt}, as a spiral as a function of time.

gets rid of the high frequencies and passes low frequencies. This is expressed using the Fourier transform as *the convolution theorem* that states:

Convolution in the time domain is equal to multiplication in the transformed (frequency) domain. The converse is also true.

A1.6 A FOURIER TRANSFORM EXAMPLE: THE SINGLE PULSE

Figure A1.2 shows a single rectangular pulse of length τ seconds and amplitude of $1/\tau$, which is defined mathematically as follows:

$$f(t) = \begin{cases} \dfrac{1}{\tau}, & \dfrac{-\tau}{2} < t < \dfrac{\tau}{2} \\ 0, & elsewhere \end{cases} \qquad (A1.11)$$

Note that irrespective of the value of τ the area of the pulse is constant and equal to one.

To find the Fourier transform of this we need to use equation A1.11 as $f(t)$ in Equation A1.9. Fortunately, as this function is zero over most of the range and the integral of zero is zero, this results in a solvable definite integral with limits $\pm(\tau/2)$ that is shown below in Equation A1.12a.

$$F(\omega) = \int_{-\frac{\tau}{2}}^{\frac{\tau}{2}} \frac{1}{\tau} e^{-j\omega t}\, dt \qquad (A1.12a)$$

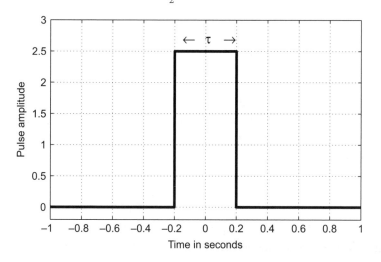

FIGURE A1.2

A single rectangular pulse of length τ, and amplitude of $1/\tau$.

This gives the following:

$$F(\omega) = \frac{-1}{j\omega}\left[\frac{1}{\tau}e^{-j\omega t}\right]_{-\frac{\tau}{2}}^{\frac{\tau}{2}} \qquad (A1.12b)$$

which, after evaluating the limits becomes:

$$F(\omega) = \frac{-1}{j\omega\tau}\left(e^{\frac{-j\omega t}{2}} - e^{\frac{j\omega t}{2}}\right) \qquad (A1.12c)$$

Expanding out the complex exponentials into $\cos\omega\tau + j\sin\omega\tau$ gives:

$$F(\omega) = \frac{-1}{j\omega\tau}\left(\cos\left(-\frac{\omega\tau}{2}\right) + j\sin\left(-\frac{\omega\tau}{2}\right) - \cos\left(\frac{\omega\tau}{2}\right) - j\sin\left(\frac{\omega\tau}{2}\right)\right) \qquad (A1.12d)$$

$$F(\omega) = \frac{-1}{j\omega\tau}\left(\cos\left(\frac{\omega\tau}{2}\right) - \cos\left(\frac{\omega\tau}{2}\right) - j\sin\left(\frac{\omega\tau}{2}\right) - j\sin\left(\frac{\omega\tau}{2}\right)\right) \qquad (A1.12e)$$

which simplifies to:

$$F(\omega) = \frac{2}{\omega\tau}\sin\left(\frac{\omega\tau}{2}\right) \qquad (A1.12f)$$

This can be expressed as:

$$F(\omega) = \frac{\sin\left(\frac{\omega\tau}{2}\right)}{\left(\frac{\omega\tau}{2}\right)} \qquad (A1.12g)$$

> If you remember that:
>
> $$\sin(\theta) = \frac{1}{2j}(e^{\theta} - e^{-\theta}),$$
>
> and recognize that equation A1.12c can be rearranged to be:
>
> $$F(\omega) = \frac{2}{\omega\tau} \cdot \frac{1}{2j}\left(e^{\frac{j\omega\tau}{2}} - e^{\frac{-j\omega\tau}{2}}\right)$$
>
> then you can cut straight to Equation A1.12e, instead of taking the "scenic route!"

The function $\sin(x)/x$ appears so often that it has its own name: $sinc(x)$. So we can say the Fourier transform of the rectangular pulse as described in Equation A1.11, is:

$$F(\omega) = \operatorname{sinc}\left(\frac{\omega\tau}{2}\right) \qquad (A1.13)$$

Note that in this case $F(\omega)$ turns out to be real; that is, it has no imaginary part. This is a consequence of our defining $f(t)$ as an even function.

$F(\omega)$ is plotted in Figure A1.3 for several values of τ. (Note that as τ gets smaller the frequency extent of the spectrum gets much wider.) It is generally true that a waveform that changes rapidly over a narrow time extent results in a very wide Fourier transform. The converse is also true. In fact if we reduce τ to zero, giving a pulse of infinite amplitude but still with an area equal to one, the spectrum becomes uniform for all frequencies. This infinitely small pulse is called a Dirac delta function and has a uniform, also known as a white, spectrum. The only other waveform to have a white spectrum is random noise.

A1.7 THE DISCRETE FOURIER TRANSFORM

With digital audio signals we can calculate something called the discrete Fourier transform, which is given by:

$$F_k = \sum_{n=0}^{N-1} f_n e^{\frac{-j2\pi nk}{N}} \qquad (A1.14)$$

This is a mixture of the continuous Fourier transform defined in Equation A1.9 and the Fourier series in regard to the following:

- The continuous function of time $f(t)$ is replaced by the discrete time sequence x_n.

- Continuous time t is replaced by a time index (or sample number) n that takes values $0, 1, 2, \ldots N - 1$.

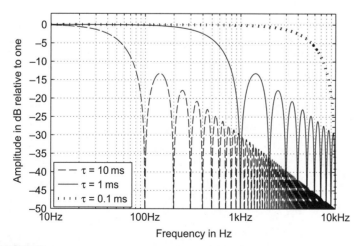

FIGURE A1.3 *The spectrum of a single rectangular pulse of length τ.*

- The length of the signal (or of the chunk being transformed) is N samples and not infinite.

- The integral is replaced by a summation.

- The continuous function of frequency $X(\omega)$ is replaced by the frequency sequence X_k.

- Continuous frequency f is replaced by frequency index k, which takes values $0, 1, 2, \ldots N - 1$. This means that the transform values only exist at a finite number of discrete frequencies. This is a bit like the harmonics in a Fourier series except that there are a finite number of them.

- N discrete input time samples are transformed into N discrete frequency samples whose spacing is proportional to the sampling frequency divided by the number of samples.

There is a corresponding *inverse* discrete Fourier transform that takes a frequency spectrum and turns it back into a time signal:

$$x_k = \frac{1}{N} \sum_{m=0}^{N-1} X_m e^{\frac{j2\pi mk}{N}} \tag{A1.15}$$

The Fourier transform is a powerful tool for the analysis and processing of acoustic signals.

Appendix 2: Solving the ERB Equation

To find the center frequency of the auditory filter whose critical bandwidth or ERB is equal to a given value, the ERB equation (Equation 2.6), which relates the critical bandwidth to the center frequency of the filter (f_c), can be rearranged as follows.

Equation 2.6 states the following.

$$\text{ERB} = \{24.7 \times [(4.37 \times f_c) + 1]\} \text{ Hz} \qquad (A2.1)$$

where f_c = the filter center frequency in kHz
ERB = the equivalent rectangular bandwidth in Hz
equation valid for (100 Hz < f_c < 10000 Hz)

Rearranging step-by-step to find f_c as a function of ERB:

$$\frac{\text{ERB}}{24.7} = (4.37 \times f_c) + 1$$

Therefore:

$$\left(\frac{\text{ERB}}{24.7}\right) - 1 = (4.37 \times f_c)$$

Therefore:

$$\left[\frac{\left(\dfrac{\text{ERB}}{24.7}\right) - 1}{4.37}\right] = f_c(\text{kHz})$$

So that:

$$f_c(\text{kHz}) = \left[\frac{\left(\dfrac{\text{ERB}}{24.7}\right) - 1}{4.37}\right] \qquad (A2.2)$$

Appendix 3: Converting between Frequency Ratios and Cents

The cent is defined as one hundredth of an equal tempered semitone, which is equivalent to one twelve-hundredth of an octave since there are 12 semitones to the octave. Thus one cent can be expressed as:

$$\sqrt[1200]{2} \text{ or } 2^{\left[\frac{1}{1200}\right]}$$

The frequency ratio of any interval (F1/F2) can therefore be calculated from that interval in cents (c) as follows:

$$\frac{F1}{F2} = 2^{\left[\frac{c}{1200}\right]}$$

and the number of cents can be calculated from the frequency ratio by rearranging to give:

$$\log_2\left[\frac{F1}{F2}\right] = \left[\frac{c}{1200}\right]$$

Therefore:

$$c = 1200 \, \log_2\left[\frac{F1}{F2}\right] \qquad (A3.1)$$

For calculation convenience, a logarithm to base 2 can be expressed as a logarithm to base 10. Suppose:

$$\log_2[x] = y \qquad (A3.2)$$

Then by definition:

$$x = 2^y$$

Acoustics and Psychoacoustics

Taking logarithms to base 10:

$$\log_{10}[x] = \log_{10}[2^y] = y\ \log_{10}[2]$$

Substituting in Equation A3.2 for y:

$$\log_{10}[x] = \log_2[x]\ \log_{10}[2]$$

Rearranging:

$$\log_2[x] = \left[\frac{\log_{10}[x]}{\log_{10}[2]}\right] \tag{A3.3}$$

Substituting Equation A3.3 into Equation A3.1:

$$c = 1200\left\{\frac{\log_{10}\left[\frac{F1}{F1}\right]}{\log_{10}[2]}\right\} = \left[\frac{1200}{\log_{10}[2]}\right]\log_{10}\left[\frac{F1}{F2}\right]$$

Evaluating the constants to give the equation for calculating the cents value of a frequency ratio:

$$c = 3986.3137^{\log_{10}\left[\frac{F1}{F2}\right]} \tag{A3.4}$$

In semitones (s), this is equivalent to:

$$s = \left[\frac{c}{100}\right] = 39.863137\ \log_{10}\left[\frac{F1}{F2}\right] \tag{A3.5}$$

Rearranging Equation A3.4 to give the equation for calculating the frequency ratio from a cent value:

$$\left[\frac{F1}{F2}\right] = 10^{\left[\frac{c}{3986.3137}\right]} \tag{A3.6}$$

Appendix 4: Deriving the Reverberation Time Equation

Clearly the length of time that it takes for sound to die is a function not only of the absorption of the surfaces in a room but also a function of the length of time between interactions with the surfaces of the room. We can use these facts to derive an equation for the reverberation time in a room. The first thing to determine is the average length of time that a sound wave will travel between interactions with the surfaces of the room. This can be found from the mean free path of the room which is a measure of the average distances between surfaces, assuming all possible angles of incidence and position. For an approximately rectangular box the mean free path is given by the following equation:

$$MFP = \frac{4V}{S} \tag{A4.1}$$

where MFP = the mean free path (in m)
V = the volume (in m^3)
and S = the surface area (in m^2)

The time between surface interactions may be simply calculated from A4.1 by dividing it by the speed of sound to give:

$$\tau = \frac{4V}{Sc} \tag{A4.2}$$

where τ = the time between reflections (in s)
and c = the speed of sound (in ms^{-1}, or meters per second)

Equation A4.2 gives us the time between surface interactions and at each of these interactions α is the proportion of the energy absorbed, where α is the average absorption coefficient discussed earlier. If α of the energy is absorbed at the surface, then $(1 - \alpha)$ is the proportion of the energy reflected back to interact with further surfaces. At each surface a further proportion, α, of energy will be removed so the proportion of the original sound energy that is reflected back will go as follows:

$$\text{Energy}_{\text{After one reflection}} = \text{Energy}_{\text{Before reflection}} (1 - \alpha)$$

$$\text{Energy}_{\text{After two reflections}} = \text{Energy}_{\text{Before reflections}} (1 - \alpha)^2 \qquad (\text{A4.3})$$

$$\text{Energy}_{\text{After three reflections}} = \text{Energy}_{\text{Before reflections}} (1 - \alpha)^3$$

$$\vdots$$

$$\text{Energy}_{\text{After } n \text{ reflections}} = \text{Energy}_{\text{Before reflections}} (1 - \alpha)^n$$

As α is less than 1, $(1 - \alpha)$ will be also. Thus Equation A4.3 shows that the sound energy decays away in an exponential manner. We are interested in the time it takes the sound to decay by a fixed proportion and so need to calculate the number of reflections that have occurred in a given time interval. This is easily calculated by dividing the time interval by the mean time between reflections, calculated using Equation A4.2, to give:

$$n = \frac{t}{\left(\dfrac{4V}{Sc}\right)} = t\left(\frac{Sc}{4V}\right) \qquad (\text{A4.4})$$

where t = the time interval (in s)

By substituting Equation A3.4 into Equation A4.3 we can get an expression for the remaining energy in the sound after a given time period as:

$$\text{Energy}_{\text{After a time interval}} = \text{Energy}_{\text{Initial}} (1 - \alpha)^{t\left(\frac{Sc}{4V}\right)} \qquad (\text{A4.5})$$

and therefore the ratio that the sound energy has decayed by at that time as:

$$\frac{\text{Energy}_{\text{After } n \text{ reflections}}}{\text{Energy}_{\text{Before reflections}}} = (1 - \alpha)^{t\left(\frac{Sc}{4V}\right)} \qquad (\text{A4.6})$$

In order to find the time that it takes for the sound to decay by a given ratio we must take logarithms, to the base $(1 - \alpha)$, on both sides of Equation A4.6 to give:

$$\log_{(1-\alpha)}\left(\frac{\text{Energy}_{\text{After } n \text{ reflections}}}{\text{Energy}_{\text{Before reflections}}}\right) = t\left(\frac{Sc}{4V}\right)$$

which can be rearranged to give the time required for a given ratio of sound energy decay as:

$$t = \left(\frac{4V}{Sc}\right) \log_{(1-\alpha)} \left(\frac{\text{Energy}_{\text{After } n \text{ reflections}}}{\text{Energy}_{\text{Before reflections}}}\right) \qquad (A4.7)$$

Unfortunately Equation A4.7 requires that we take a logarithm to the base $(1 - \alpha)$! However, we can get round this by remembering that this can be calculated using natural logarithms as:

$$\log_{(1-\alpha)} \left(\frac{\text{Energy}_{\text{After } n \text{ reflections}}}{\text{Energy}_{\text{Before reflections}}}\right) = \frac{\ln\left(\dfrac{\text{Energy}_{\text{After } n \text{ reflections}}}{\text{Energy}_{\text{Before reflections}}}\right)}{\ln(1 - \alpha)}$$

So Equation A4.7 becomes:

$$t = \left(\frac{4V}{Sc}\right) \frac{\ln\left(\dfrac{\text{Energy}_{\text{After } n \text{ reflections}}}{\text{Energy}_{\text{Before reflections}}}\right)}{\ln(1 - \alpha)} \qquad (A4.8)$$

Equation A4.8 gives a relationship between the ratio of sound energy decay and the time it takes, and so can be used to calculate this time. There are an infinite number of possible ratios that could be used. However, the most commonly used ratio is that which corresponds to a decrease in sound energy of 60 dB, or 106. When this ratio is substituted into Equation A4.8 we get an equation for the 60 dB reverberation time, known as T_{60}, which is:

$$T_{60} = \left(\frac{4V}{Sc}\right) \frac{\ln(10^{-6})}{\ln(1 - \alpha)} = \left(\frac{V}{S \ln(1 - \alpha)}\right) \frac{4 \times (-13.82)}{344 \text{ ms}^{-1}}$$
$$= \frac{-0.161V}{S \ln(1 - \alpha)} \qquad (A4.9)$$

where T_{60} = the 60 dB reverberation time (in s)

Thus the reverberation time is given by:

$$T_{60} = \frac{-0.161V}{S \ln(1 - \alpha)} \qquad (A4.10)$$

where T_{60} = the 60 dB reverberation time (in s)

Equation A4.10 is known as the "Norris–Eyring reverberation formula" and the negative sign in the numerator compensates for the negative sign arising from the natural logarithm, resulting in a reverberation time that is positive. Note that it is possible to calculate the reverberation time for other ratios of decay, and that the only difference between these and Equation A4.10 would be the value of the constant. The argument behind the derivation of reverberation time is a statistical one and so there are some important assumptions behind Equation A4.10. These assumptions are:

- that the sound visits all surfaces with equal probability, and at all possible angles of incidence; that is, the sound field is diffuse. This is required in order to invoke the concept of an average absorption coefficient for the room. Note that this is a desirable acoustic goal for subjective reasons as well; we prefer to listen to, and perform, music in rooms with a diffuse field.

- that the concept of a mean free path is valid. Again, this is required in order to have an average absorption coefficient but in addition it means that the room's shape must not be too extreme. This means that this analysis is not valid for rooms which resemble long tunnels; however, most real rooms are not too deviant and the mean free path equation is applicable.

Appendix 5: Deriving the Reverberation Time Equation for Different Frequencies and Surfaces

In real rooms we must also allow for the presence of a variety of different materials, as well as accounting for their variation of absorption as a function of frequency. This is complicated by the fact that there will be different areas of material, with different absorption coefficients, and these will have to be combined in a way that accurately reflects their relative contribution. For example, a large area of a material with a low value of absorption coefficient may well have more influence than a small area of material with more absorption.

In the Sabine equation this is easily done by multiplying the absorption coefficient of the material by its total area and then adding up the contributions from all the surfaces in the room. This resulted in a figure which Sabine called the equivalent open window area as he assumed, and experimentally verified, that the absorption coefficient of an open window was equal to one. It is therefore easy to incorporate the effects of different materials by simply substituting the total open window area for different materials, calculated using the method described above for the open window area. This gives a modified equation which allows for a variety of frequency-dependent materials in the room:

$$T_{60(\alpha<0.3)} = \frac{0.161V}{\displaystyle\sum_{\text{All surfaces } S_i} S_i\alpha_i(f)} \tag{A5.1}$$

where $\alpha_i(f)$ = the absorption coefficient for a given material
and S_i = its area

For the Norris–Eyring reverberation time equation the situation is a little more complicated because the equation does not use the open window area directly. There are two possible approaches. The first is to calculate a weighted average absorption coefficient by calculating the effective open window area, as done in the Sabine equation, and then dividing the result by the total surface area. This gives the following equation for the average absorption coefficient:

$$\alpha_{\text{weighed average}}(f) = \frac{\displaystyle\sum_{\text{All surfaces } S_i} S_i \alpha_i(f)}{S}$$

which can be substituted for α in the Norris–Eyring reverberation time equation to give a modified equation, which allows for different materials in the room:

$$T_{60} = \frac{-0.161V}{S \ln\left(1 - \dfrac{\displaystyle\sum_{\text{All surfaces } S_i} S_i \alpha_i(f)}{S}\right)} \tag{A5.2}$$

Equation A5.2 can be used to calculate the effect of a variety of frequency-dependent materials in the room. However, there is an alternative way of looking at the problem which is more in the spirit of the reasoning behind the Norris–Eyring reverberation time equation. This second approach can be derived by considering the effect on the sound energy amplitude of successive reflections which hit surfaces of differing absorption coefficients. In this case the proportion of the original sound energy that is reflected back will vary with each reflection as follows:

$$\text{Energy}_{\text{After one reflection}} = \text{Energy}_{\text{Before reflection}} (1 - \alpha_1)$$
$$\text{Energy}_{\text{After two reflections}} = \text{Energy}_{\text{Before reflections}} (1 - \alpha_1)(1 - \alpha_2)$$
$$\text{Energy}_{\text{After three reflections}} = \text{Energy}_{\text{Before reflections}} (1 - \alpha_1)(1 - \alpha_2)(1 - \alpha_3)$$
$$\vdots$$
$$\text{Energy}_{\text{After } n \text{ reflections}} = \text{Energy}_{\text{Before reflections}} (1 - \alpha_1)(1 - \alpha_2)(1 - \alpha_3)$$
$$\times \cdots \times (1 - \alpha_n)$$

This can be couched in terms of an average α by taking the geometric mean of the different reflection coefficients $(1 - \alpha)$. For example, after two reflections the energy is at a level which would be the same as if there had been two reflections from a material whose reflected energy was given by:

$$(1 - \alpha)_{\text{average}} = \sqrt{(1 - \alpha_1)(1 - \alpha_2)}$$

After three reflections the average reflection coefficient would be given by:

$$(1 - \alpha)_{average} = \sqrt[3]{(1 - \alpha_1)(1 - \alpha_2)(1 - \alpha_3)}$$

And after n reflections the average reflection coefficient would be given by:

$$(1 - \alpha)_{average} = \sqrt[n]{(1 - \alpha_1)(1 - \alpha_2)(1 - \alpha_3) \times \cdots \times (1 - \alpha_n)}$$

Because there are only a finite number of different materials in the room, but of differing areas, it is necessary only to consider an average based on just the number of different materials but weighted to allow for their differing surface areas. Because logarithms convert products to additions, this weighted geometric mean can be simply expressed as a sum of the individual absorption terms and so the Norris–Eyring reverberation time equation can be rewritten in a modified form, which allows for the variation in material absorption due to both nature and frequency, as:

$$T_{60} = \frac{-0.161V}{\sum\limits_{\text{All surfaces } S_i} S_i \ \ln(1 - \alpha_i(f))} \tag{A5.3}$$

Equation A5.3 is also known as the "Millington–Sette equation." Although Equation A5.3 can be used irrespective of the absorption level it is still more complicated than the Sabine equation, and if the absorption coefficient is less than 0.3 it can be approximated very effectively by it, as discussed previously. Thus in many contexts the Sabine equation, Equation A5.1, is preferred.

Appendix 6: The Effect of Speaker Size on its Polar Pattern

In order to understand how the properties of sequences, or the size of a loudspeaker, affect polar performance we must first look at some theory behind array polar patterns.

A6.1 AN ARRAY OF POINT SOURCES

Consider an evenly spaced, linear array of perfect point source radiators, as shown in Figure A6.1, with complex amplitudes $A_0 \ldots A_{N-1}$. This corresponds to the radiated sound from an array of speakers, and the amplitudes represent the *illumination of the surface*. If we are an infinite, or at least very large, distance away, we can make the following approximations:

1. The wavefronts are planar and therefore all the radiators will have the same angle of incidence (θ) to the far off point.
2. The differences in path lengths are so small that only the initial phase difference, due to θ affects the received amplitude.

These approximations are known as the far-field assumptions and, in theory, will be satisfied provided one is a reasonable distance from the array.

Assuming, for the moment, that the far-field assumptions are satisfied, we can say the following about our linear array of ideal point sources:

1. The far-field response will be given by the sum of the individual point sources with an additional phase delay/advance due to due to θ, which is the angle from the normal, as shown in Figure A6.1.

2. The phase delay due to θ will be given by:

$$Phase\ delay = nd \sin \theta \qquad (A6.1)$$

where n is proportional to the point source number, as shown in Figure A6.1.

For the example shown in Figure A6.1, this results in an equation for the far-field polar response, at a frequency whose wavenumber is k, which is:

$$P(\theta_k) = A_0 e^{-j(0)kd \sin \theta} + A_1 e^{-j(1)kd \sin \theta}$$
$$+ A_2 e^{-j(2)kd \sin \theta} + \cdots + A_{N-1} e^{-j(N-1)kd \sin \theta} \qquad (A6.2)$$

where the wave number k is given by:

$$k = \frac{2\pi}{\lambda} = \frac{\omega}{c} = \frac{2\pi f}{c} \qquad (A6.3)$$

This can be rewritten as:

$$P(\theta_k) = \sum_{n=0}^{N-1} A_n e^{-jnkd \sin \theta} \qquad (A6.4)$$

If we make $\Omega = kd \sin\theta$ then Equation A6.4 can be rewritten as:

$$P(\theta_k) = \sum_{n=0}^{N-1} A_n e^{-jn\Omega} \qquad (A6.5)$$

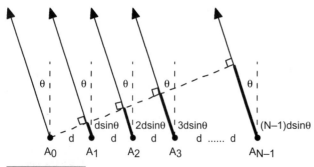

FIGURE A6.1   *A linear array of n point sources.*

Equation A6.5 is in fact a discrete Fourier transform (DFT) in which $\Omega =$ $kd \sin\theta$. This means that the far-field polar pattern of an array of point sources is related to the illumination of the surface by a Fourier transform relationship. Therefore all the theorems that apply to the Fourier transform apply to an array of point sources. In particular, these are:

1. *Linearity and superposition*: Weighted addition in the spatial domain is equivalent to addition in the transformed polar pattern domain.

2. *The convolution theorem*: This theorem states that convolution in the spatial domain is equivalent to multiplication in the transformed polar pattern domain. The converse is also true.

3. *The Wiener–Khinchin theorem*: The Wiener–Khinchin theorem states that the squared Fourier transform magnitude of a sequence, in the spatial domain (that is, its polar pattern) is equal to the Fourier transform of its autocovariance (or autocorrelation function).

4. *The shift theorem*: A shift in the spatial domain leads to a linear (progressive) phase change in the transformed polar pattern domain and vice versa.

As we shall see later these have some important consequences.

A6.1.1 The visible region

Although, in theory, the variable in Equation A6.5 can range from $-\infty$ to $+\infty$, in reality it cannot. In fact, because $\sin\theta$ cannot exceed ± 1, there is only a limited range that makes any physical sense. This region is known as the "visible region" and, because $\Omega = kd \sin\theta$, the visible region corresponds to $-kd \leq \Omega \leq + kd$. The visible region corresponds to the angles between $\pm 90°$ of the normal direction.

This is shown in Figure A6.2 for a 10-element array of points, with the elements spaced 4.3 cm apart, at 1 khz ($kd = 0.79$). If we double the frequency to 2 kHz then kd doubles ($kd = 1.58$) and the visible region also doubles, as shown in Figure A6.3.

As the visible region corresponds to the angles between $\pm 90°$ of the normal direction, the effect of doubling the visible region also implies a narrowing of the main lobe—if its shape does not change as the visible region increases, as in our examples.

A6.1.2 The effect of sampling

When the frequency gets high enough so that the spacing between the point sources becomes greater than half a wavelength, the array becomes

The visible region of an array of points in Ω space (kd = 0.79).

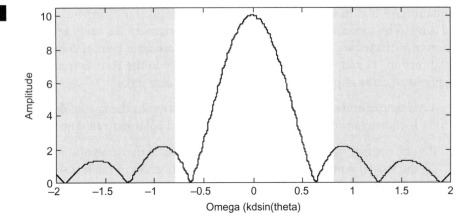

The visible region of an array of points in Ω space (kd = 1.58).

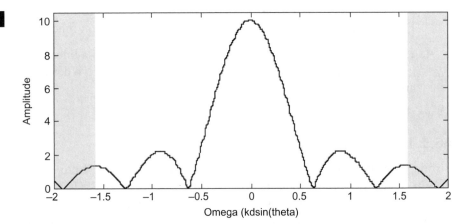

under-sampled. Under these conditions one gets *spatial aliasing*, which results in multiple main lobes. Figures A6.4, A6.5 and A6.6 illustrate this. Figure A6.4 shows the 1 kHz example with the scale expanded. The first thing to note is that the visible region still covers the same region as that of Figure A6.2. The second thing to note is that the expanded scale reveals the multiple peaks that indicate the possibility of spatial aliasing.

Figure A6.5 shows the visible region when the frequency equals 7 kHz ($kd = 5.5$). Here we can see that although the aliased main lobe is not visible, there is an increase in side-lobe levels due to the spatial aliasing. Figure A6.6 shows the visible region when the frequency equals 10 kHz ($kd = 7.85$). Here we can see that the aliased main lobe is now visible, and there is a large increase in the side-lobe levels due to the spatial aliasing.

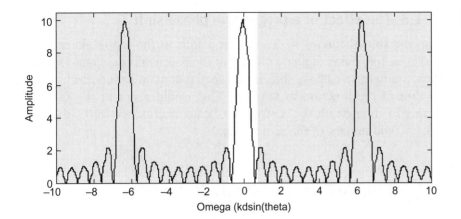

FIGURE A6.4

The visible region of an array of points in a larger Ω space (kd = 0.79).

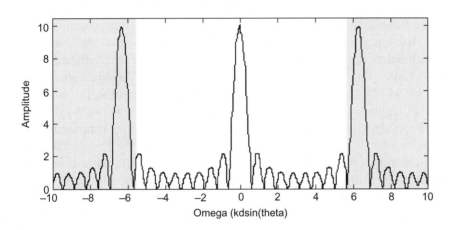

FIGURE A6.5

The visible region of an array of points in a larger Ω space (kd = 5.5).

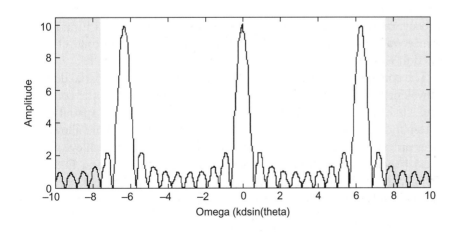

FIGURE A6.6

The visible region of an array of points in a larger Ω space (kd = 7.85).

A6.1.3 The effect of a progressive phase shift

From the shift theorem, we know that a shift in the spatial domain leads to a linear (progressive) phase change in the Fourier domain and vice versa. Thus, a progressive phase shift in the spatial domain would result in a linear shift of the function in Ω space. This would result in the main lobe moving to an angle off the central axis: beam steering. However, the visible region would remain in the same place.

A6.2 APPLICATION TO DIFFUSER DESIGN

For a diffuser, we wish to have a $P(\theta_k)$ that is uniform with respect to angle. Which, in turn, corresponds to a pattern of coefficients. This corresponds to a diffuser structure that has a constant Fourier transform magnitude over the visible region. The effect of an obliquely incident wavefront is to add an additional progressive phase shift across the diffuser's re-radiated sound. This causes the visible region to be shifted in angle. Therefore, in addition, we would like the Fourier transform magnitude to be uniform outside the visible region as well to cover oblique incidence; and so we need to find diffusion structures that have uniform magnitude Fourier transforms, such as Schroeder (Schroeder, 1975) diffusers. We can use the Fourier transform relationship between the far-field polar pattern and the pattern of amplitudes at the diffuser's surface to help us choose appropriate sequences for diffusion structures.

The Wiener–Khinchin theorem states that the squared Fourier transform magnitude of a sequence is equal to the Fourier transform of its autocovariance (or autocorrelation function). Therefore sequences whose autocovariance is either a delta function, or close to a delta function, will form good diffusers, because the Fourier transform magnitude of a delta function is uniform.

The convolution theorem states that convolution in the spatial domain is equivalent to multiplication in the Fourier domain and that the converse is true. This means that multiplication, or modulation, in the spatial domain corresponds to convolution in the polar pattern domain. This allows us to use a variety of modulation techniques on short diffusers to achieve good diffusers (Angus and McManmon, 1998; Angus, 2000; Cox and D'Antonio, 2009) without the lobe narrowing that results from a repeated set of short diffusers. In fact, the Fourier relationship allows one to develop new diffusion structures, such as "Binary Amplitude" (Angus, 2000; Cox and D'Antonio, 2009) and "Ternary" (Cox et al., 2005) diffusers.

A6.3 APPLICATION TO ARRAY LOUDSPEAKERS

An early example of an array loudspeaker was the column loudspeaker. In this arrangement a number of small loudspeakers were arranged in a closely spaced line. Because of the extended length of the source in one plane, directivity control was achieved in that plane. However, the beam pattern would get progressively more directive with frequency, as predicted by the Fourier transform. Techniques were developed to reduce this behavior, usually by applying the necessary frequency-dependent weighting, tapering, or windowing using simple electrical circuits—a direct application of the convolution theorem. Methods of steering these line speakers were also developed, either by using simple analog delay techniques or by using the inherent phase shifts in the filters used to taper the array. Again, this is a direct application of the shift theorems of the Fourier transform.

Array loudspeakers can also exhibit unwanted side lobes at higher frequencies, due to aliasing, which reduce their utility; that is, above some frequencies the spacing between the drivers is greater than half the wavelength of the sound being produced. This results in spatial aliasing and in a loss of control of the beam pattern.

To avoid spatial aliasing requires a huge number of small loudspeakers, which results in a prohibitive cost for the array. For example, ideally we want pattern control over the entire audio frequency range. However, even if we make the speaker spacing 4.3 cm, which is unfeasibly small because we would need a large number to achieve low-frequency pattern control, we still have significant aliasing at 10 kHz.

A6.3.1 Acoustic spatial filtering

One way of reducing the effect of spatial aliasing is to use directive loudspeakers, instead of point sources, as the array elements. If one uses directive sources then their polar patterns will act as a form of spatial filter; that is, the off-axis side lobes will be reduced by the axis reduction in sound level that a directive source affords. Figure A6.7 shows an array response at 10 kHz ($kd = 7.85$) with the response of a continuous line source (of length equal to the element spacing) superimposed upon it. Of particular note is that the zeros of the continuous line source fall on the aliased main lobes from the point source array. Because the far-field polar pattern of an array of point sources is related to the applied signals by a Fourier transform relationship, all the theorems that apply to the discrete Fourier transform apply to the array loudspeaker. This means that the theorem that convolution in one domain is equal to multiplication in the other domain applies to this

FIGURE A6.7

An array speaker and a continuous source equal to the spacing (kd = 7.85).

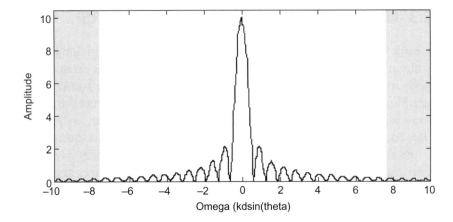

FIGURE A6.8

An array speaker made of continuous sources equal to the spacing (kd = 7.85).

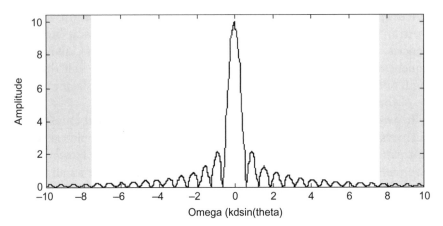

situation. Replacing each of the point sources with a continuous line source is equivalent to convolving it with the point array. Therefore, the effect of replacing the point source with the continuous sources is to multiply their far-field patterns together.

This pattern multiplication is well known and the effect for our example is shown in Figure A6.8. One can see that the aliased main lobes have been eliminated. In fact, the response has become equivalent to a continuous line source of the same extent as the array. Clearly, using directional sources such as constant directivity horns can also be used to achieve similar effects. It is this that results in the success of large arrays based on constant directivity horns, providing the horns have directivity control before spatial aliasing occurs. Once the directivity of the individual elements is considered, the need for curved arrays also becomes apparent as the spatial filtering effect of the sources must also be factored in.

A6.4 APPLICATION TO CONSTANT DIRECTIVITY HORNS

Keele (2000, 2002, 2003a & b) extended the work of Van Buren (Rogers and Van Buren, 1978; Van Buren et al., 1983) to develop the constant beamwidth theory (CBT) arrays. In his papers, the transducer is a circular spherical cap of arbitrary half-angle with Legendre function shading. It provides a constant beam pattern and directivity with extremely low side lobes for all frequencies above a certain cut-off frequency.

To maintain constant beamwidth behavior, CBT circular-arc loudspeaker line arrays require that the individual transducer drive levels be set according to a continuous Legendre shading function. This shading gradually tapers the drive levels from maximum at the center of the array to zero at the outside edges of the array. Keele developed approximations to the Legendre shading that both discretise the levels and truncate the extent of the shading so that practical CBT arrays can be implemented. He determined by simulation that a 3 dB stepped approximation to the shading maintained out to −12 dB did not significantly alter the excellent pattern control of a CBT line array.

Conventional CBT arrays require a driver configuration that conforms to either a spherical-cap curved surface or a circular arc. Keele also showed how CBT arrays can be implemented in flat-panel or straight-line array configurations using signal delays and Legendre function shading of the driver amplitudes. CBT arrays do not require any signal processing except for simple frequency-independent shifts in loudspeaker level. This is in contrast with conventional constant-beamwidth flat-panel and straight-line designs, which require strongly frequency-dependent signal processing.

These results are important because they also provide a link between array loudspeakers and constant directivity horns. Figure A6.9, reproduced from Keele (2002), shows how the delays for a planar CBT array effectively move the driver from its position on a flat surface to a point on a circular arc; that is, it provides a delay that makes a wavefront at the planar array that is spherical. According to the Fourier theory described earlier, the Fourier transform of the combination of phase shifts—due to a spherical wavefront and Legendre weighting—results in a frequency-independent constant beamwidth above a certain cut-off frequency. Furthermore, the cut-off frequency is a function of both the required directivity angle and the length of the array.

If we compare this to a constant directivity horn we observe many similarities:

1. Firstly, the conical flare of such horns results in a spherical wavefront at the horn's mouth.

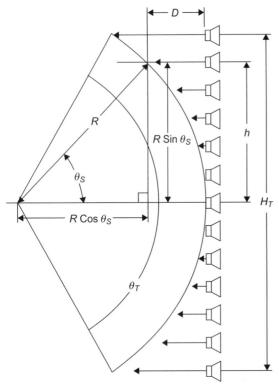

FIGURE A6.9 *Relationships required to calculate the delays for a planar CBT array (from Keele, 2002).*

2. Secondly, the projection of the spherical wavefront's intensity onto the planar front of the horn results in an intensity that has a cosine roll-off from the center of the horn. This approximates a Legendre weighting at the center of the planar front of the horn, as shown in Figure A6.10. Unfortunately, the weighting is not enough at the edges of the horn aperture. However, practical constant directivity horns have an additional, more extreme flare at the mouth. This would have the effect of more rapidly reducing the amplitude at the edge of the horn's mouth; thus more closely approaching Legendre weighting.

Thus, constant directivity horns can be seen as a simple approximation to a CBT array!

A6.5 THE EFFECT OF MOUTH, OR ARRAY, SIZE ON BEAMWIDTH

How does the size of the horn mouth, or the array size, affect the lowest frequency for a given beamwidth?

The normalized polar pattern for a linear array of N equally driven point sources is given by:

$$P_N(\theta) = \frac{\sin\left(Nk\frac{d}{2}\sin\theta\right)}{N\sin\left(k\frac{d}{2}\sin\theta\right)} \qquad (A6.6)$$

where N = the number of sources
k = the wavenumber
d = the distance between the sources
θ = the angle from the normal

and its $-6\,dB$ angle occurs when:

$$k\frac{d}{2}\sin\theta_{-6dB} \approx \frac{2}{N} \qquad (A6.7)$$

This can be rewritten as:

$$\theta_{-6dB} \approx \sin^{-1}\left(\frac{4}{kNd}\right) \qquad (A6.8)$$

But:

$$k = \frac{2\pi}{\lambda} = \frac{\omega}{c} = \frac{2\pi f}{c}$$

So:

$$\theta_{-6dB} \approx \sin^{-1}\left(\frac{4}{\frac{2\pi}{\lambda}Nd}\right) \approx \sin^{-1}\left(\frac{2\lambda}{\pi Nd}\right) \quad (A6.9)$$

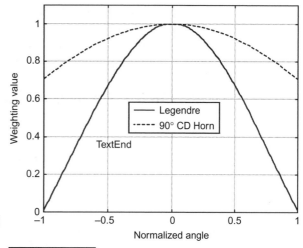

FIGURE A6.10 *Legendre versus cosine weighting.*

which, if we combine all the constants, becomes:

$$\theta_{-6dB} \approx \sin^{-1}\left(0.64\frac{\lambda}{Nd}\right) \qquad (A6.10)$$

Equation A6.10 shows that the $-6\,dB$ angle depends on the ratio of the wavelength, and the number of sources times the distance between them. The number of sources times the spacing between them is simply the length of the array "L." So equation A6.10 becomes:

$$\theta_{-6dB} \approx \sin^{-1}\left(0.64\frac{\lambda}{L}\right) \qquad (A6.11)$$

Where: L = The size of the source in meters

or:

$$\theta_{-6dB} \approx \sin^{-1}\left(0.64\frac{c}{fL}\right) \Rightarrow \theta_{-6dB} \approx \sin^{-1}\left(\frac{219}{fL}\right) \qquad (A6.12)$$

as a function of frequency.

Because of the convolution theorem discussed earlier, this equation applies to continuous sources, e.g., horn mouths, as well as speaker arrays. This means that the directivity of a uniformly driven speaker array, or a uniformly illuminated horn mouth, is a function of the number of wavelengths that fit into the size "L."

For example, if exactly one wavelength fits across the speaker length, then the $-6\,dB$ angle will be $\pm40°$, a beamwidth of $80°$, and for two wavelengths the beamwidth will be 37°, and so on, the beamwidth, approximately halving for every doubling of frequency. On the other hand if the wavelength to length ratio is $(\lambda / L) \geq (1/0.64) \geq 1.57$ then the speaker has no directivity at all – it becomes omnidirectional!

A6.5.1 The minimum beamwidth frequency as a function of size

It is sometimes useful to be able to calculate the minimum frequency that a speaker of a given size can achieve at a particular coverage angle. We can do this by rearranging Equation A6.7:

$$k_{min} \frac{d}{2} \sin\theta_{-6dB} \approx \frac{2}{N} \Rightarrow \frac{2\pi f_{min}}{c}$$

$$\approx \frac{4}{Nd \sin\theta_{-6dB}} \Rightarrow f_{min} \approx \frac{4c}{2\pi L \sin\theta_{-6dB}} \tag{A6.13}$$

which, if we combine all the constants becomes:

$$f_{min} \approx \frac{219}{L \sin\theta_{-6dB}} \tag{A6.14}$$

Equation A6.13 gives a simple relationship between the size of a speaker array, or a constant directivity horn mouth, and the minimum frequency that sustains the desired beamwidth.

This appendix has presented the basic Fourier relationship between the far-field polar response and the near-field illumination of the aperture. It has demonstrated its utility in a variety of electroacoustic applications. In particular it is possible to derive some useful relationships between the size of the speaker, with respect to wavelength, and its directivity performance versus frequency. In the case of directivity, size really does matter!

REFERENCES

Angus, J.A.S., McManmon, C.I., 1998. Orthogonal sequence modulated phase reflection gratings for wide-band diffusion. Journal of the Audio Engineering Society. 46 (12), 1109–1118.

Angus, J.A.S., 2000. Using grating modulation to achieve wideband large area diffusers. Applied Acoustics. 60 (2), 143–165.

Cox, T.J., D'Antonio, P., 2009. Acoustic Absorbers and Diffusers: Theory, Design and Application. Spon Press.

Cox, T.J., Angus, J.A.S., D'Antonio, P., 2005. Ternary sequence diffusers. Forum Acusticum, Budapest paper 501.0.

Keele D.B. Jr., 2000. The application of broadband constant beamwidth transducer (CBT) theory to loudspeaker arrays. 109th Conv. Audio Eng. Soc. Preprint, 5216 (Sept.).

Keele, D.B. Jr., 2002. Implementation of straight-line and flat-panel constant beamwidth transducer (CBT) loudspeaker arrays using signal delays. Presented at the 113th Conv. Audio Eng. Soc., Oct., Preprint 5653.

Keele, D.B. Jr. 2003. Practical implementation of constant beamwidth transducer (CBT) loudspeaker circular-Arc Line Arrays. Presented at the 115th Conv. Audio Eng. Soc., Oct., Preprint 5863.

Rogers, P.H., Van Buren, A.L., July 1978. New approach to a constant beamwidth transducer. Journal of the Acoustical Society of America. 64 (1), 38–43.

Schroeder, M.R., 1975. Diffuse sound reflection by maximum-length sequences. Journal of the Acoustical Society of America. 57 (1), 149–150.

Van Buren, A.L., Luker, L.D., Jevnager, M.D., Tims, A.C., June 1983. Experimental constant beamwidth transducer. Journal of the Acoustical Society of America. 73 (6), 2200–2209.

FURTHER READING

Angus, J.A.S. and D'Antonio, P. 1999. Two-Dimensional Binary Amplitude Diffusers. 107th Convention of the Audio Engineering Society, Preprint 5061.

Keele D.B. Jr., July/August 2003a. The full-sphere sound field of constant beamwidth transducer (CBT) loudspeaker line arrays. Journal of the Audio Engineering Society 51 (7/8), 611–624.

Trevelyan, J., 1994. Boundary Elements for Engineers: Theory and Applications. Computational Mechanics Publications, Southampton.

Appendix 7: Track Listing for the Audio Compact Disc

The compact disc (CD) which accompanies this book contains a number of audio examples that are provided for listening and experimentation in relation to discussions provided in the text, to support learning and further understanding. The purchase of this book and the audio CD grants the owner the right to use the audio material provided solely for this purpose. Copyright © in these recordings belongs to David M. Howard. Any use of this copyright material for commercial gain is not permitted unless prior agreement has been gained from the copyright © holder.

The Acoustics and Psychoacoustics CD is
© David M. Howard, York, 2009

The tracks on the CD are listed below along with a brief description and an indication of the section in the text that is supported by the audio material. Some tracks are provided in support of psychoacoustic aspects of the text, and these should be listened to at a comfortable volume level, using only headphones where indicated. Other tracks provide anechoic recordings of acoustic musical instruments as source material to allow readers to carry out their own analyses. For example, these sounds can be analyzed in terms of their waveforms and spectra to enable comparisons to be made with figures provided in the text such as Figures 4.11, 4.17, 4.22, 4.29, and 5.1–5.5. A number of freeware programs are available for this purpose which can readily be found via Internet search engines.

The acoustic recordings were made in the acoustic anechoic room at the Department of Electronics, University of York, UK. A Sennheiser MKH20 omnidirectional microphone, an RME quad microphone amplifier and an Edirol R4 hard disk recorder (44 kHz sampling rate, 24-bit resolution) were used to make the master recordings. Please note that no attempt has been made to remove extraneous sounds in the live anechoic recordings, such as breath noises, instrument key clicks and page rustling; these recordings are presented "in the raw" as source material intended for use as learning

Acoustics and Psychoacoustics
Copyright © 2009 Elsevier Ltd. All rights of reproduction in any form reserved.

exercises. One such exercise for readers might involve practicing removing these extraneous sounds.

The track number is given as "No." and the "Section" entry indicates where in the text the track is most relevant. If the track links specifically to an aspect of the text, such as a figure, this is indicated under "Track contents."

Table A7	Acoustics and psychoacoustics audio compact disc track listing	
No.	**Track contents**	**Section**
1	*Critical bands by sine waves*: Two sine waves are heard—one at a constant frequency (F_1) of 440 Hz and the second starting at 440 Hz, moving to around 660 Hz and back to 440 Hz. The changes described in Section 2.2 and illustrated in Figure 2.6 should be audible.	2.2
2	*Average hearing change with age*: The first few section of the chorale prelude used in Track 73 has been filtered to allow the average effect of aging, or presbycusis, to be illustrated acoustically for men and women aged (in years) 20 (no difference), 60 and 80 based on the data plotted in Figure 2.11. Track has five snippets: (1) male or female aged 20; (2) female aged 40; (3) male aged 40; (4) female aged 60; (5) male aged 60.	2.3
3	*Loudness doubling*: A doubling in loudness requires, on average, a 10 dB increase in sound pressure level as indicated in Example 2.4. This example uses an anechoic recording set to the following levels: 0 dB, +3 dB, +6 dB and +10 dB to illustrate the effect of doubling intensity (+3 dB), doubling pressure (+6 dB) and average doubling in loudness (+10 dB).	2.4
4	*Pitch demo*: Three pitch demonstrations are presented: (A) the first five harmonics with a fundamental of 200 Hz; (B) as (A) but without the fundamental (the "missing fundamental"); (C) a sound consisting of components at 1800 Hz, 2000 Hz and 2200 Hz for which the perceived pitch is usually 200 Hz.(D) a sound consisting of components at 1840 Hz, 2040 Hz and 2240 Hz for which the perceived pitch is usually reported as 207 Hz but with ambiguity. For explanations of (C) and (D), see comments in Section 3.2.1, Schouten (1940) and Moore (1982) — references listed at the end of Chapter 3.	3.2.1, 3.2.2 & 3.2.3
5	*Prime partials*: Synthesized sounds whose frequency components are set to prime numbers (557 Hz, 1381 Hz, 1663 Hz, 1993 Hz and 2371 Hz); thus they are not harmonics of a common fundamental.	3.2.2 & 5.3.1
6	*Residue pitch*: An arpeggio and a glissando are played using noise added to a delayed version of itself as described in the final paragraph of Section 3.2.6.	3.2.6
7	*Tuning systems*: A chord sequence (tonic—I, sub-dominant—IV, dominant—V, tonic—I) is played in just temperament (rooted in the key of C) in all 12 keys starting and ending in C major. Notice the difference between the consonant tuning in C major and the rather more dissonant tuning in keys such as F# and B major.	3.4
8	*Half-size violin arpeggio*: G major arpeggio (G3 to G5) on a half-size violin.	4.2
9	*Half-size violin music*: Brief music snippet on half-size violin.	4.2
10	*Violin arpeggio*: G major arpeggio (G3 to G7) on a violin.	4.2
11	*Flute arpeggio*: G major arpeggio (D4 to B6) on a flute.	4.3.3
12	*Flute music*: Brief music snippet on a flute.	4.3.3
13	*Piccolo arpeggio*: G major arpeggio (D5–B7) on a piccolo.	4.3.3
14	*Piccolo music*: Brief music snippet on a piccolo.	4.3.3
15	*Bass recorder arpeggio*: G major arpeggio plus extreme (F2, G2 to G4) on a bass recorder.	4.3.3
16	*Bass recorder music*: Brief music snippet on a bass recorder.	4.3.3

(Continued)

No.	Track contents	Section
	Table A7　Continued	
17	*Tenor recorder arpeggio*: G major arpeggio plus extremes (C3, D3 to B4, C5) on a tenor recorder.	4.3.3
18	*Tenor recorder music*: Brief music snippet on a tenor recorder.	4.3.3
19	*Treble recorder arpeggio*: G major arpeggio plus extreme (F3, G3 to G5) on a treble recorder.	4.3.3
20	*Treble recorder music*: Brief music snippet on a treble recorder.	4.3.3
21	*Descant recorder arpeggio*: G major arpeggio plus extremes (C4, D4 to B5, C6) on a descant recorder.	4.3.3
22	*Descant recorder music*: Brief music snippet on a descant recorder.	4.3.3
23	*Sopranino recorder arpeggio*: G major arpeggio plus extreme (F4, G4 to D6) on a sopranino recorder.	4.3.3
24	*Sopranino recorder music*: Brief music snippet on a sopranino recorder.	4.3.3
25	*Swanee whistle*: Fast and slow sweeps on a swanee whistle.	4.3.3
26	*Bassoon arpeggio*: G major arpeggio plus extreme (Bb1, B1 to B4) on a bassoon.	4.3.6
27	*Bassoon music*: Brief music snippet on a bassoon.	4.3.6
28	*Oboe arpeggio*: G major arpeggio (B3 to D6) on an oboe.	4.3.6
29	*Oboe music*: Brief music snippet on an oboe.	4.3.6
30	*Bagpipe steady chord*: Drone and steady note on bagpipes.	4.3.6
31	*Bagpipe music*: Brief music snippet on bagpipes.	4.3.6
32	*Bass clarinet arpeggio*: G major arpeggio (D2 to G4) on a bass clarinet.	4.3.6
33	*Bass clarinet music*: Brief music snippet on a bass clarinet.	4.3.6
34	*Clarinet arpeggio*: G major arpeggio (D3 to G5) on a clarinet.	4.3.6
35	*Clarinet music*: Brief music snippet on a clarinet.	4.3.6
36	*Tuba arpeggio*: G major arpeggio (G1 to G3) on a tuba.	4.3.7
37	*Tuba music*: Brief music snippet on a tuba.	4.3.7
38	*Bird call*: cuckoo (3 times).	4.3.3
39	*Bird all*: duck (8 times).	4.3.6
40	*Bird call*: goose (7 times).	4.3.6
41	*Bird call*: nightingale (5 times).	4.3.3
42	*Bird call*: quail (8 times).	4.3.3
43	*Adult female speech*: Read passage (both channels: microphone).	4.5
44	*Adult female speech:* Read passage (right: microphone; left: electrolaryngograph[*]).	4.5
45	*Adult female arpeggio*: G major arpeggio from B3 to B5 (both channels: microphone).	4.5
46	*Adult female arpeggio*: G major arpeggio from B3 to B5 (right: microphone; left: electrolaryngograph[*]).	4.5
47	*Adult female singing*: Song snippet (both channels: microphone).	4.5
48	*Adult female singing*: Song snippet (right: microphone; left: electrolaryngograph[*]).	4.5
49	*Adult male speech*: Read passage (both channels: microphone).	4.5
50	*Adult male speech*: Read passage (right: microphone; left: electrolaryngograph[*]).	4.5
51	*Adult male arpeggio*: G major arpeggio from G2 to B4 (both channels: microphone).	4.5

(Continued)

Table A7	Continued	
No.	**Track contents**	**Section**
52	*Adult male arpeggio*: G major arpeggio from G2 to B4 (right: microphone; left: electrolaryngograph*).	4.5
53	*Adult male singing*: Four-part multi-tracked male voice barbershop (TTBB) snippet (stereophonic presentation of microphone outputs).	4.5
54	*Adult male singing*: Four-part multi-tracked male voice barbershop (TTBB) snippet (stereophonic presentation of electrolaryngograph outputs).	4.5
55	*Adult male singing*: Four-part multi-tracked male voice barbershop (TTBB) snippet (stereophonic presentation of the upper part from the 1st tenor part accompanied by the electrolaryngograph outputs of the three other parts).	4.5
56	*Girl (9years old) singing*: Hymn snippet (both channels: microphone).	4.5
57	*Girl (9years old) singing*: Hymn snippet (right: microphone; left: electrolaryngograph*).	4.5
58	*Boy (12years old) speech*: Read passage (both channels: microphone).	4.5
59	*Boy (12years old) speech*: Read passage (right: microphone; left: electrolaryngograph*).	4.5
60	*Boy (12years old) singing*: Hymn snippet (both channels: microphone).	4.5
61	*Boy (12years old) singing*: Hymn snippet (right: microphone; left: electrolaryngograph*).	4.5
62	*Timbre demonstration material*: Four notes are provided—one each from a violin, flute, bassoon and oboe. The notes have been pitch shifted to G4, amplitude ramped at the start and finish, and the steady-state portions are equalized in level. Any difference that is heard between any of these sounds results from a change in timbre, since their pitch, loudness and durations are the same (see the definition of timbre given in Section 5.1).	5.1
63	*Helmholtz timbre rules*: A series of sounds to illustrate each of the four timbre "rules" described in Section 5.3.2: (A) sine wave at 200 Hz (rule 1—simple tones); (B) harmonics at 200, 400, 600, 800, and 1000 Hz (rule 2—musical tones); (C) harmonics at 200, 600, 1000, 1400, 1800, 2200, 2400, 2600, 3000, 3400 and 3800 Hz (rule 3—uneven partials); (D) harmonics 1 to 20 at 200 (rule 4—distinct partials above the sixth or seventh).	5.3.2
64	*Organ stops reinforcing all harmonics*: The stops that reinforce the first nine harmonics ($1f_0$, $2f_0$, $3f_0$, $4f_0$, $5f_0$, $6f_0$, $7f_0$, $8f_0$, $9f_0$) are drawn one after the other whilst the note G4(392Hz) is held. The stops on this instrument are: chimney flute 8', open flute 4', nazard 2 2/3', block flute 2', tierce 1 3/5, larigot 1 1/3, septième 1 1/7', octavin 1' and none 8/9'. This enables the effect of each stop to be heard as it contributes to the timbre of the sound: a form of "analytic" listening. Then the note is repeated a few times and an arpeggio is played, and now the notes are usually heard as a whole: a form of "holistic" listening.	5.4
65	*Organ stops reinforcing odd harmonics*: The stops that reinforce the first five odd harmonics ($1f_0$, $3f_0$, $5f_0$, $7f_0$, $9f_0$) are drawn while the note G4(392Hz) is held. The stops on this instrument are: chimney flute 8', nazard 2 2/3', tierce 1 3/5, septième 1 1/7' and none 8/9'. This enables the effect of each stop to be heard as it contributes to the timbre of the sound: a form of "analytic" listening. Then the note is repeated a few times and a short tune is played; now the timbre of the notes is usually heard as a whole: a form of "holistic" listening. Notice that the final effect is somewhat clarinet-like due to the spectrum consisting of the odd harmonic series (see Section 4.3.6 which discusses the acoustics of the clarinet).	5.4
66	*Frequency proximity streaming*: The score snippet from the Preludio from Partita III in E major by J.S. Bach shown in Figure 5.13 is synthesized to enable the streaming effect to be heard.	5.5.3
67	*Grouping and frequency proximity streaming*: The score snippet from the final movement of Tchaikovsky's sixth symphony shown in Figure 5.14 is synthesized to enable the streaming effect to be heard as follows: (A) full orchestra; (B) 1st violins and violas; (C) 2nd violins and cellos.	5.5.3

(Continued)

Table A7	Continued	
No.	**Track contents**	**Section**
68	*Shepherd tone illusion*: A demonstration of the Shepherd tone illusion which is based on the spectra shown in Figure 5.18.	5.5.4
69	*Shepherd tone element*: A demonstration of the circular nature of the contribution made to the Shepherd tone illusion by each component in the previous track. Figure 5.18 illustrates how each component is controlled.	5.5.4
70	*Continuous scale pitch illusion 1*: The snippet from the Fantasia in G minor (BWV 542) by J.S. Bach shown in Figure 5.19 is played using a registration with no reed stops in the manual parts. Follow the score and try to identify whether the upward seventh leaps are ambiguous in terms of which octave the note following the leap is in.	5.5.4
71	*Continuous scale pitch illusion 2*: The snippet from the Fantasia in G minor (BWV 542) by J.S. Bach shown in Figure 5.19 is played using a registration with reed stops in the manual parts. Follow the score and try to identify whether the upward seventh leaps are ambiguous in terms of which octave the note following the leap is in.	5.5.4
72	*Continuous scale pitch faked*: The snippet from the Fantasia in G minor (BWV 542) by J.S. Bach shown in Figure 5.19 is played using registration 1 (no reed stops) for the manual parts, but stop changes are made during the snippet to produce a pedal line that is continually descending using reed stops. This is achieved by using a schalmei 4' for the first descending octave, then a trumpet 8' for the second descending octave, an ophecleide 16' for the third descending octave and a contra bombarde 32' for the final descending seventh. It should be noted that these reeds have different timbres and intensities so the stop changes are clearly audible. Nevertheless, the long descending scale is well illustrated.	5.5.4
73	*Virtual pitch—chorale played normally*: Following the suggestion from Roederer quoted towards the end of Section 5.5.4, this demonstration uses the Chorale Prelude "Ich ruf' zu dir, Herr Jesu Christ" from the Orgelbüchlein by J.S. Bach (BWV 639), which is listed in the Novello Edition (edited by Ivor Atkins and revised by Walter Emery) as chorale number 41. Chorale number 40, which is suggested by Roederer, is, according to the Novello Edition: "Es ist das Heil uns kommen her" (BWV 638), which has no solo chorale melody line and thus is not suitable for this demonstration. In this demonstration, a snippet from the chorale is performed using the stops that reinforce the first nine harmonics (chimney flute 8', open flute 4', nazard 2 2/3', block flute 2', tierce 1 3/5, larigot 1 1/3, septième 1 1/7' octavin 1', and none 8/9') for the melody (a registration that would probably not be used in a performance, but one that enhances this demonstration).	5.5.4 & 3.2.2
74	*Virtual pitch organ chorale*: In this demonstration, the snippet from the Chorale Prelude used in Track 73 starts using the same stops that reinforce the first nine harmonics for the melody, but the stops are pushed in the following order: chimney flute 8', open flute 4', nazard 2 2/3', block flute 2', tierce 1 3/5, larigot 1 1/3, septième 1 1/7' and octavin 1' until only the none 8/9' (reinforcing the ninth harmonic) remains. Then the accompaniment is removed and later brought back in. The tune is still perceivable when the stops reinforcing the low harmonics are removed, even when only the one harmonic (the ninth) remains, and it is a mutation (not a whole number of octaves away from the fundamental). When the accompaniment is removed, the pitch of the melody is somewhat ambiguous.	5.5.4 & 3.2.2
75	*Short impulse response*: This is the impulse response of a small church provided courtesy of Dr Damian Murphy to enable reverberation convolution for a short T_{60}.	6.1.4
76	*Medium impulse response*: This is the impulse response of a medium sized space provided courtesy of Dr Damian Murphy to enable reverberation convolution for a medium T_{60}.	6.1.4
77	*Long impulse response*: This is the impulse response of a large space provided courtesy of Dr Damian Murphy to enable reverberation convolution for a long T_{60}.	6.1.4

(*Continued*)

Table A7	Continued	
No.	**Track contents**	**Section**
78	*Mono, stereo and surround sound*: This demonstration is for headphone listening. It illustrates how sound can be placed outside the head and moved around by tracking the change from monophonic to stereophonic to surround sound listening. The authors thank the creators of this demonstration for permission to use it on this CD. It was originally produced for The Royal Society's Summer Science Exhibition "Surrounded by Sound" in July 2001 by members of The Music Technology Research Group, Department of Electronics, University of York.	
79	*Tone used in the "mosquito" or "teen deterrent"*: The sound used for a mosquito or teen deterrent is a 16.8 kHz sine wave. In this demonstration, four pure tones rising in octaves to the mosquito average frequency of 16.8 kHz are presented as follows: 2.1 kHz, 4.2 kHz, 8.4 kHz and 16.8 kHz. PLEASE NOTE: many youngsters will find the 16.8 kHz tone unpleasant—be cautious with the listening level.	7.7

The electrolaryngograph (or electroglottograph) is a device that monitors vocal fold vibration by measuring the electrical impedance between two electrodes that are placed externally on the neck at the level of the larynx. Listening to the output waveform from the electrolaryngograph (Lx) gives an appreciation of the sound source during speech and singing. Please note that there are high-frequency artifacts associated with the Lx waveform which are particularly obvious when listening to it. Whilst these recordings can be low-pass filtered to remove these artifacts (this exercise is left to interested readers), any filtering is likely to alter the detailed shape of the Lx waveform due to the phase response of the filter. The Lx output is therefore presented in its "raw" form in these recordings to enable further analysis of the waveshape if desired. Measurement of the fundamental frequency (f_0) of the Lx waveform, usually based on measuring the fundamental period (T_0) and finding its reciprocal as $f_0 = (1/T_0)$, provides a reliable and accurate experimental method for finding the f_0 in speech or singing. More details can be found in Baken (1987), Howard (1995, 1999), and Howard et al. (1990)—these references are listed at the end of Chapter 4.

The authors would like to thank the following for their contributions to the recording of this audio CD (names are given in alphabetical order):

Chris Bouchard, Jude Brereton, Helena Daffern, Annie Howard, Clare Howard, David Howard, Joey Howard, Ross Kirk, Damian Murphy, Tom Whalley, Rachel van Besouw, Jez Wells, and members of the Audio Laboratory of the Intelligent Systems Research Group (formerly The Music Technology and Media Engineering Research Groups) of the Department of Electronics, at the University of York, UK.

Production mastering: Jez Wells
Copyright: This CD and its contents are © David M. Howard, York 2009

Index

479